Seguin's
COMPUTER
Applications

with Microsoft® Office 2016

Second Edition

Denise Seguin
Fanshawe College, London, Ontario

PARADIGM
EDUCATION SOLUTIONS

St. Paul

Senior Vice President	Linda Hein
Editor in Chief	Christine Hurney
Senior Editor	Cheryl Drivdahl
Developmental Editor, Digital and Print	Tamborah Moore
Assistant Developmental Editors	Mamie Clark and Katie Werdick
Contributing Writer	Janet Blum
Testers	Jeff Johnson and Desiree Carvel
Director of Production	Timothy W. Larson
Production Editors	Rachel Kats and Carrie Rogers
Cover and Text Designer, Senior Production Specialist	Jaana Bykonich
Copy Editor	Heidi Hogg
Indexer	Terry Casey
Vice President Sales and Marketing	Scott Burns
Director of Marketing	Lara Weber McLellan
Vice President Information Technology	Chuck Bratton
Digital Projects Manager	Tom Modl
Digital Learning Manager	Troy Weets
Digital Production Manager	Aaron Esnough
Senior Digital Production Specialist	Julie Johnston

Care has been taken to verify the accuracy of information presented in this book. However, the authors, editors, and publisher cannot accept responsibility for Web, e-mail, newsgroup, or chat room subject matter or content, or for consequences from application of the information in this book, and make no warranty, expressed or implied, with respect to its content.

Trademarks: Microsoft is a trademark or registered trademark of Microsoft Corporation in the United States and/or other countries. Some of the product names and company names included in this book have been used for identification purposes only and may be trademarks or registered trade names of their respective manufacturers and sellers. The authors, editors, and publisher disclaim any affiliation, association, or connection with, or sponsorship or endorsement by, such owners.

Image Credits: See page 391

We have made every effort to trace the ownership of all copyrighted material and to secure permission from copyright holders. In the event of any question arising as to the use of any material, we will be pleased to make the necessary corrections in future printings. Thanks are due to the aforementioned authors, publishers, and agents for permission to use the materials indicated.

Paradigm Publishing, Inc., is independent from Microsoft Corporation and not affiliated with Microsoft in any manner.

ISBN 978-0-76386-885-7 (print)
ISBN 978-0-76386-888-8 (digital)

© 2017 by Paradigm Publishing, Inc.
875 Montreal Way
St. Paul, MN 55102
Email: educate@emcp.com
Website: www.emcp.com

Printed in the United States of America

25 24 23 22 21 20 19 18 17 16   1 2 3 4 5 6 7 8 9 10

Brief Contents

Preface ix

Chapter 1
Using Windows 10 and Managing Files 3

Chapter 2
Navigating and Searching the Web.. 39

Chapter 3
Exploring Microsoft Office 2016 Essentials.. 63

Chapter 4
Organizing and Managing Class Notes Using OneNote 93

Chapter 5
Communicating and Scheduling Using Outlook 111

Chapter 6
Creating, Editing, and Formatting Word Documents................................. 135

Chapter 7
Enhancing a Word Document with Special Features 157

Chapter 8
Creating, Editing, and Formatting Excel Worksheets 185

Chapter 9
Working with Functions, Charts, Tables, and Page Layout Options in Excel 213

Chapter 10
Creating, Editing, and Formatting a PowerPoint Presentation.................. 239

Chapter 11
Enhancing a Presentation with Multimedia and Animation Effects 263

Chapter 12
Using and Querying an Access Database 287

Chapter 13
Creating a Table, Form, and Report in Access.. 315

Chapter 14
Integrating Word, Excel, PowerPoint, and Access Content...................... 339

Chapter 15
Using OneDrive and Other Cloud Computing Technologies....................... 355

Glossary/Index.. 377

Contents

Preface ix

Chapter 1

Using Windows 10 and Managing Files 3

Topic 1.1   Using Touch, Mouse, and Keyboard Input to Navigate Windows 10 4

Topic 1.2   Starting Windows 10 and Exploring Apps 8

Topic 1.3   Using the Search Text Box and Working with Windows 12

Topic 1.4   Using the Notification Area 14

Topic 1.5   Locking the Screen, Signing Out, and Shutting Down Windows 10 16

Topic 1.6   Using and Customizing the Start Menu 18

Topic 1.7   Personalizing the Lock Screen and the Desktop 20

Topic 1.8   Understanding Files and Folders and Downloading and Extracting Data Files 22

Topic 1.9   Browsing Files with File Explorer.............................24

Topic 1.10   Creating Folders and Copying Files and Folders 26

Topic 1.11   Moving, Renaming, and Deleting Files and Folders and Ejecting a USB Flash Drive 30

Chapter 2

Navigating and Searching the Web............ 39

Topic 2.1   Introduction to the Internet and the World Wide Web 40

Topic 2.2   Navigating the Web Using Microsoft Edge 42

Topic 2.3   Navigating the Web Using Chrome....................................46

Topic 2.4   Navigating the Web Using Mozilla Firefox 50

Topic 2.5   Searching for Information and Printing Web Pages 54

Topic 2.6   Downloading Content from a Web Page..............................58

Chapter 3

Exploring Microsoft Office 2016 Essentials 63

Topic 3.1   Starting and Switching Programs, Starting a New Presentation, and Exploring the Ribbon Interface 64

Topic 3.2   Using the Backstage Area to Manage Documents 68

Topic 3.3   Customizing and Using the Quick Access Toolbar 72

Topic 3.4   Selecting Text or Objects, Using the Ribbon and Mini Toolbar, and Selecting Options in Dialog Boxes 74

Topic 3.5   Using the Office Clipboard and Formatting Using a Task Pane...78

Topic 3.6   Finding Help or Options Using the Tell Me Feature 82

Topic 3.7   Using OneDrive for Storage, Scrolling in Documents, and Using Undo.............................84

Topic 3.8   Changing the Zoom Option and Screen Resolution................88

Chapter 4

Organizing and Managing Class Notes Using OneNote... 93

Topic 4.1   Opening a Notebook and Adding Notes, Sections, and Pages 94

Topic 4.2   Inserting Web Content into a Notebook...................................98

Topic 4.3   Inserting Files into a Notebook 100

Topic 4.4   Tagging Notes, Viewing Tags, and Jumping to a Tagged Note.. 102

Topic 4.5   Searching Notes and Closing a Notebook 104

Topic 4.6   Creating a New Notebook and Sharing a Notebook........... 106

Chapter 5

Communicating and Scheduling Using Outlook 111

Topic 5.1   Using Outlook to Send Email ... 112

Topic 5.2   Attaching a File to a Message and Deleting a Message 116

Topic 5.3   Previewing File Attachments and Using File Attachment Tools 118

Topic 5.4   Scheduling Appointments and Events in Calendar 120

Topic 5.5   Scheduling a Recurring Appointment and Editing an Appointment 122

Topic 5.6   Scheduling a Meeting 124

Topic 5.7   Adding and Editing Contacts in People 126

Topic 5.8   Adding and Editing Tasks 128

Topic 5.9   Searching Outlook Items 130

Chapter 6

Creating, Editing, and Formatting Word Documents 135

Topic 6.1   Creating and Editing a New Document 136

Topic 6.2   Inserting Symbols and Completing a Spelling and Grammar Check 140

Topic 6.3   Finding and Replacing Text 142

Topic 6.4   Moving Text and Inserting Bullets and Numbering 144

Topic 6.5   Formatting Text with Font and Paragraph Alignment Options 146

Topic 6.6   Indenting Text and Changing Line and Paragraph Spacing 148

Topic 6.7   Formatting Using Styles 150

Topic 6.8   Creating a New Document from a Template 152

Chapter 7

Enhancing a Word Document with Special Features 157

Topic 7.1   Inserting, Editing, and Labeling Pictures in a Document 158

Topic 7.2   Adding Borders and Shading, and Inserting a Text Box 162

Topic 7.3   Inserting a Table 164

Topic 7.4   Formatting and Modifying a Table 166

Topic 7.5   Changing Layout Options 170

Topic 7.6   Adding Text and Page Numbers in a Header for a Research Paper 172

Topic 7.7   Inserting and Editing Citations 174

Topic 7.8   Creating a Works Cited Page and Using Word Views 176

Topic 7.9   Inserting and Replying to Comments 178

Topic 7.10   Creating a Resume and Cover Letter from Templates 180

Chapter 8

Creating, Editing, and Formatting Excel Worksheets 185

Topic 8.1   Creating and Editing a New Worksheet 186

Topic 8.2   Formatting Cells 190

Topic 8.3   Adjusting Column Width and Row Height, and Changing Cell Alignment 194

Topic 8.4   Entering or Copying Data with the Fill Command and Using AutoSum 196

Topic 8.5   Inserting and Deleting Rows and Columns 200

Topic 8.6   Sorting Data and Applying Cell Styles 202

Topic 8.7   Changing Orientation and Scaling, and Displaying Cell Formulas 204

Topic 8.8   Inserting and Renaming a Worksheet, Copying Cells, and Indenting Cell Contents 206

Topic 8.9   Using Go To; Freezing Panes; and Shading, Wrapping, and Rotating Cell Entries 208

Chapter 9

Working with Functions, Charts, Tables, and Page Layout Options in Excel 213

Topic 9.1   Using Absolute Addresses and Range Names in Formulas 214

Topic 9.2 Entering Formulas Using Statistical Functions.................216

Topic 9.3 Entering, Formatting, and Calculating Dates....................218

Topic 9.4 Using the IF Function 220

Topic 9.5 Using the PMT Function.........222

Topic 9.6 Creating and Modifying a Pie Chart 224

Topic 9.7 Creating and Modifying a Column Chart......................226

Topic 9.8 Creating and Modifying a Line Chart 228

Topic 9.9 Using Page Layout View, Adding a Header, and Changing Margins..................230

Topic 9.10 Creating and Modifying Sparklines and Inserting Comments..........................232

Topic 9.11 Working with Tables 234

Chapter 10

Creating, Editing, and Formatting a PowerPoint Presentation 239

Topic 10.1 Creating a New Presentation and Inserting Slides..................240

Topic 10.2 Changing the Theme and Inserting and Modifying a Table....................................244

Topic 10.3 Formatting Text with Font and Paragraph Options.............246

Topic 10.4 Selecting, Resizing, Aligning, and Moving Placeholders..........248

Topic 10.5 Using Slide Sorter View and Moving, Duplicating, and Deleting Slides 250

Topic 10.6 Modifying the Slide Master.......252

Topic 10.7 Adding Notes and Comments...254

Topic 10.8 Displaying a Slide Show 256

Topic 10.9 Preparing Audience Handouts and Speaker Notes 258

Chapter 11

Enhancing a Presentation with Multimedia and Animation Effects 263

Topic 11.1 Inserting Graphic Images from Picture Collections 264

Topic 11.2 Inserting a SmartArt Graphic....266

Topic 11.3 Converting Text to SmartArt and Inserting WordArt.............268

Topic 11.4 Creating a Chart on a Slide 270

Topic 11.5 Drawing a Shape and Adding a Text Box 272

Topic 11.6 Adding Video to a Presentation 274

Topic 11.7 Adding Sound to a Presentation 276

Topic 11.8 Adding Transitions and Animation Effects to Slides for a Slide Show 278

Topic 11.9 Setting Up a Self-Running Presentation 282

Chapter 12

Using and Querying an Access Database 287

Topic 12.1 Understanding Database Objects and Terminology..........288

Topic 12.2 Adding Records to a Table Using a Datasheet 292

Topic 12.3 Editing and Deleting Records Using a Datasheet 294

Topic 12.4 Adding, Editing, and Deleting Records Using a Form 296

Topic 12.5 Finding and Replacing Data and Adjusting Column Widths in a Datasheet.........................298

Topic 12.6 Sorting and Filtering Records...300

Topic 12.7 Creating a Query Using the Simple Query Wizard.........302

Topic 12.8 Creating a Query Using Design View 304

Topic 12.9 Entering Criteria to Select Records in a Query.................306

Topic 12.10 Entering Multiple Criteria to Select Records and Sorting a Query...................................308

Topic 12.11 Inserting and Deleting Columns, Creating a Calculated Field in a Query, and Previewing a Datasheet 310

Chapter 13

Creating a Table, Form, and Report in Access 315

Topic 13.1   Creating a New Database File and Understanding Table Design Guidelines 316

Topic 13.2   Creating a New Table Using a Datasheet 318

Topic 13.3   Creating a New Table Using Design View and Assigning a Primary Key 320

Topic 13.4   Adding Fields to an Existing Table 322

Topic 13.5   Modifying Field Properties Using Design View 324

Topic 13.6   Creating a Lookup List 326

Topic 13.7   Displaying and Editing a Relationship 328

Topic 13.8   Creating and Editing a Form 330

Topic 13.9   Creating, Editing, and Viewing a Report 332

Topic 13.10   Compacting, Repairing, and Backing Up a Database 334

Chapter 14

Integrating Word, Excel, PowerPoint, and Access Content 339

Topic 14.1   Importing Excel Worksheet Data into Access 340

Topic 14.2   Exporting an Access Query to Excel 344

Topic 14.3   Embedding an Excel Chart into a Word Document 346

Topic 14.4   Embedding Excel Data into PowerPoint and Editing the Embedded Data 348

Topic 14.5   Linking an Excel Chart with a Presentation and Updating the Link 350

Chapter 15

Using OneDrive and Other Cloud Computing Technologies 355

Topic 15.1   Creating a Document Using Word Online 356

Topic 15.2   Creating a Worksheet Using Excel Online 360

Topic 15.3   Creating a Presentation Using PowerPoint Online 362

Topic 15.4   Editing a Presentation in PowerPoint Online 364

Topic 15.5   Downloading Files from and Uploading Files to OneDrive 366

Topic 15.6   Sharing a File on OneDrive 368

Topic 15.7   Creating a Document Using Google Docs 370

Glossary/Index 377

Preface

Course Overview

Today's students arrive in the classroom with more confidence in using technology than any generation before them. You have grown up with technology as a part of your life and the Internet as a source of information, entertainment, and communication. Chances are you have used a word processor and a presentation program for several years to prepare materials for school projects. You may have learned your way around a computer application by trial and error. However, to be efficient and successful, you need to learn how to use software applications in a way that saves you time and makes the best use of the available feature set. To that end, *Seguin's COMPUTER Applications with Microsoft® Office 2016*, Second Edition, provides the tools you need to succeed immediately in your academic and personal lives as well as prepare yourself for success in your future career. In this book, you will learn skills you can apply immediately to accomplishing projects and assignments for school and to organizing, scheduling, recording, planning, and budgeting for your personal needs. You will find the work done in this course to be relevant and useful, with the content presented in a straightforward approach.

Along with well-designed textbook pedagogy, practice and problem solving will help you learn and apply computer topics and skills. Technology provides opportunities for interactive learning as well as excellent ways to quickly and accurately assess your performance. To this end, this textbook is supported by SNAP, Paradigm Education Solution's web-based training and assessment learning management system. Details about SNAP as well as additional student and instructor resources appear on page xiv.

Course Goals

According to Microsoft, more than 1.2 billion people use Microsoft Office—that represents one in seven people on the planet. Time.com reported that in June 2015, the Microsoft Windows operating system was running on 91 percent of desktop computers. With the prevalence of Microsoft products in the workplace, to be successful in your career requires competence with the Microsoft Windows operating system, Microsoft Edge web browser, and applications within the Office suite. In this course, you will learn to navigate and operate these software programs at a standard expected of an entry-level employee.

Seguin's COMPUTER Applications offers instruction that will guide you to achieve entry-level competence with the latest editions of Microsoft Windows, web browsers, and the Microsoft Office productivity suite, including OneNote, Outlook, Word, Excel, PowerPoint, and Access. You will also be introduced to cloud computing alternatives to the traditional desktop suite. No prior experience with these software programs is required. Even those with some technological savvy can benefit from completing the course by learning new ways to perform tasks or reinforce skills. After completing a course that uses this textbook, you will be able:

- to navigate the Windows operating system and manage files and folders.
- to use web browsers such as Microsoft Edge, Google Chrome, or Mozilla Firefox to navigate and search the web, as well as download content to a PC or mobile device.
- to use navigation, file management, commands, and features within the Microsoft Office suite that are standard across all applications.
- to organize and manage class notes in OneNote.
- to communicate and manage personal information in Outlook.
- to create, edit, format, and enhance documents in Word.
- to create, edit, analyze, format, and enhance workbooks in Excel.
- to create, edit, format, and enhance slides and set up a slideshow in PowerPoint.
- to create and edit tables, forms, queries, and reports in Access.
- to integrate information among the applications within the Microsoft Office suite.
- to use cloud computing technologies to create, edit, store, and share documents.

Reading time is minimized; you will learn just what you need to know to succeed within these programs. You will practice features with step-by-step instruction interspersed with text that explains why a feature is used or how the feature can be beneficial to you. You should work through each chapter at a PC or with a tablet, so that you can complete the steps as you learn.

Textbook Organization and Methodology

This textbook is divided into 15 chapters that are best completed in sequence; however, after assigning the essential skills learned in Chapters 1 through 5, your instructor may opt to assign Word, Excel, PowerPoint, Access, integration, and cloud computing technologies in the order of his or her choice.

Each chapter begins with a brief introduction to the chapter content along with a precheck and a list of learning objectives. Following the chapter opener, each chapter topic begins with a list of skills to be mastered. The topics are presented in two or four pages. A variety of marginal notes and other features expand or clarify the content. You will gain experience with topic features by working through hands-on exercises, which consist of step-by-step instructions and illustrative screen shots.

At the end of each chapter, you will have a chance to review a summary of the topics, and then take a Recheck quiz to see how much you have learned and which topics require more attention. The student ebook accompanying this textbook includes workbook pages that offer additional study tools, interactive review exercises, and assessments that will reinforce and expand the knowledge you've gained.

What Makes This Textbook Different from Others?

Many textbooks that teach computer applications were designed and organized for software that was in effect one or two decades ago. As software evolves and becomes more flexible and streamlined, so too should software textbooks. With this mandate, this textbook has been designed and organized with a fresh look at the skills a student should know to be successful in today's world. The author has chosen and placed in a logical sequence those skills that are considered essential for today's student. Consider this book a "software survival kit for school and life." Nothing more, nothing less!

Many of the student data files in this textbook are based on files created by students for projects or assignments in courses similar to those students may be enrolled in now. Students will open and manipulate real work completed by someone just like them. Other files include practical examples of documents that students can readily relate to their school and personal experiences.

This Book Is Green!

Instructions to print results have been intentionally omitted for all exercises and assessments. This approach is consistent with a green computing initiative to minimize wasteful printing for nongraded topics or assessment work and also provides instructors with maximum flexibility in designing their course structure.

Course Features

The following guide shows how this textbook and its digital resources use a visual approach combined with hands-on activities to help you learn and master key skills and topics.

SNAP Resources

SNAP icons in the margins of the textbook are accompanied by blue text listing exercises and assessments that are available in SNAP. If you are a SNAP user, go to your SNAP Assignments page to complete the activities.

Interactive Resources

Arrow icons in the margins of the textbook indicate interactive resources that are available through the links menu in your student ebook and, in some cases, in SNAP.

- **Precheck quizzes** test your knowledge of the chapter content before you study the material. Use the results to help focus your study on the topics and skills you need to learn. SNAP users should go to their SNAP Assignments page to complete these quizzes.

- **Tutorials** provide hands-on guided skills practice. SNAP users should go to their SNAP Assignments page to complete these tutorials.
- **View Model Answer** features link to completed model answers that you can compare with your completed work.
- The **Data Files** needed for the exercises and assessments are downloaded as a zipped file in chapter 1. In the workbook section of the student ebook you have the option of downloading the individual data files as you need them to complete each assessment.
- The **Audio Files** needed for some assessments are included in the zipped file of student data files downloaded in Chapter 1 and may also be accessed from the workbook sections of the student ebook.
- A **Recheck quiz** at the end of each chapter enables you to recheck your understanding of the chapter content. You may recheck your understanding at any time and as many times as you wish. SNAP users should go to their SNAP Assignments page to complete these quizzes.
- The workbook sections of the student ebook provides study resources, review exercises, and assessments to help you reinforce and demonstrate your understanding of the topics covered in each chapter.

Textbook Elements

Topics are presented in two or four pages.

A **Skills** list presents the skills that will be learned by completing the steps in the topic.

Tutorials reinforce and supplement the skills being taught.

Step-by-step instructional text accompanied with frequent illustrative screen shots presents instructions for completing the exercises in simple, easy-to-follow steps.

Screen captures with step numbers provide visual confirmation and guidance. Text to be typed is set in **red font** to stand out from the instructional text.

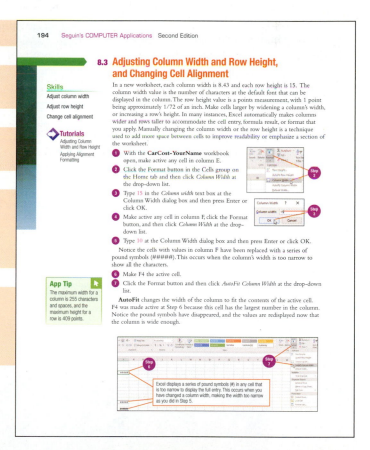

App Tips extend or add to your knowledge of a feature.

Quick Steps briefly summarize the steps to complete major tasks, for quick reference and review.

Quick Steps

Enter a Formula
1. Activate formula cell.
2. Type =.
3. Type first cell address.
4. Type operator symbol.
5. Type next cell address.
6. Continue Steps 4–5 until finished.

Check This Out features provide links to websites at which you can explore more information.

> **Check This Out** ✔
>
> **http://CA2.Paradigm College.net/Chrome**
>
> Go here to download and install Chrome on your PC or mobile device.

Good to Know features highlight interesting or fun facts or trivia about the topic.

> **Good to Know**
>
> Many businesses that operate globally have adopted the International Standards Organization (ISO) date format YYYY-MM-DD to avoid confusion

Oops! hints anticipate common challenges and provide solutions to help you succeed with the topic.

> **Oops!** !
>
> No Cell Styles gallery? On smaller displays, the gallery is accessed by clicking the Cell Styles

Alternative Method boxes present different ways to accomplish the task learned in the topic.

> **Alternative Method** **Changing Print Options Using the Page Layout Tab**
>
> You can change some print options on the Page Layout tab, using the Margins and Orientation buttons in the Page Setup group and the Width, Height, and Scale options in the Scale to Fit group.

Beyond Basics boxes provide additional information about the feature that extends the skills described in the topic.

> **Beyond Basics**   **Order of Operations in Formulas**
>
> If you combine operations in a formula, Excel will automatically calculate exponentiation, multiplication, and division before addition and subtraction. You can tell Excel to perform a particular operation first by using parentheses around that part of the formula. For example, in the formula $=(A1+A2)*A3$, Excel adds the values in A1 and A2 first and then multiplies the result by the value in A3.

Security Alert boxes offer tips for security and privacy to help you learn safe computing practices.

> **Security Alert**   **Update Links with Caution**
>
> At Step 23, you clicked the Update Links button because you knew the linked file was from a trusted source. Be aware before you click Update Links when opening a document, workbook, presentation, or database that you know where the linked object originates and that it is from a trusted source. In 2015, Sophos (an IT security company) reported a resurgence of malware circulated in Microsoft Office documents. Sophos reported the malware is targeted more towards Word and Excel files and the files are usually attached to email messages.

Tables and Figures organize information in a streamlined format to minimize your reading load.

Table 8.1

Excel Features

Feature	Description
Active cell	Location in which the next typed data will be stored and that will be affected by the next command. Make a cell active by clicking it or by moving to it using the Arrow
	when you are able to select cells with the mouse by clicking or
	e with no mouse attached, tap a cell to display selection handles the top left and bottom right corners.
	contents stored in the active cell and is also used to create
	w parts of a worksheet not shown in the current viewing area.
	the address or name of the active cell.

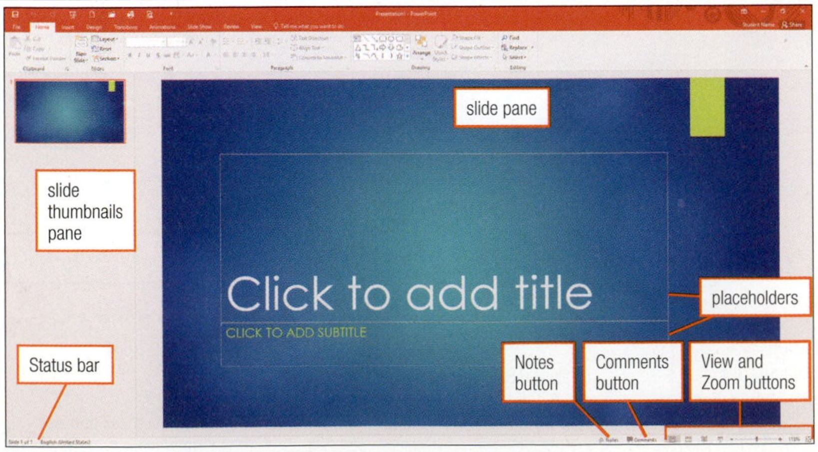

Figure 10.1

A new PowerPoint presentation with Ion theme and blue color variant in the default Normal view is shown above. See Table 10.1 for a description of screen elements.

A **Topics Review** chart at the end of each chapter summarizes the main chapter content and key words learned.

Topics Review

Topic	Key Concepts	Key Terms
8.1 Creating and Editing a New Worksheet	A spreadsheet is an application in which data is created, analyzed, and presented in a grid-like structure of columns and rows.	spreadsheet application
	A workbook is an Excel file that consists of a collection of individual worksheets.	workbook
	A new workbook opens with a blank worksheet into which you add text, insert values, and create formulas.	worksheet
	The intersection of a column and a row is called a *cell*.	cell
	The active cell is indicated with a green border, and is the location into which the next data typed will be stored, or the next command will be acted upon.	active cell
	Create a worksheet by making a cell active and typing text, a value, or a formula.	formula
	A formula is used to perform mathematical operations on values.	Clear button
	Formula entries begin with an equals sign and are followed by cell references with operators between the references.	
	Edit a cell by typing new data to overwrite existing data, by double-clicking to open the cell for editing, or by inserting or deleting characters in the Formula bar.	

Workbook Elements

The workbook sections of the student ebook provide a variety of materials you can use to review, reinforce, and demonstrate your understanding of the skills and topics covered in the textbook. For each chapter, you will find the following workbook elements.

Interactive **study tools**, including a presentation with audio support, help reinforce your understanding of the topics. These resources are accessed from the links menu.

Interactive **review exercises**, including multiple-choice, matching, and completion questions, give you an opportunity to practice and review your understanding of the skills and topics. These resources are accessed from the links menu and are also available in SNAP. SNAP Users should go to their SNAP Assignments page to complete these exercises.

SNAP Exercises offer additional practice for SNAP users. SNAP Users should go to their SNAP Assignments page to complete these exercises.

Three to seven **assessments** allow you to apply and demonstrate comprehension of the major skills learned in the chapter. In general, assessments increase in complexity from the first to the last. Most assessments are intended to be completed individually; however, your instructor may opt to assign some to pairs or teams of students.

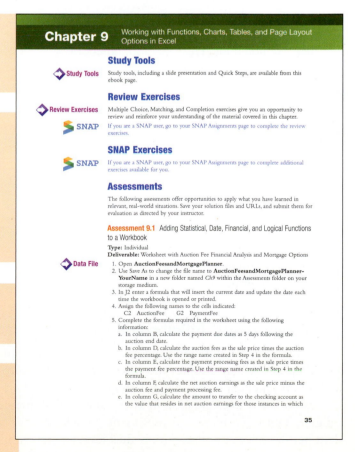

Beginning with Chapter 3, most chapters include a culminating **Visual assessment** in which you are to create a document similar to the example shown. The visual project requires that you notice details and problem solve to create a deliverable with less direction than is provided in other assessments. In some cases, you are required to do some Internet research and composition to complete the visual assessment.

Assessment 10.5 Visual—Creating a Graduation Party Planning Presentation

Type: Individual
Deliverable: Presentation on college graduation party planning

1. Create a presentation similar to the one shown in the Assessment 10.5 Graduation Party Planning Presentation on the next page with the following additional information:
   a. The theme is *Integral* with one of the variants selected.
   b. The bullet symbols have been changed on the slide master.
   c. The font color for the slide titles has been changed on the slide master. Use your best judgment to choose a similar color.
   d. Use your best judgment to determine other formatting, placeholder size, and alignment options.
2. Save the presentation in the Ch10 folder within the Assessments folder as **GradParty-YourName**.
3. Submit the assessment to your instructor in the manner she or he has requested.
4. Close the presentation.

View
Model Answer
Compare your completed file with the model answer.

In Chapters 6-11, an **Audio assessment** asks you to listen to instructions provided in an **Audio File** and then to compose the document, workbook, or presentation as instructed.

Assessment 10.6 Audio—Internet Research and Composing a New Presentation

Type: Individual or Pairs
Deliverable: Presentation about US or Canadian historical figure

You have been asked to help the president of the school's history society with a presentation for a guest speaker at a new activity called History Conversations. Research a US or Canadian historical figure and prepare a six-slide presentation with the main facts about the person and his or her significance in US or Canadian history.

1. Listen to the audio file named *HistoricalFigure_Instructions*.
2. Complete the research and compose the presentation as instructed.
3. Save the presentation in the Ch10 folder within the Assessments folder as **HistoricalFigure-YourName**.

Audio File

Beginning with Chapter 4, a **OneNote assessment** instructs you to send output to a OneNote notebook. The OneNote notebook is a repository for all your work. Your instructor may opt to have you share your OneNote notebook with him or her in a OneDrive folder. The instructor then has the option to check all your work in one place.

Assessment 10.7 OneNote—Sending Assessment Work to OneNote Notebook

Type: Individual
Deliverable: New page in Shared OneNote notebook

1. Start OneNote and open the MyAssessments notebook created in Chapter 4, Assessment 4.4.
2. Make PowerPoint the active section and then add a new page titled *Chapter 10 Assessments*.
3. Switch to PowerPoint. For each assessment that you completed, open the presentation, send the slides formatted as handouts with six slides horizontal per page, and with your name in a header to OneNote 2016, selecting the *Chapter 10 Assessments* page in the *PowerPoint* section in the MyAssessments notebook, then close the presentation, saving changes if prompted to do so.
4. Close your MyAssessments notebook in OneNote and then close OneNote.
5. Close PowerPoint.
6. Submit the assessment to your instructor in the manner she or he has requested.

Course Components

The *Seguin's COMPUTER Applications with Microsoft® Office 2016*, Second Edition, textbook contains the essential content you will need to master the key skills and topics covered. Additional resources are provided by the following digital components.

SNAP Web-Based Training and Assessment for Microsoft® Office 2016   SNAP

SNAP is a web-based training and assessment program and learning management system (LMS) for computer skills and topics and Microsoft Office 2016. SNAP offers rich content, a sophisticated grade book, and robust scheduling and analytics tools. SNAP includes a quiz and exam for each chapter, plus an item bank that can be used to create custom assessments. SNAP provides automatic scoring and detailed feedback on the program's many exercises and assessments to help identify areas where additional support is needed, evaluating student performance at both the individual level and the course level. The *Seguin's COMPUTER Applications* SNAP course content is also available to export into any LMS system that supports LTI tools. Paradigm Education Solutions provides technical support for SNAP through 24-7 chat at ParadigmCollege.com. In addition, an online user guide and other training tools for SNAP are available.

Student eBook

The student ebook provides access to all program content from any device (desktop, tablet, and smartphone) anywhere, through a live Internet connection. The versatile ebook platform features dynamic navigation tools including a linked table of contents and the ability to jump to specific pages, search for terms, bookmark, highlight, and take notes. The student ebook offers live links to the interactive content and resources that support the textbook, including the Precheck and Recheck quizzes, the tutorials, View Model Answer images, student data files, and the workbook sections.

By completing all Chapter 1 topics and assessments, students download and extract the student data files and set up a folder structure for storing completed work in the course. Student data files are downloaded and extracted to a USB flash drive from a link in the ebook. As an alternative, instructors may choose to load student data files on a network, in which case alternative instructions may need to be provided.

Integrated into the student ebook, the workbook sections include access to study tools such as chapter-based presentations with audio support, interactive end-of-chapter review exercises, and end-of-chapter assessments. The student ebook is accessed through SNAP or online at Paradigm.bookshelf.emcp.com.

SNAP users should go to their SNAP Assignments page to complete all interactive quizzes, tutorials, and exercises in the student ebook.

Instructor eResources

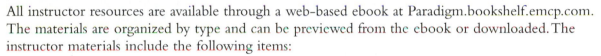

All instructor resources are available through a web-based ebook at Paradigm.bookshelf.emcp.com. The materials are organized by type and can be previewed from the ebook or downloaded. The instructor materials include the following items:

- Grading rubrics for evaluating responses to chapter assessments
- Lesson blueprints with teaching hints, lecture tips, and discussion questions
- Syllabus suggestions and course planning resources
- Chapter-based PowerPoint presentations with lecture notes
- Chapter-based quizzes and exams
- Annotated Model Answers for the end-of-chapter assessments

Acknowledgments

The author and editors would like to thank the following students for their contributions to and feedback on the textbook: Patti Ann Reynolds, Toni McBride, and Nicole Oke, Fanshawe College; and Michael Seguin, University of Windsor, Windsor, Ontario.

About the Author

Denise Seguin has served on the Faculty of Business at Fanshawe College of Applied Arts and Technology in London, Ontario, from 1986 until her retirement from full-time teaching in December 2012. She developed curriculum and taught a variety of office technology, software applications, and accounting courses to students in postsecondary Information Technology diploma programs and Continuing Education courses. Seguin served as Program Coordinator for Computer Systems Technician, Computer Systems Technology, Office Administration, and Law Clerk programs and was acting Chair of the School of Information Technology in 2001. Along with authoring *Seguin's COMPUTER Concepts*, First and Second Editions, and *Seguin's COMPUTER Applications with Microsoft® Office 2013*, First Edition, and *with Microsoft® Office 2016*, Second Edition, she has also authored Paradigm Education Solution's *Microsoft Outlook* 2000 to 2016 editions and co-authored *Our Digital World* First, Second, Third and Fourth Editions, *Benchmark Series Microsoft® Excel®*, 2007, 2010, and 2013, *Benchmark Series Microsoft® Access®* 2007, 2010, and 2013, *Marquee Series Microsoft® Office*, 2000 to 2013, and *Using Computers in the Medical Office*, 2003 to 2010.

In 2007, Seguin earned her Masters in Business Administration specializing in Technology Management, choosing to take her degree at an online university. She has an appreciation for those who are juggling work and life responsibilities while furthering their education, and she has taken her online student experiences into account when designing instruction and assessment activities for this textbook.

Seguin's
COMPUTER
Applications

with Microsoft® Office 2016

Second Edition

Chapter 1

Using Windows 10 and Managing Files

Precheck
Check your understanding of the topics covered in this chapter.

Windows 10 is the operating system (OS) software published by Microsoft Corporation. An OS provides the user interface that allows you to work with a computer or mobile device. The OS also manages all the hardware resources, routes data between devices and applications, and provides the tools for managing files and application programs. Every computing device requires an OS; without one your computer would not function. Think of the OS as the data and device manager that ensures data flows to and from each device and application. When you touch the screen, click the mouse, or type words on a keyboard, the OS recognizes the input and sends the data to the application or device that needs it. If a piece of hardware is not working, the OS senses the problem and displays a message to you. For example, if a printer is not turned on, the OS communicates that the printer is offline. When you power on a computer or mobile device, the OS loads automatically into memory and displays the user interface when the computer is ready.

Computers and mobile devices have an OS preloaded and ready to use. Some tasks require you to interact with the OS directly, such as when you launch an app, program, or system setting window, switch windows, and manage your files and folders. In this chapter, you will learn to navigate in Windows 10 and manage files on a storage medium.

Note: You will need a removable storage medium (USB flash drive) with enough space to download a copy of the student data files for this textbook.

Learning Objectives

1.1 Navigate Windows 10 using touch, a mouse, or a keyboard

1.2 Sign in to Windows 10, launch an app, and switch between apps

1.3 Use the search text box and minimize, maximize, close, and snap windows

1.4 Use icons in the notification area to view the status of a setting and view customization options for the notification area

1.5 Lock the screen, sign out, and shut down Windows 10

1.6 Use and customize the Start menu

1.7 Personalize the Lock screen, desktop background, and color scheme

1.8 Describe file, folder, download, and extract student data files

1.9 Browse files with File Explorer

1.10 Create new folders and copy files and folders

1.11 Move, rename, and delete files and folders and safely eject a USB flash drive

1.1 Using Touch, Mouse, and Keyboard Input to Navigate Windows 10

Skills

Describe Windows 10 touch gestures

Describe basic mouse actions

List common keyboard commands for Windows 10

Windows 10, released in July 2015, is the latest edition of the Microsoft OS for personal computers (PCs) and mobile devices. The Windows 10 **user interface (UI)** incorporates the familiar desktop and Start menu from Windows 7, with the tiles and enhanced touchscreen functionality from Windows 8.1. The UI uses icons, a Start menu, and other means to interact with users. The **Start menu** is a pop-up menu that displays when you click the **Start button** located at the bottom left corner. In the left pane of the Start menu are links to your most-used applications; options to launch File Explorer, Settings, the All apps menu; and a Power button. The right pane contains tiles. A **tile** is a square or rectangle that, when clicked, starts an app, application, or other Windows feature. An **app** is a smaller program used on a smartphone, tablet, or PC that is designed usually for one purpose, to work on a touchscreen, to use less power than an application, and to use fewer network resources. An **application**, which is more powerful than an app, is a program used to get a task done with a wide range of functions and features. Applications make use of a full-size keyboard and a mouse, and they are designed for a larger screen and more powerful computing resources.

Windows 10 works with touchscreen input, mouse input, and keyboard input devices. Becoming familiar with input actions using touch, the mouse, and the keyboard will help you navigate the UI.

Using Touch to Navigate Windows 10

App Tip

You can adjust settings for some touch actions in the Pen and Touch dialog box accessed from the Control Panel.

On a touchscreen, you perform actions using gestures. A **gesture** is an action or motion you perform with one or two fingers, a thumb, or a stylus. Table 1.1 (page 5) explains the gestures used with the Windows interface.

On a touchscreen, when a task requires typed characters, such as an email address, message text, or web address, the **touch keyboard** shown in Figure 1.1 is used. Tapping the area that requires typed input

In Windows, touch actions are called *gestures*.

generally causes the touch keyboard to appear. The touch keyboard is also available in thumb keyboard mode (Figure 1.2, page 5) and in handwriting mode (Figure 1.3, page 6). The touch keyboard changes some buttons depending on the application in use. For example, in the Mail app the @ symbol is located to the right of the spacebar, while in Microsoft Edge the slash (/) displays in that keyboard location.

Figure 1.1

The full touch keyboard in Windows appears when typed characters are expected.

Table 1.1

Windows Touch Gestures

Gesture	Description and Mouse Equivalent	What It Does	What It Looks Like
Tap or Double-tap	One finger touches the screen and immediately lifts off the screen once or twice in succession. **Mouse:** Click or double-click left mouse button.	• Launches an app or application • Follows a link • Performs a command or selection from a button or icon	
Press and hold	One finger touches the screen and stays in place for a few seconds. **Mouse:** Point or right-click.	• Shows a context menu • Shows pop-up information or details	
Slide	Move one or two fingers in the same direction. **Mouse:** Drag (may need to drag or scroll using scroll bars).	• Used to drag, pan, or scroll through lists or pages	
Swipe	Move one finger in from the left or right a short distance. **Mouse:** Click the Notification icon on the taskbar. **Mouse:** Click the Task View button on the taskbar.	• In from the right reveals the Action Center panel with Notifications and quick action tiles at the bottom • In from the left shows thumbnails of open windows so you can switch to another window	
Zoom Out	Two fingers touch the screen apart from each other and move closer together. **Mouse:** Ctrl + scroll mouse wheel toward you.	• Shrinks the size of text, an item, or tiles on the screen	
Zoom In	Two fingers touch the screen together and move farther apart. **Mouse:** Ctrl + mouse scroll wheel away from you.	• Expands the size of text, an item, or tiles on the screen	

Figure 1.2

The Windows touch keyboard in thumb mode has the keys split on either side of the screen.

Figure 1.3

The Windows touch keyboard in handwriting mode is shown.

For some purposes, you still use a mouse and traditional keyboard with a touch-enabled device. For example, typing an essay for a school project is easier using a traditional keyboard, and doing precise graphics editing is easier with a mouse. Connect a universal serial bus (USB) or wireless mouse and/or keyboard for intensive work like this.

Using a Mouse to Navigate Windows 10

For traditional desktops, laptops, or notebook PCs, navigating the Windows user interface requires the use of a **mouse**, **trackball**, **touchpad**, or other pointing device. These devices are used to move and manipulate a **pointer** (displayed as) on the screen; however, the white arrow pointer can change appearance depending on the action being performed.

To operate the mouse, move the device up, down, right, or left on the desk surface to move the pointer on the screen in the corresponding direction. A scroll wheel on the top of the mouse can be used to scroll up or down. Newer mice include the ability to push on the left or right side of the scroll wheel to scroll right or left. Left and right buttons on the top of the mouse are used to perform actions when the pointer is resting on an item. Table 1.2 (page 7) provides a list and description of mouse actions.

To use a touchpad, move your finger across the surface of the touchpad in the direction required. Tap the touchpad or a button below it to perform an action.

A mouse, trackball, or touchpad can be used as a pointing device to navigate Windows.

Using a Keyboard to Navigate Windows 10

Most actions in Windows are performed using touch gestures or a mouse; however, some people like to use a **keyboard command** (also called a *keyboard shortcut*) because it is fast and easy to use. The Windows logo key is positioned at the bottom left of a keyboard between the Ctrl or Function key (labeled Fn) and the Alt key. Many keyboard shortcuts use the Windows logo key. For example, press the Windows logo key at any time to bring up the Start menu. Useful keyboard commands are described in Table 1.3 (page 7).

If you prefer using keyboard commands, search the web for articles that give other Windows 10 keyboard navigational commands.

Press the Windows logo key with a letter to perform an action.

Note: Instructions in this textbook are written with mouse actions. If necessary, check with your instructor for the equivalent touchpad or other pointing device action. If you prefer to use touch gestures, refer to the gestures with mouse equivalent actions provided in Table 1.1 on page 5.

Table 1.2

Mouse Movements and Actions

Term or Action	Description
Point	Move the mouse in the direction required to rest the white arrow pointer on a button, icon, option, tab, link, or other screen item.
Click	Quickly tap the left mouse button once while the pointer is resting on a button, icon, option, tab, link, or other screen item.
Double-click	Quickly tap the left mouse button twice. On the desktop, a program is launched by double-clicking the program's icon.
Right-click	Quickly tap the right mouse button. Within a software application such as Word or Excel, a right-click causes a shortcut menu to appear. Shortcut menus in software applications are context-sensitive, meaning that the menu that appears varies depending on the item the pointer is resting upon when the right-click occurs.
Drag	Hold down the left mouse button, move the mouse up, down, left, or right, and then release the mouse button. Dragging is an action often used to move or resize an object.
Scroll	Use the scroll wheel on the mouse to scroll in a window. If the pointing device you are using does not include a scroll wheel, click the scroll arrows on a horizontal or vertical scroll bar at the right or bottom of a window, or drag the scroll box in the middle of the scroll bar up, down, left, or right.

Table 1.3

Keyboard Commands or Shortcuts

Keyboard Shortcut	What It Does
Windows logo key	Displays Start menu
Windows logo key + a	Opens the Action Center panel at the right side of the screen
Windows logo key + d	Returns to the desktop from an app or application window
Windows logo key + e	Opens a File Explorer window
Windows logo key + l	Locks the screen
Windows logo key + q	Starts a search for apps, files, settings, and web links
Windows logo key + Tab	Displays Task View to switch to another window
Alt + F4	Closes app, application, or other active window
Alt + Tab	Switches between open apps
Up, Down, Left, or Right Arrow keys	Moves selection on Start menu to an app or application name, setting, folder name, or tile; pressing the Enter key launches the selection

App Tip

To use a keyboard command, hold down the Windows logo key or Alt, press and release the letter or function key, and then release the Windows logo key or Alt.

Beyond Basics   Windows 10 is Software as a Service

With the release of Windows 10, Microsoft adopted a new model called *Windows as a service*. This means that new features and technologies will be released periodically as updates rather than Microsoft accumulating new features and issuing a new version every few years. The default setting is for automatic updates to download and install as they become available. Expect that the Windows 10 you see in this textbook may vary slightly as quarterly or semiannual updates will become routine.

1.2 Starting Windows 10 and Exploring Apps

Skills

Start and sign in to
Windows 10

Launch an app

Switch apps

If you are turning on your PC from a no power state, the **Lock screen** shown in Figure 1.4 appears. The Lock screen also appears if you resume computer use after the system has gone into sleep mode. Depending on your PC or mobile device, turning on or resuming system use from sleep mode involves pressing the Power button or moving a mouse.

Each person who uses a PC or mobile device will have his or her own **user account**. A user account includes a user name and a password or personal identification number (PIN). Windows stores program and settings information for each user's account so that each person can customize options without conflicting with the settings for other people who use the computer. In Windows, Microsoft offers two types of user accounts at sign-in: a Microsoft account or a local account.

Signing In with a Microsoft Account

A user account set up as a **Microsoft account** means you sign in to Windows using an email address from hotmail.com, live.com, or outlook.com (referred to as a Windows Live ID). A Microsoft account means you can download new apps from the Windows store, see live updates from messaging and social media services in tiles on the Start menu, and sync your Windows and browser settings online so that they are the same across all devices.

Signing In with a Local Account

A user account set up as a **local account** means Windows and browser settings on a PC or mobile device cannot be shared with other devices. Automatic connections to messaging and social media services also do not work with a local account.

Note: The screens shown throughout this textbook show Windows signed in with a Microsoft account.

1. Turn on the computer or mobile device, or resume system use from sleep mode.

2. At the Lock screen shown in Figure 1.4, click anywhere on the screen, or press any key on the keyboard to reveal the sign-in screen.

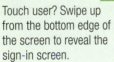

App Tip

Signing in with a Microsoft account is considered connecting to the cloud. Your apps, games, music, photos, files, and settings are stored online and can be used or viewed from any other Windows device.

App Tip

Touch user? Swipe up from the bottom edge of the screen to reveal the sign-in screen.

You can customize the image that appears on the Lock screen. Your lock screen may show a different picture.

Icons that show power and network connectivity status and notifications for Mail and Calendar (if any are available) appear on the lock screen.

Your date and time will vary.

10:29
Friday, July 10

Figure 1.4
The Windows 10 lock screen.

③ Depending on the system, the next step will vary. Complete the sign in by following the steps in 3a, 3b, or 3c that match the configuration of your device:

   a. Type your password in the *Password* text box below your Microsoft account name and email address and press Enter or click the Submit button (right-pointing arrow).

   b. Type your PIN in the text box below your Microsoft account name and email address, or perform the touch gestures over your account's picture password.

A picture password uses a picture from your Pictures library that you perform three touch gestures over. A PIN code lets you sign in faster using a numeric code.

Step 3b

For an account with multiple sign-in options, click this link to show the sign-in icons if you want to select a different sign-in method.

   c. If the active account is not your account, click the account name, picture, or silhouette icon at the bottom left of the sign-in screen, type your password in the *Password* text box, and then press Enter or click the Submit button (right-pointing arrow).

The Desktop

Once signed in, the desktop, similar to the one shown in Figure 1.5, appears. The **desktop** displays icons that are used to launch programs or open other windows and

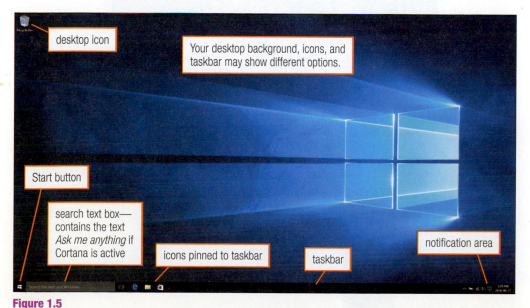

desktop icon

Your desktop background, icons, and taskbar may show different options.

Start button

search text box—contains the text *Ask me anything* if Cortana is active

icons pinned to taskbar

taskbar

notification area

Figure 1.5
The Windows 10 desktop displays icons and a taskbar at the bottom.

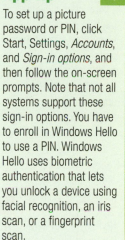

App Tip

To set up a picture password or PIN, click Start, Settings, *Accounts*, and *Sign-in options*, and then follow the on-screen prompts. Note that not all systems support these sign-in options. You have to enroll in Windows Hello to use a PIN. Windows Hello uses biometric authentication that lets you unlock a device using facial recognition, an iris scan, or a fingerprint scan.

App Tip

Signing in to Windows 10 on a tablet or other touchscreen device converts the display to *tablet mode*, in which the Start screen displays tiles for common apps, a menu button, a power button, and a back button to navigate the UI.

a taskbar along the bottom of the screen. The **taskbar** has the Start button, the search text box, icons to start programs, and the notification area. The **search text box** is used to locate apps, programs, settings, files on your PC, or find information on the web. The **notification area** at the right end of the taskbar shows the current date and time, and icons to view or change the status of system settings. When you sign in, Windows opens the Start menu on the desktop for you.

Launching an App

Built-in apps for Windows 10 include Photos, Maps, Mail, Calendar, Groove, and Movies & TV, to name a few. The Windows 10 family of apps is designed to look and operate consistently on all devices. To launch an app, click the Start button and then click the tile for the app in the right pane.

Note: If you are using a tablet, turn off tablet mode by swiping in from the right edge and then tapping the Tablet mode quick action tile.

4 If necessary, click the Start button. Click the Photos tile in the right pane on the Start menu to launch the Photos app.

The **Photos app** shows thumbnails of the photos stored in the Pictures library on a PC, tablet, smartphone, and in OneDrive, grouped by date. Photos are set to auto-enhance and can be easily shared with others from the app.

Step 4

Photos

Your Photos tile may be larger and will vary if the tile is live.

Photos on your screen will be grouped by date and will vary from this example.

5 Click the Start button, and then click the Calendar tile in the right pane. If this is the first time you have started the app, click the Get started button, and then the Ready to go button to set up the app with your Microsoft account.

The **Calendar app** shows appointments and reminders for events stored in your Microsoft account. The calendar for the current month loads. Birthday reminders for those accounts connected to Facebook are shown in the calendar.

Step 5

Calendar

Your Calendar tile may be larger and will vary if the tile is live.

In Windows 10 apps, the Menu button shows the app commands, settings, and views.

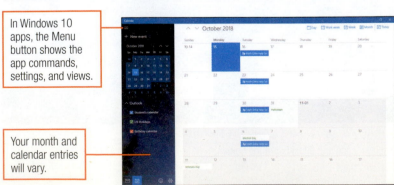

Your month and calendar entries will vary.

6   Click the Store icon on the taskbar.

In the **Store app**, you can search for and download new apps for your PC or mobile device.

Step 6

Switching between Apps

You can switch between apps by clicking the button on the taskbar for the app, by viewing open apps in Task View, or by using the keyboard shortcut Alt + Tab. Each app, program, or other window that is opened has a button on the taskbar. Pointing to a button on the taskbar causes a thumbnail to appear above the button with a preview of the window and its active contents. The preview is helpful when you have multiple documents open in the same program to select the correct window.

7   Click the Photos button on the taskbar.

The Photos app moves to the foreground.

8   Click the Task View icon on the taskbar.

Step 7

Step 8

9   Click the Calendar thumbnail.

Task View is a new view in Windows 10 that shows all open apps in the middle of the screen with larger thumbnails to easily see the active contents in each window. Click the thumbnail for the desired app to switch to another window.

Step 9

Alternative Method | **Switching Apps Using the Keyboard**

You can switch to another app by holding down the Alt key and pressing the Tab key. This brings up thumbnails for open apps in the middle of the screen. Continue to hold down the Alt key while pressing and releasing Tab until the desired app is selected (outlined with a white box). Release Alt to switch to the highlighted window.

Beyond Basics | **Other Windows 10 Apps to Explore**

In this topic, you briefly viewed the Photos app, Calendar app, and Windows Store. These are just three of several universal apps installed with Windows 10. Other useful apps worth exploring include Alarms & Clock, Calculator, Groove Music (requires an XBox profile to run), Maps, Money, Movies & TV, News, People, Sports, Voice Recorder, and Weather. Click All apps in the left pane of the Start menu to view the full list.

1.3 Using the Search Text Box and Working with Windows

Skills

Use the search text box to start an app

Minimize, maximize, restore down, close, and snap windows

The search text box is a quick way to start an app, launch an application, locate a file or setting, find information from the web, or find information on how to use a Windows feature. Click in the search text box and then start typing the first few letters of the app or program name, file name, or other topic. The search text box is the fastest way to locate help on anything in Windows 10. The search text box contains the dimmed text *Ask me anything* or *Search the web and Windows* depending on whether Cortana is active. Cortana is the new personal assistant included with Windows 10.

Oops! !

No Search text box on taskbar? Right-click the taskbar, point to *Search*, and then click *Show search box*. If the *Show search box* option does not appear, then the taskbar is set to show small buttons only. In that case, right-click the taskbar, click *Properties*, then click the *Use small taskbar buttons* check box to remove the check mark. Click OK.

1 Click in the search text box and then type al.

Watch the entries appear in the results list above the search text box as you type. Each character you add in the search text box further refines the list. Notice that options appear in the *Settings* section and web links appear in the *Web* section of the results list.

2 Click *Alarms & Clock* in the *Best match* section of the results list.

You should now have four windows overlapping each other on your desktop.

Each app, application, setting, or other feature opens in a window, which allows you to view and work on multiple tasks at the same time. Each window contains standard features for moving, resizing, and closing a window so you can arrange items on your desktop to suit your work preferences. Figure 1.6 identifies the standard window features.

Oops! !

Don't have four windows open? If you closed one or all three of the windows that were opened in the previous topic, return to Topic 1.2 and reopen them.

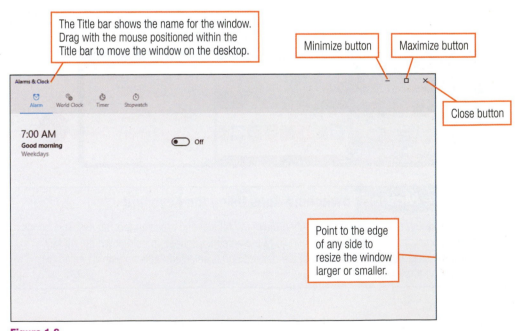

The Title bar shows the name for the window. Drag with the mouse positioned within the Title bar to move the window on the desktop.

Minimize button   Maximize button

Close button

Point to the edge of any side to resize the window larger or smaller.

App Tip

Cortana is the personal assistant included in Windows 10 that operates from the search text box on the taskbar. Cortana responds to text typed in the search text box or to voice commands if you use the microphone icon. Cortana personalizes content the more you use the feature. Some things Cortana will do include helping you find a file on your PC, find a restaurant on the web, track your flight status, and give you a reminder from your calendar.

Figure 1.6
Shown above, is the Alarms & Clock window displaying standard features.

3 Click the Minimize button located at the top right of the Alarms & Clock window.

The **Minimize** button is used to reduce the window to a button on the taskbar.

④ With the Calendar window now active, click the Close button to close the Calendar app.

Step 4

The **Close** button is used to close the window. In a window that contains an open document, the document is also closed, and you are prompted to save changes before closing if changes have been made since the last save operation.

⑤ With the Photos window now active, click the Maximize button.

Step 5

Clicking the **Maximize** button expands the window size to fill the entire desktop, and the Maximize button changes to the Restore Down button. **Restore Down** returns the window to its previous size after a window has been maximized.

⑥ Click the Restore Down button on the Photos window title bar.

Step 6

⑦ Position the mouse pointer on the Photos window Title bar and then drag the window up, down, left, or right a few inches to practice moving the window on the desktop.

Photos   — ☐ ✕

≡   COLLECTION   ⟳ ⁙ ⎘ …

Step 7

⑧ Click the Alarms & Clock button on the taskbar to restore the window to the desktop.

Step 8

The **Snap** feature in Windows lets you dock a window to one side of the screen without moving and then resizing the window manually. To snap a window to the left or right side of the screen, drag the title bar to the desired edge, then release the mouse when you see the window outline fill the right or left half. In Windows 10, you can snap a window to one of four quadrants by dragging it to a corner. The **Snap Assist** feature pops up when a window is snapped, leaving part of the screen empty and other windows open. In the empty portion of the screen, Snap Assist shows thumbnails of the remaining windows. Click the thumbnail for the window you want to fill the remainder of the screen.

⑨ Drag the Alarms & Clock Title bar to the left edge of the screen until the outline of a window in the left half of the screen appears, and then release the mouse.

The Alarms & Clock window now fills the left half of the desktop, and Snap Assist shows you thumbnails for the Photos and Store windows in the empty space on the right.

⑩ Click the Photos window thumbnail.

The screen is now split in half, with the Alarms & Clock window in the left half and the Photos window in the right half.

⑪ Close the Photos, Alarms & Clock, and Store windows.

Step 10

Alarms & Clock window is snapped to the left half of screen at Step 9.

Quick Steps

Search for an App, Application, Setting, File, or Web Link
1. Click in the search text box.
2. Type first few letters of app, application, setting, or topic.
3. Click desired item in results list.

App Tip

You can also maximize a window by double-clicking the Title bar or by dragging the window's title bar to the top edge of the screen until the window maximizes.

1.4 Using the Notification Area

The notification area at the right end of the taskbar shows the current date and time, the Notification icon that opens the Action Center panel, the Speaker icon used to adjust the volume, and the Network icon, which shows network connectivity. A Power icon appears next to the network icon if the device uses a battery. Additional icons may appear depending on the applications installed on your PC or mobile device. You use the notification area to view or manage the status of a setting, or other feature.

1 Click the current date and time in the notification area at the right end of the taskbar.

A calendar appears above the date and time displaying the current month. Change the month using the up and down pointing arrows, or change the system date by clicking <u>Date and time settings</u> to open the Settings app.

2 Click in an unused area on the desktop to remove the pop-up calendar.

3 Click the Speaker icon in the notification area on the taskbar.

The volume control slider opens. Take note of the current value for the volume level.

4 Drag the slider right to adjust the volume higher.

A chime sounds when you release the mouse at the new volume level. If you are in a computer lab classroom without speakers and do not have headphones plugged in, you will not hear the chime.

5 Drag the slider left to restore the volume control to the original level.

6 Click in an unused area on the desktop to remove the volume control slider.

7 Click the Network icon in the notification area on the taskbar.

A list of Wi-Fi networks within range are shown with the active network shown as *Connected*. At the bottom of the panel in the *Network settings* section are quick action tiles. A **quick action tile** is a link to a setting. Clicking the quick action tile for a wireless network you are connected to turns the Wi-Fi connection off. Click the Airplane mode quick action tile to turn on the airplane mode setting, which turns off all wireless communications for your device. You use the Network panel to manage your network connectivity by disconnecting from your current Wi-Fi network or by connecting to another one. To connect to another Wi-Fi network, click the network name and then follow the prompts that appear.

8 Click in an unused area on the desktop to remove the Network panel.

9 Click the Power icon in the notification area. Skip to Step 11 if you do not see a Power icon similar to the one shown because you are using a PC that does not use a battery.

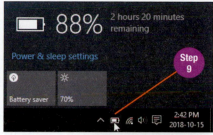

Quick Steps

View the Status of or Change a System Setting Using the Notification Area
Point at or click the icon in the notification area at the right end of the taskbar.

The Power & sleep settings link opens the Settings app with the Power & sleep panel active, where you can adjust the time periods when the screen turns off or sleep mode is activated. The Battery saver quick action tile turns on the feature, which limits background activity to conserve the battery. Cycle through various screen brightness settings using the Brightness quick action tile. The brightness setting affects the battery life.

10 Click in an unused area on the desktop to remove the pop-up Power panel.

11 Click the Notifications icon to open the Action Center panel.

In the top section of the **Action Center** panel, notifications appear from settings, apps, applications, and your connected accounts. For example, you will see messages regarding installed updates, new mail, or a birthday reminder from the Calendar app. At the bottom of the panel are rows of quick action tiles. The tiles shown vary depending on the type of device in use.

12 Click the All settings quick action tile in the Action Center panel to start the Settings app.

13 Click System (first option) in the Settings app window.

14 Click Notifications & actions in the left pane of System Settings.

15 Scroll down and review the options for customizing Notifications.

Drag to slide a switch from On to Off or vice versa to change any of the Notifications options.

16 Click the Select which icons appear on the taskbar link and then review the options.

17 Click the Back button and then click the Turn system icons on or off link.

18 Review the system icon options that can be turned on or off and then close the Settings app window.

Your panel will vary with notifications specific to your device and account.

App Tip

On a touchscreen, swipe in from the right edge of the screen to display the Action Center panel.

App Tip

The Notifications icon is white when new notifications are available.

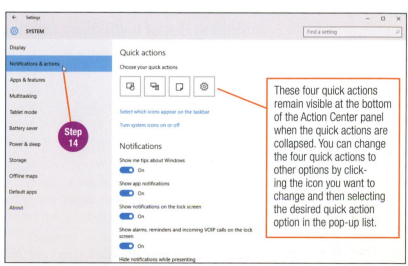

These four quick actions remain visible at the bottom of the Action Center panel when the quick actions are collapsed. You can change the four quick actions to other options by clicking the icon you want to change and then selecting the desired quick action option in the pop-up list.

1.5 Locking the Screen, Signing Out, and Shutting Down Windows 10

Skills

Lock the Screen

Sign out of Windows 10

Shut down Windows 10

If you need to leave your PC or mobile device for a short period of time and do not want to close all your apps and documents, you can lock the device. Locking the system causes the lock screen image to appear full screen so someone else cannot see your work. All your apps, applications, and documents are left open in the background, and you can resume work right away once you unlock the device with your password, PIN, or picture password.

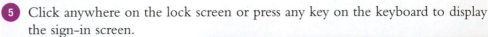

1. Click the Start button, type *photos* in the search text box and press Enter, or click the Photos tile to launch the Photos app.

2. Click the Start button.

3. Click your account name at the top of the left pane on the Start menu.

4. Click *Lock* at the drop-down list.

The Windows lock screen appears with the current date and time and notifications, if any are available.

5. Click anywhere on the lock screen or press any key on the keyboard to display the sign-in screen.

6. Type your password in the *Password* text box and press Enter, type your PIN, or perform the touch gestures on your picture password.

Your lock screen at Step 5 may vary.

Your date and time will vary.

Notice when you are back at the desktop that the Photos app remained open while the device was locked.

Signing Out of a Windows Session

When you are finished with a Windows session, you should **sign out** of the PC or mobile device. Signing out is also referred to as *logging off*. Signing out closes all apps, applications, and files. If a computer or mobile device is shared with other people, signing out is expected so that other users can sign in to their accounts. Signing out and locking also provide security for your device because someone would need to know your password, PIN, or touch gestures to access programs or files.

7. Click the Start button.

8. Click your account name at the top of the left pane on the Start menu, and then click *Sign out* at the drop-down list.

The Windows lock screen appears; however, this time Windows closes the Photos app automatically.

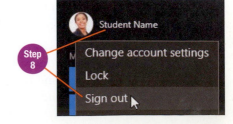

9 Display the sign-in screen, type your password and press Enter, type your PIN, or perform the touch gestures on your picture password.

Notice the Photos app is not on the desktop because it was closed when you signed out of the previous Windows session.

Shutting Down the PC or Mobile Device

If you want to turn off the power to your computer or mobile device, perform a **shut down**. Shutting down the system ensures that all Windows files are properly closed. The power will turn off automatically when shut down is complete.

10 If necessary, click the Start button.

11 Click Power near the bottom of the left pane on the Start menu.

Note: Check with your instructor before proceeding to Step 12 because some schools do not allow students to shut down computers. If necessary, click in an unused area on the desktop to remove the Start menu and proceed to the next topic.

12 Click *Shut down* at the pop-up list.

If someone else is signed in on the device you are shutting down, a message will display informing you the other user could lose unsaved work if you proceed. Click *Shut down anyway* to proceed with turning off the device, or click in an unused area on the desktop to remove the message and cancel the shutdown operation.

The system will perform the shutdown operation, and in a few seconds the power will turn off. In some instances, system updates will be installed during a shutdown operation.

13 Wait a moment or two and then press the Power button to turn the PC or mobile device back on.

14 When the lock screen appears, display the sign-in screen and then sign back in to Windows.

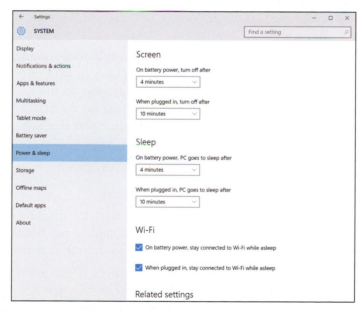

Figure 1.7

The Power & sleep panel is shown in the Settings app.

Beyond Basics   **Other Power Options**

The *Sleep* option from the Power pop-up list is used when you are away from the computer and want to leave all programs and documents open but use less power. A system goes into sleep mode after a period of time elapses with no activity. See Figure 1.7 to view the Power & sleep panel in the Settings app where you can change the options. Choose *Restart* to shut down Windows and immediately start it up again without turning off the power. A restart is used when a system is not performing correctly or is otherwise not responding.

1.6 Using and Customizing the Start Menu

Skills

Navigate the Start menu

Resize the Start menu

Pin/unpin a tile to/from the Start menu

Move a tile

Resize a tile

Turn on and off live updates

The Start menu is divided into two panes. The left pane contains links to the apps or applications most used or recently added at the top of the pane, and File Explorer, Settings, Power, and All apps are at the bottom of the pane. The right pane, as you learned in Topic 1.2, contains tiles to apps, applications, or other features. The Start menu can be customized to suit your preferences by adding or removing tiles, rearranging tiles, resizing a tile, and by turning a live tile on or off to enable or stop notifications or other content from appearing on the tile.

Note: In some school computer labs, the ability to change Windows settings is disabled. If necessary, complete this topic and the next topic on your PC or mobile device at home.

1. Double-click the Recycle Bin icon on the desktop to open the Recycle Bin window.

2. Click the Start button, and then click Settings to open the Settings app.

3. Click the Start button, click All apps, scroll down the apps list in the left pane, and then click Money to launch the Money app.

4. Click the Start button, click All apps, then click Back to return the left pane to the previous list.

5. Click in the search text box, type calc, and then press Enter to launch the Calculator app.

6. Close the Calculator, Money, Settings, and Recycle Bin windows.

Whether you choose to start an app, application, setting, or open another window by double-clicking an icon on the desktop, clicking an option in the left pane on the Start menu, clicking a tile in the right pane on the Start menu, or by typing a name using the search text box is a personal preference. Over time, you will develop your own style for using the Start menu and desktop.

Resizing the Start Menu

You can make the Start menu wider or taller by dragging the top or right border of the Start menu upward or to the right. Drag the top border downward or the right border to the left to make the Start menu shorter or narrower.

7. Click the Start button, point at the right border of the Start menu, and then drag to the right when the pointer changes to a left- and right-pointing arrow to make the menu wider.

8. With the Start menu still open, point at the top border and then drag upward when the pointer changes to an up- and down-pointing arrow to make the menu taller.

App Tip

Right-click the Start button to open the Quick Access menu, which provides access to advanced system tools, such as Device Manager, Computer Management, Task Manager, and Control Panel to name a few.

Pinning and Unpinning Tiles to and from the Start Menu

Right-click a tile to reveal the shortcut menu with the options Unpin from Start, Resize, or More, which is used to access the options Turn live tile on or Turn live tile off (depending on its current state), and Pin to taskbar. Click **Unpin from Start** to remove the tile from the Start menu. A tile can be added to the Start menu by right-clicking the app or application name in the All apps list on the Start menu, and then clicking **Pin to Start**. The shortcut menu offers more or fewer choices depending on the app or application name that you right-click on the Start menu.

9 With the Start menu open, right-click the Photos tile and then click *Unpin from Start* at the shortcut menu.

The tile is removed, and the Start menu remains open.

10 Click All apps in the left pane on the Start menu, scroll down the All apps menu, and then right-click Photos.

11 Click *Pin to Start* at the shortcut menu.

The Photos tile is added back to the Start menu, and the Start menu remains open. Notice that the tile is not added back to the same location from which it originated.

Rearranging Tiles

Tiles can be moved to a new location on the Start menu by dragging the tile to the desired location.

12 Drag the Photos tile back to its original location in the right pane on the Start menu.

As you drag a tile around on the Start menu, the tiles are dimmed, and you will notice other tiles shifting around to make room for the tile.

Resizing a Tile and Turning Live Updates Off or On

Tiles are either square or rectangular in shape and can be sized Small, Medium, Wide, or Large using the Resize option on the shortcut menu. Some tiles do not offer all size options. Some tiles display live updates, with notifications, headlines, or pictures displayed on the tile. Change the live update status using the **Turn live tile off** or **Turn live tile on** option at the shortcut menu.

13 With the Start menu open, right-click the Calendar tile, point to *Resize*, and then click *Large*.

14 Right-click the Calendar tile, point to *Resize*, and then click *Medium*. If necessary, drag the Mail or other tile back upward to fill in the space next to the Calendar tile.

15 Right-click the Mail tile, point to *More*, and then click *Turn live tile off*.

16 Right-click the Mail tile, point to *More*, and then click *Turn live tile on*.

17 Resize the Start menu back to its original height and width if desired.

18 If necessary, rearrange tiles or make other changes as needed to restore the Start menu to its original settings and then click in an unused area of the desktop to close the Start menu.

Quick Steps

Remove a Tile from the Start Menu
1. Right-click tile.
2. Click *Unpin from Start*.

Add a Tile to the Start Menu
1. Click Start button and then click All apps.
2. Scroll down All apps list to locate desired app or application name.
3. Right-click app or application name.
4. Click *Pin to Start*.

Rearrange Tiles
Drag tile to desired location.

Resize a Tile
1. Right-click tile.
2. Point to *Resize*.
3. Click desired size.

Turn Off/On Live Updates
1. Right-click tile, then point to *More*.
2. Click *Turn live tile off* or *Turn live tile on*.

1.7 Personalizing the Lock Screen and the Desktop

Skills

Change the Lock screen
background image

Add an app to the Lock
screen notifications

Change the desktop
background image

Change the color scheme

Most people like to put a personal stamp on their PC or mobile device. In Windows 10, you can personalize the Lock screen by changing the picture that displays and the apps that provide notifications when the screen locks. The desktop background can be personalized to a different image and color scheme. A theme changes the background image, color, and sound scheme in one step. The Start menu by default shows the most used and recently added apps or applications in the left pane. You can turn one or both of these categories off.

Open the Settings app and choose *Personalization* to make changes to the lock screen, desktop background, colors, themes, and Start menu.

1 Click the Start button and then click Settings.

2 Click *Personalization* in the Settings app window.

3 Click Lock screen in the left pane of the Settings app window.

The current Lock screen image is shown above five thumbnails for other background pictures in the right pane, as shown in Figure 1.8. Click a different picture to preview the image or click the Browse button to choose a picture on your PC or mobile device to use as the Lock screen image. You can also opt to show a slideshow for the lock screen by changing the Background option from Picture to Slideshow.

Figure 1.8

The Lock screen pane is shown in the Settings app.

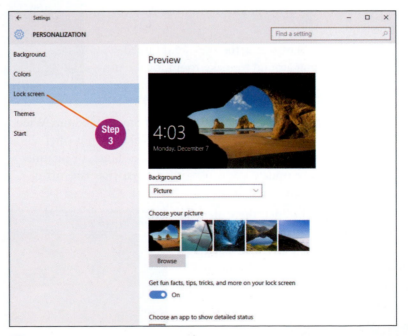

Oops! !

No background thumbnails? The Background option below the Preview is at a different setting (such as *Windows spotlight*). Click the *Background* option box arrow and then click *Picture*.

4 Click one of the five background thumbnails below the current Lock screen image. You determine the background image you want to view.

The app buttons below the background thumbnails control which apps give notifications on the lock screen image.

5 If necessary, scroll down to view the Lock screen app buttons. Click the first button with a plus symbol in the *Choose apps to show quick status* section and then click *Weather* at the pop-up list.

The Weather app will now provide notifications when the screen is locked.

6 Click Background in the left pane of the Settings app window.

7 Click one of the five background thumbnails in the *Choose your picture* section. You determine which picture you want for your background. If the *Choose your picture* thumbnails are not visible, click the Background option box arrow and then click *Picture*.

8 Click Colors in the left pane.

9 Click a color square in the *Choose your accent color* section and then close the Settings app window. You determine the accent color you want to use for tiles and window accents.

The new background image is shown on the desktop.

10 Click the Start button and look at the new accent color used on tiles.

11 Lock the screen and notice the new picture on your lock screen.

12 Sign back in to unlock the system.

13 If necessary, click the Start button. Click Settings, then repeat Step 2 to Step 9 restoring the lock screen, background, and color scheme to their original settings. At Step 5, click *None* in the Weather button pop-up list to remove the app on the lock screen.

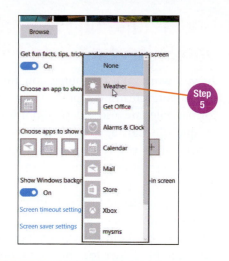

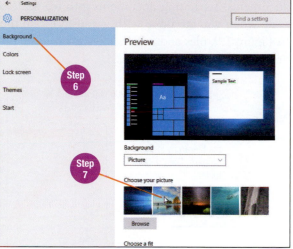

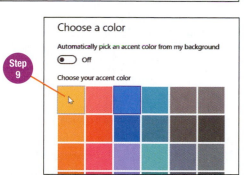

Beyond Basics   **Adding a Picture for Your Account**

To add a picture of yourself to your Microsoft account so that your picture shows in place of the silhouette icon, begin by clicking the Start button, clicking your account name at the top of the left pane, and then clicking *Change account settings*. In the Settings app window with *Your account* active in the left pane, click the Browse button below the silhouette, navigate to the image you want to use, and then double-click the file. You can also use the Camera button to take a new picture of yourself using your PC or mobile device webcam.

1.8 Understanding Files and Folders and Downloading and Extracting Data Files

Skills

Define file, file name, and folder

Download and extract student data files to a USB flash drive

As you work on a computer, you are creating and modifying files. A **file** is a document, spreadsheet, presentation, picture, or any other data that you save in digital format. Files are also videos and music that you play on a device. Each file has a unique **file name**, which is a series of characters you assign to the file when you save it that allows you to identify and retrieve the file later.

A system of organizing files so that you can find and retrieve a specific document or photo when you need it is necessary. Folders are created on a storage device to organize electronic files. A **folder** is a name assigned to a placeholder or container in which you store a group of related files. Think of a folder on the computer in the same way you consider a paper file folder in a desk drawer at your home. You might have one file folder for your household bills and another file folder for your school documents. Separating documents into file folders makes it easy to put a document away when you are done with it and to locate the document when you need it again. In Figure 1.9, files and folders are shown to help you understand how digital data is organized on a storage device.

Figure 1.9

Digital data is stored in files, which are organized into folders.

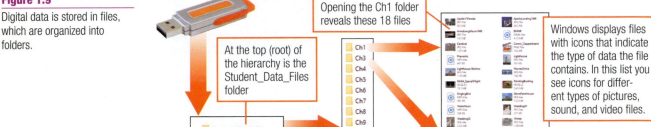

Opening the Ch1 folder reveals these 18 files

At the top (root) of the hierarchy is the Student_Data_Files folder

Windows displays files with icons that indicate the type of data the file contains. In this list you see icons for different types of pictures, sound, and video files.

Opening the Student_Data_Files folder reveals the individual chapter folders Ch1, Ch3, and so on to Ch15

Opening the Ch3 folder reveals these 4 files

In this list the file icons indicate the folder contains a Word document, an Excel worksheet, a picture, and a PowerPoint presentation.

To complete the tasks in the rest of this textbook, a set of student data files must be copied to a USB flash drive or other storage medium. Check with your instructor for his or her preference on the storage medium to use for completing the work in your course. You may be instructed to save files to a network folder at your school or in your OneDrive account. Substitute another storage medium if necessary when the USB flash drive is used in the remaining topics.

The student data files are copied from the accompanying ebook. The files are packaged so that you need to download only one file, called an *archive file*, or a ZIP file. A **ZIP file** assembles and compresses a group of files under one name that looks like a folder name with a .zip file extension. The files are compressed so that the ZIP file size is as small as possible, meaning the file will copy faster over a network. A ZIP file should be uncompressed before the files stored within it are used so that each file is restored to its original size and put back in the folder from which the file originated. To restore the files within a ZIP file to their original size and folder structure, you perform a process called *extracting*.

Note: You will need a USB flash drive to complete the steps in this topic and the next three topics. Check with your instructor for alternative instructions if he or she prefers that you save files somewhere else, such as in a network folder on a school server.

App Tip

You can recognize a ZIP file in a file list because Windows displays the icon as a file folder with a zipper.

1. Insert a USB flash drive into an available USB port on your PC. Skip this step if you are saving files to another storage medium.

2. Go to your ebook for this textbook and navigate to this page in the ebook.

3. Click the Ancillary Links button as shown in the image at the right. The button may appear at the top or along the side of the window depending on the window size.

Step 3

4. Click the Student_Data_Files hyperlink that appears in the Ancillary Links dialog box as shown in the image at the right.

5. Click the Download hyperlink at the top of the OneDrive window.

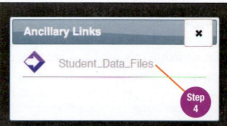

Step 4

The ZIP archive file will start copying to the Downloads folder. The download will take a few moments depending on the speed of your network connection. Watch the progress of the download at the bottom of the browser window.

6. Click the Open button that appears when the message displays that Student_Data_Files.zip finished downloading.

A File Explorer window opens. The Compressed Folder Tools Extract tab appears because the file opened is a ZIP archive file. **File Explorer** is the utility used to browse files and folders on storage devices and perform file management routines. You will learn more about File Explorer in the next three topics.

Step 6

Student_Data_Files.zip finished downloading.   Open   Open folder   View downloads   ×

7. If necessary, click the Compressed Folder Tools Extract tab.

8. Click the Extract all button.

9. Click the Browse button in the Extract Compressed (Zipped) Folders dialog box.

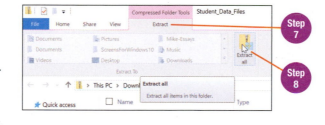

Step 7

Step 8

10. Click your USB flash drive in the left pane of the Select a destination dialog box, and then click the Select Folder button.

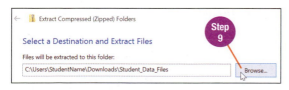

Step 9

Step 10

Your drive letter may vary.

11. Click Extract in the Extract Compressed (Zipped) Folders dialog box.

A progress message displays the status of the operation. When finished a new File Explorer window opens with the Student_Data_Files folder shown.

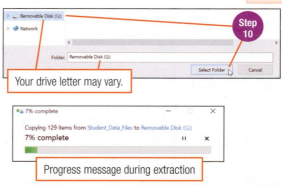

Progress message during extraction

12. Close all the open windows including the OneDrive browser window.

Data File

Student_Data_Files.

Quick Steps

Download and Extract Student Data Files

1. Navigate to this page in ebook.
2. Click Ancillary links button, then Student_Data_Files hyperlink.
3. Click Download hyperlink in OneDrive.
4. Click Open.
5. If necessary, click Compressed Folder Tools Extract tab.
6. Click Extract all button.
7. Click Browse button.
8. Click USB flash drive in left pane, then Select Folder.
9. Click Extract.

Oops!

A window other than File Explorer opens at Step 6? If there is another ZIP archive program installed on your device (such as 7-Zip), your file may open in the archive program window. In that case, close the window that opened. Click the File Explorer icon on the taskbar. Click Downloads in the left pane. Right-click the file Student_Data_Files.zip in the right pane, point to *Open with*, click *Windows Explorer*, and then proceed to Step 7.

1.9 **Browsing Files with File Explorer**

To view or manage the files on a storage device, open a File Explorer window by clicking the File Explorer icon on the taskbar or near the bottom of the left pane on the Start menu. In File Explorer, you perform file maintenance tasks, such as creating new folders to organize files, renaming files, copying files, moving files, and deleting files. To begin, you will learn how to navigate the File Explorer window (see Figure 1.10).

1 Click the File Explorer icon on the taskbar.

The File Explorer window is divided into two panes. The Navigation pane at the left displays the list of places associated with the device. The right pane, called the Content pane, displays the files and folders stored on the selected item. The window opens by default in Windows 10 in a new view called **Quick access**, which shows a list of frequently used and recently used files. Each frequent folders list contains at least four folders: Desktop, Downloads, Documents, and Pictures.

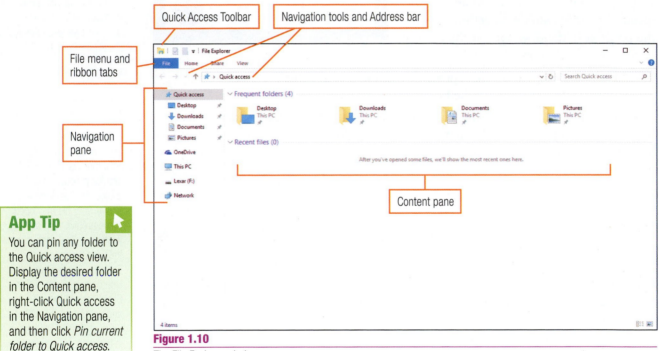

Figure 1.10

The File Explorer window

2 Click This PC in the Navigation pane.

A This PC window displays all the available storage options on the computer.

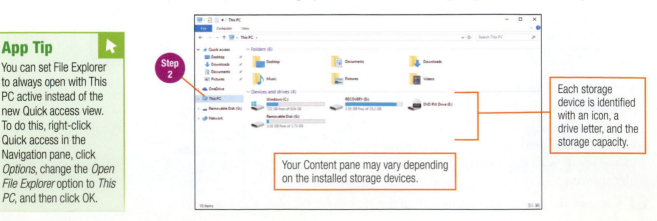

3 If necessary, insert your USB flash drive into an available USB port. If a new window opens when you insert the USB flash drive, close the window.

4 Double-click the icon for the Removable Disk representing your USB flash drive in the *Devices and drives* section to view the contents in the Content pane.

The Content pane shows the name of the folder you extracted from the ZIP file in the previous topic. Your content pane may show additional files if you did not use a blank USB flash drive.

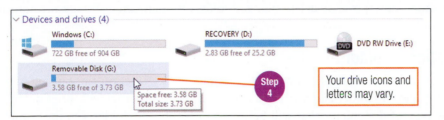

Step 4

Your drive icons and letters may vary.

5 Double-click the folder *Student_Data_Files*.

A list of folders stored within the Student_Data_Files folder appears. A folder within another folder is sometimes referred to as a **subfolder**.

Folder icon

Step 5

6 Double-click the folder *Ch1*.

A list of files stored in the Ch1 subfolder appears in the Content pane.

7 Click the View tab.

8 Click the *Tiles* option in the Layout group.

Use the Layout options to view files in the Content pane as small, medium, large, and extra large icons; in a list format with or without file details, such as the date and time the file was created or last updated and the file size; in a list with the file type and file properties such as author (Content); or as tiles (shown below).

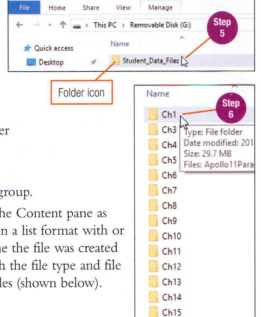

Step 6

Type: File folder
Date modified: 201
Size: 29.7 MB
Files: Apollo11Para

Step 7

Step 8

Your list will vary if this check box has a check mark in it. File extensions add a period and three or four characters after the file name. The extension indicates the type of file. For example, .png, .jpg, and .jpeg all indicate picture files.

9 Click Desktop in the Navigation pane.

Browse content in File Explorer by clicking names in the Navigation pane, or by double-clicking device or folder names in the Content pane.

10 Close the File Explorer window.

Note: Leave the USB flash drive in the device for the next topics.

Quick Steps

Browse Files in File Explorer
1. Click File Explorer icon on taskbar.
2. If necessary, insert a removable storage medium such as a USB flash drive.
3. Click desired location or device in Navigation pane, *or* double-click a folder or device name in Content pane.

App Tip

Double-clicking a file name opens the program associated with the file and automatically opens/plays the document, image, video, or sound.

App Tip

Changing the view to tiles or icons lets you easily locate a file you are looking for when browsing pictures or videos.

1.10 Creating Folders and Copying Files and Folders

Skills

Create a folder and subfolder

Copy files and folders

As you work with software applications, such as Word or Excel, you will create many files that are documents or spreadsheets. You may also download files from a digital camera or a website and receive other files from email or text messages. Storing the files in an organized manner with recognizable names will mean you can easily locate the document or picture later. Creating folders in advance of creating files means having an organizational structure already in place. From time to time, you also need to rename, copy, move, or delete files and folders to maintain a storage medium in good order.

Creating a Folder

Creating a folder on a computer is like placing a sticky label on the outside of an empty paper file folder and writing a title on it. The title provides a brief description of the type of documents that will be stored inside the file folder. On the computer, in File Explorer, click the **New folder** button in the New group on the Home tab, and then type a name for the folder to set up the electronic equivalent of a paper filing system.

1 Click the File Explorer icon on the taskbar.

2 Click This PC in the Navigation pane of the File Explorer window.

3 Double-click the Removable Disk representing your USB flash drive in the *Devices and drives* section.

The number of drives and devices and their corresponding letters may vary.

4 If necessary, click the Home tab to display the ribbon interface.

App Tip

Tabs and/or buttons on the ribbon change and become available or unavailable depending on what is selected in the window.

The **ribbon** provides the buttons you need to perform file management tasks. Buttons are organized into the tabs Home, Share, View, and Manage. Within each tab, buttons are further organized into groups, such as Clipboard, Organize, New, Open, and Select (on the Home tab).

5 Click the New folder button in the New group on the Home tab.

Tabs

Step 4

Step 5

A group of related buttons

App Tip

In the folder structure you are creating, no spaces between words are used. Windows allows the use of spaces; however, a good practice is to avoid spaces in folder names.

6 Type ComputerCourse and then press Enter.

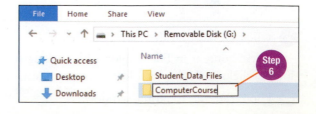

Step 6

7 Click the View tab and then click the *List* option in the Layout group.

List view displays names with small icons representing each file or folder and without details such as the date or time the file or folder was created or modified, the file type, and the file size. Each folder can have a different view.

8 Click the Home tab. If the ribbon is not currently expanded, a pushpin icon displays below the Close button near the top right of the window. Click the pushpin to expand the ribbon; otherwise, proceed to Step 9.

The pushpin expands the ribbon permanently in the window. The pushpin icon changes to an up-pointing arrow (like a caret ^ symbol), which is the Minimize the Ribbon icon.

9 Double-click the *ComputerCourse* folder name in the Content pane.

You have now opened the ComputerCourse folder. File management tasks performed next will occur inside this folder.

10 With the Home tab active, click the New folder button, type ChapterTopicsWork, and then press Enter.

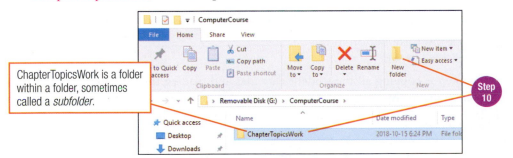

ChapterTopicsWork is a folder within a folder, sometimes called a *subfolder*.

11 Click the New folder button, type Assessments, and then press Enter.

You now have two folders (or subfolders) within the ComputerCourse folder.

12 Click the Up arrow button at the left of the Address bar.

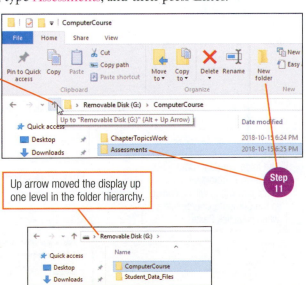

Up arrow moved the display up one level in the folder hierarchy.

Notice you now see only the ComputerCourse folder and the Student_Data_Files folder. The two folders created in Steps 10 and 11 are no longer visible because they are *inside* the ComputerCourse folder.

Quick Steps

Create a Folder
1. Click File Explorer icon on taskbar.
2. Navigate to desired storage medium and/or folder.
3. If necessary, click Home tab.
4. Click New folder button.
5. Type folder name.
6. Press Enter.

Oops!

No pushpin icon below the Close button? Then your ribbon is already pinned in the window. Skip Step 8.

App Tip

The terms *folder* and *subfolder* are used interchangeably when a subfolder exists. Windows does not distinguish a subfolder from a folder; however, some people prefer to use *subfolder* to differentiate a folder within the hierarchy on a disk.

App Tip

The Back button (left-pointing arrow) returns the display to the previous folder list.

Copying Files and Folders

Copying a file or folder from one storage medium to another makes an exact copy of a document, spreadsheet, presentation, picture, video, music file, or other object. Copying is one way of making a backup copy of an important file on another storage medium. Use the Copy button and the Paste button in the Clipboard group on the Home tab to copy a selected file or folder.

13 Click to select the folder *Student_Data_Files*.

Before you can copy, you must first select a file or folder—a single-click selects a file. A selected file or folder displays with a blue background in the Content pane. On touchscreen devices, a check box also displays next to folder and file names; a check mark in the check box indicates the file is selected.

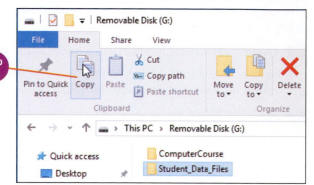

> **App Tip**
> Ctrl + C is the universal keyboard shortcut to Copy.

14 Click the Copy button in the Clipboard group on the Home tab.

15 Double-click the folder *ComputerCourse* in the Content pane.

16 Click the Paste button in the Clipboard group.

> **App Tip**
> Ctrl + V is the universal keyboard shortcut to Paste.

This step is copying all the student data files and placing the copy of the folder and all its contents in the ComputerCourse folder. As the copying takes place, Windows displays a progress message. When the message disappears, the copy is complete.

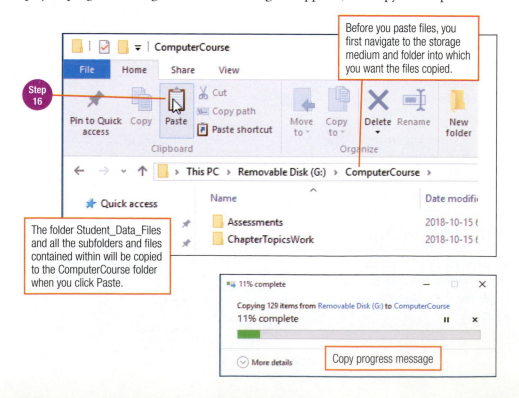

Before you paste files, you first navigate to the storage medium and folder into which you want the files copied.

The folder Student_Data_Files and all the subfolders and files contained within will be copied to the ComputerCourse folder when you click Paste.

Copy progress message

17 Double-click the *Student_Data_Files* folder name.

18 Double-click the *Ch1* folder name.

In the next steps, you will copy individual files to another folder.

19 If necessary, click the View tab and then click List in the Layout group to show only the file names in the Ch1 folder. Skip this step if List view is already active.

20 Click to select *Lighthouse*, hold down the Shift key, and then click to select *Winter*.

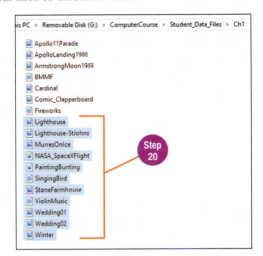

This action selects all the files, starting with **Lighthouse** and ending with **Winter**. Holding down the Shift key while clicking a file name instructs Windows to select all files from the file name selected first to the file name selected second.

Another way to select multiple files is to hold down the Ctrl key while clicking a file name. Use the Ctrl key if you do not want all the files in between to be selected. In other words, Ctrl + click is used to select multiple files that are not next to each other.

21 If necessary, click the Home tab.

22 Click the Copy button.

23 Click *ComputerCourse* in the Address bar.

Clicking a device or folder name in the Address bar is another way to navigate to devices or folders.

24 Double-click *ChapterTopicsWork* in the Content pane.

25 Click the Paste button.

The selected files are copied into the ChapterTopicsWork folder.

26 Click the Close button to close the File Explorer window.

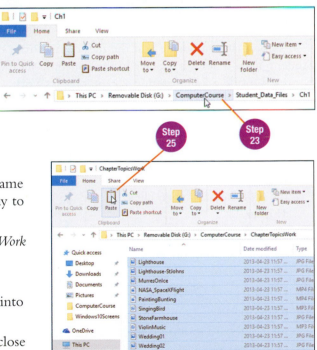

> **Quick Steps**
>
> **Copy Files or Folders**
> 1. Open File Explorer.
> 2. Navigate to source data storage medium and/or folder.
> 3. Select file(s) or folder(s) to be copied.
> 4. If necessary, click Home tab.
> 5. Click Copy button.
> 6. Navigate to destination storage medium and/or folder.
> 7. Click Paste button.

Alternative Method | **Copying Files or Folders in File Explorer Using Other Methods**

- Select files, choose the Copy to button in the Organize group on the Home tab, and then choose the destination folder.
- Select files, right-click inside selection, click Copy, navigate to the destination folder, right-click in Content pane, and then click Paste.
- Drag and drop folders/files from the source location to the destination location.

1.11 Moving, Renaming, and Deleting Files and Folders and Ejecting a USB Flash Drive

Skills

Move a file or folder

Rename a file or folder

Delete a file or folder

Empty the Recycle Bin

Eject a USB flash drive

File Explorer is also used to move, rename, and delete files and folders. Sometimes you will copy a file or save a file in a folder and later decide you want to move it elsewhere. You may also assign a file name to a file or folder and later decide you want to change the name. Files or folders no longer needed can be deleted to clean up the disk. When you are finished using a USB flash drive, you should properly eject the drive to avoid problems that can occur when files are not properly closed.

Moving Files

Files are moved using a process similar to copying. Begin by selecting files and choosing **Cut** in the Clipboard group on the Home tab in File Explorer. Navigate to the destination location and choose Paste. When files are cut they are removed from the source location.

1. Click the Start button and then click File Explorer in the left pane on the Start menu.

2. Click This PC in the Navigation pane and then double-click the Removable Disk for your USB flash drive in the *Devices and drives* section.

3. Double-click *ComputerCourse* in the Content pane.

4. Double-click *ChapterTopicsWork* in the Content pane.

Assume that you decide that the files copied from the Student_Data_Files Ch1 folder in the last topic should be stored inside a folder within ChapterTopicsWork.

5. Click the New folder button in the New group on the Home tab, type Ch1, and then press Enter.

6. Click *Lighthouse*, hold down the Shift key, and then click *Winter*.

7. Click the Cut button in the Clipboard group on the Home tab.

8. Double-click *Ch1*.

9. Click the Paste button in the Clipboard group on the Home tab.

App Tip

Ctrl + X is the universal keyboard shortcut to Cut.

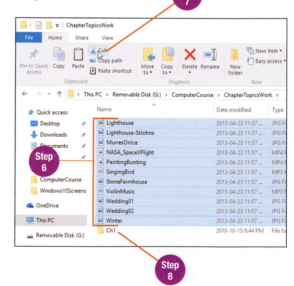

The Address bar shows the destination folder is Ch1 within ChapterTopicsWork.

The files are removed from the ChapterTopicsWork folder and placed within the Ch1 folder.

10 Click the Back button to return to the previous list.

Notice the files are no longer in the ChapterTopicsWork folder.

11 Click the Forward button (right-pointing arrow next to Back button) to return to the Ch1 folder.

Files are moved to Ch1 folder.

Renaming Files and Folders

At times you will receive a file from someone else and decide you want to rename the file to something more meaningful to you, or you may decide upon a new name for a file or folder after the file or folder was created. Change the name of a file or folder with the **Rename** button in the Organize group on the Home tab.

12 Double-click *Winter* in the Content pane. The photograph opens in the Photos app unless your system is set up to display images in another application. If a message window appears asking *How do you want to open this file?*, double-click Photos to open the picture in the Photos app window.

Assume a friend sent you this picture of a weeping birch tree laden with snow in a winter scene. You decide to rename the picture.

13 Close the Photos app window (or other picture viewing application) to return to the File Explorer window.

14 If necessary, click to select *Winter* in the Content pane.

15 Click the Rename button in the Organize group on the Home tab.

Step 15

Step 14

16 Type BirchTreeInWinter and then press Enter.

17 Click to select the file named *ViolinMusic*.

Assume this file is the song Danny Boy played on a violin and recorded by you as you heard the song at an outdoor event.

Step 16

Quick Steps

Move Files or Folders
1. Open File Explorer.
2. Navigate to source data storage medium and/or folder.
3. Select files or folder to be moved.
4. If necessary, click Home tab.
5. Click Cut button.
6. Navigate to destination storage medium and/or folder.
7. Click Paste button.

Rename Files or Folders
1. Open File Explorer.
2. Navigate to data storage medium and/or folder.
3. Select file to be renamed.
4. If necessary, click Home tab.
5. Click Rename button.
6. Type new name and press Enter.

18 Click the Rename button, type DannyBoyViolin, and then press Enter.

19 Click *ComputerCourse* in the Address bar.

You can rename a folder as well as a file.

20 If necessary, click to select the folder named *ChapterTopicsWork*.

21 Click the Rename button, type CompletedTopicsByChapter, and then press Enter.

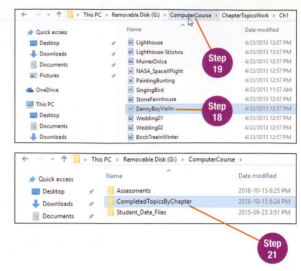

Deleting Files and Folders

Delete files or folders when you no longer need to keep them. You should also delete files or folders if you have copied them to a removable storage medium for archive purposes and want to free up space on the local disk. In File Explorer, select the files or folders to be deleted and then click the **Delete** button in the Organize group on the Home tab. Files and folders deleted from the local hard disk are moved to the Recycle Bin. While it is in the Recycle Bin, you can restore the file to its original location if you deleted the file in error. Files deleted from a USB flash drive are not sent to the Recycle Bin; therefore, exercise caution when deleting a file from a USB flash drive.

22 Double-click the folder *CompletedTopicsByChapter* and then double-click the folder *Ch1*.

23 Click to select **Lighthouse**, hold down the left mouse button and then drag the file to Desktop in the Navigation pane.

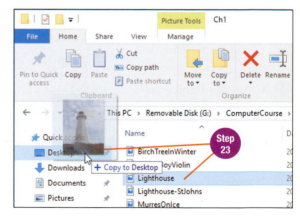

Dragging a file from a folder on one storage medium to a location on another storage medium copies the file.

24 Click Desktop in the Navigation pane.

25 Click **Lighthouse** to select the file and then click the Delete button in the Organize group on the Home tab (click the top of the button—not the down-pointing arrow).

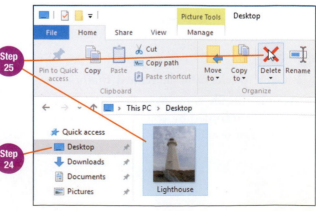

26   Click This PC in the Navigation pane and then double-click the Removable Disk for your USB flash drive in the *Devices and drives* section.

27   Double-click *ComputerCourse*, *CompletedTopicsByChapter*, and then *Ch1*.

28   Click to select **SingingBird** and then click the Delete button in the Organize group on the Home tab.

29   Click Yes at the Delete File message box that appears.

The confirmation message appears when you delete a file from a USB flash drive because the files are permanently deleted, not sent to the Recycle Bin from which the file could be restored.

30   Close the File Explorer window.

31   Double-click the Recycle Bin icon on the desktop.

32   Click the Empty Recycle Bin button in the Manage group on the Recycle Bin Tools Manage tab.

33   Click Yes to permanently delete the file at the Delete File message box.

34   Close the Recycle Bin window.

The file deleted in Step 25 was moved here.

Oops!

Other files in your Recycle Bin? Other files may appear in your Recycle Bin from previous work. To be safe, select and delete only the **Lighthouse** file in this activity.

When you are finished with a USB flash drive, you should always use the **Safely Remove Hardware and Eject Media** feature before pulling the drive out of the USB port to make sure all files are properly closed to avoid data corruption.

35   Click Show hidden icons (up-pointing arrow) in the notification area on the taskbar and then click the Safely Remove Hardware and Eject Media icon.

36   Click *Eject Storage Media* at the pop-up menu. Your menu may show *Eject Disk* or *Eject USB Flash Drive* depending on the device.

37   Remove your USB flash drive from the computer when the message appears that it is safe to do so.

Oops!

The Safely Remove Hardware and Eject Media icon is not always hidden; you may not have to reveal it with the Show hidden icons up-pointing arrow.

If multiple USBs are plugged in, click the correct drive in the list.

A message appears when the USB can be safely removed.

Topics Review

Topic	Key Concepts	Key Terms
1.1 Using Touch, Mouse, and Keyboard Input to Navigate Windows 10	Windows 10 is the latest operating system from Microsoft. The user interface is a desktop that includes icons, a Start menu accessed from the Start button, and a taskbar along the bottom of the screen.	Windows 10
		user interface (UI)
	A tile is a square or rectangle on the Start menu used to launch an app, application, or other feature.	Start menu
		Start button
	An app is a small program designed for a smartphone, tablet, or notebook with touchscreen input and is mainly for one task.	tile
		app
	An application is a more complex program that does more tasks than an app, uses full-size keyboards, uses a mouse, and requires more extensive hardware resources.	application
		gesture
	A gesture is an action or motion you perform with your finger or stylus on a touch-enabled device, such as a tap or swipe motion.	touch keyboard
		mouse
	The touch keyboard appears on-screen when typed characters are expected, such as in text messages.	trackball
		touchpad
	A mouse, trackball, touchpad, or other pointing device is used to move the white arrow pointer on the screen and perform mouse actions, such as click or double-click.	pointer
		keyboard command
	Pressing the Windows logo key brings up the Start menu.	
	A keyboard command involves pressing the Windows logo key, Ctrl, Alt, or a function key with a letter.	
1.2 Starting Windows 10 and Exploring Apps	The Lock screen appears when you start Windows or resume use from sleep mode.	Lock screen
		user account
	A user account is a user name and password used to sign in to Windows.	Microsoft account
		local account
	Signing in to Windows using a hotmail.com, live.com, or outlook.com email address is considered a Microsoft account and means that your settings are stored online and synced with other devices.	desktop
		taskbar
		search text box
	Signing in with a local account means that your settings are not shared with other devices, and you do not see live updates on some of the tiles on the Start menu.	notification area
		Photos app
	Once signed in, the desktop appears and the Start menu opens. The desktop has icons to launch programs and a taskbar along the bottom of the screen.	Calendar app
		Store app
		Task View
	On the taskbar, the Start button, search text box, and icons are pinned, which are used to start apps or applications. A notification area at the right end has icons to show system status.	
	The Photos app shows pictures stored on the local PC and from connected accounts such as OneDrive.	
	The Calendar app is used to enter appointments or events.	
	Use the Store app to search for and download new apps.	
	Switch between apps using the app button on the taskbar or by displaying Task View using the Task View icon on the taskbar to click the desired thumbnail.	

continued…

Topic	Key Concepts	Key Terms
1.3 Using the Search Text Box and Working with Windows	Click in the search text box, type the first few letters of what you are looking for, and then click the item in the search results.	Minimize
	The Minimize button reduces a window to a button on the taskbar.	Close
	Click the Close button to close a window and any content within the window.	Maximize
	The Maximize button expands a window size to fill the entire desktop.	Restore Down
	The Maximize button changes to the Restore Down button when a window is maximized. Use Restore Down to return the window to its previous size.	Snap
	The Snap feature lets you dock a window to the left or right half of the window, or to one of four quadrants by dragging the window left, right, or to a corner.	Snap Assist
	Snap Assist shows thumbnails of open windows when you snap a window and leave part of the screen empty.	
1.4 Using the Notification Area	Click the current date and time at the right end of the taskbar to view a pop-up calendar.	quick action tile
	Click the Speaker icon to adjust the volume.	Action Center
	Click the Network icon to manage network connectivity.	
	A quick action tile displays in Notification panels and is used to turn on or turn off a setting or open a setting window.	
	The Power icon displays the battery status and provides quick actions and links to manage power options.	
	Click the Notifications icon to open the Action Center panel where you can view system messages or access several quick action tiles for managing settings.	
1.5 Locking the Screen, Signing Out, and Shutting Down Windows 10	Lock the screen if you need to leave your PC or mobile device for a short period of time by clicking Start, your account name, and then Lock.	sign out
	Locking leaves all documents and apps open but unavailable to anyone but yourself.	shut down
	Click Start, your account name, and then Sign out to close all apps and documents (also referred to as *logging off*).	
	Perform a shutdown command if you want to turn off the power to the PC or mobile device.	
	Shut down is accessed from the Power option on the Start menu.	

continued…

Topic	Key Concepts	Key Terms
1.6 Using and Customizing the Start Menu	Apps or applications can be started by double-clicking an icon on the desktop, by clicking a tile on the Start menu, by scrolling the options in the left pane on the Start menu, or by typing the first few letters of the name in the search text box. Resize the Start menu by dragging the right border or the top border left, right, up, or down as desired. Right-click a tile and click *Unpin from Start* to remove the tile from the Start menu; locate an app or application in the All apps menu, right-click the name, and click *Pin to Start* to add a tile for the program to the Start menu. Move tiles to a new location on the Start menu by dragging the tile. App tiles are either square or rectangular in shape and can be resized larger or smaller. The Turn live tile off command stops a tile from displaying notifications or status updates; use Turn live tile on to restore the tile's notifications.	Unpin from Start Pin to Start Turn live tile off Turn live tile on
1.7 Personalizing the Lock Screen and the Desktop	Click the Start button, click Settings, then click *Personalization* in the Settings app to personalize the lock screen and desktop options. You can choose from five other pictures for the Lock screen or browse to a picture on your PC or mobile device. You can also change the apps that display notifications on the lock screen. Change the desktop background image and/or the color scheme in the Settings app using the Background and Colors panels.	
1.8 Understanding Files and Folders and Downloading and Extracting Data Files	A file is any document, spreadsheet, picture, or other text or image saved as digital data. When you create a file, you assign a unique file name that allows you to identify and retrieve the file. A folder is a name assigned to a placeholder where a group of related files are stored. A group of files and folders can be compressed and saved as a ZIP file, which bundles everything together in one file with a .zip file extension. When you open a ZIP file, a window opens with the Extract all button used to restore the files and folders from the zipped package, allowing you to use them in their original state and folder structure. File Explorer is the Windows utility used to browse files and perform file management tasks.	file file name folder ZIP file File Explorer

continued…

Topic	Key Concepts	Key Terms
1.9 Browsing Files with File Explorer	In Windows 10, File Explorer opens in the new Quick access view, which shows frequently used and recently used files. The left pane in File Explorer is the Navigation pane, and the right pane is the Content pane. Click This PC in the Navigation pane to view all the available storage devices. You browse content by clicking names in the Navigation pane or double-clicking names in the Content pane. Use options in the Layout group on the View tab to change how the files and folders display in the Content pane. A folder created within another folder is sometimes called a *subfolder*.	Quick access subfolder
1.10 Creating Folders and Copying Files and Folders	Click the New folder button in the New group on the Home tab in File Explorer to create a new folder. A ribbon in File Explorer provides buttons organized into tabs and groups used to carry out file management tasks. Copying a file or folder makes an exact duplicate of a document, spreadsheet, presentation, picture, video, music file, or other object in another folder and/or storage medium. Begin a copy task by first selecting the file or folder to be copied and then use the Copy and Paste buttons in the Clipboard group on the Home tab.	New folder ribbon Copy Paste
1.11 Moving, Renaming, and Deleting Files and Folders and Ejecting a USB Flash Drive	Files are moved by selecting the files or folders, choosing Cut, navigating to the new destination drive and/or folder, and choosing Paste. Click the Rename button in the Organize group on the Home tab, type a new name, and then press Enter to change the name of the selected file or folder. Files deleted from a hard disk drive are sent to the Recycle Bin and remain there until the Recycle Bin is emptied. Files deleted from a USB flash drive are not sent to the Recycle Bin. Select files or folders to be deleted and then click the Delete button in the Organize group on the Home tab. Open the Recycle Bin from the desktop to view files deleted from the hard disk drive. Emptying the Recycle Bin permanently deletes the files or folders. Eject a USB flash drive using the Safely Remove Hardware and Eject Media icon in the notification area on the taskbar.	Cut Rename Delete Safely Remove Hardware and Eject Media

Recheck

Recheck your understanding of the topics covered in this chapter.

Workbook

Chapter review and assessment resources are available in the *Workbook* ebook.

Chapter 2

Navigating and Searching the Web

Precheck

Check your understanding of the topics covered in this chapter.

For many people reading this textbook, the Internet is part of daily life, used to search for information, connect with friends and relatives, watch videos, listen to music, play games, or shop. Mobile devices, such as tablets and smartphones, allow people to browse the web anywhere at any time. Being able to effectively navigate and search the web is a requirement for all workers and consumers.

In this chapter, you will learn definitions for Internet terminology and how to navigate the web using three different web browsers. You will also learn to use search tools to find and print information, view multiple websites in a browsing session, bookmark web pages you visit often, and copy an object from a website to your computer.

Note: While the emphasis in this chapter is on using Microsoft Edge, which is included with Windows 10, feel free to work through the topic activities and projects using Google Chrome or Mozilla Firefox. In that case, be aware that for some topics, the steps provided may need to be altered to suit the Chrome or Firefox browser.

If you are using a computer with Windows 7, Internet Explorer is the web browser from Microsoft included with the operating system. The steps you complete and screens you see will vary from the ones shown in this chapter. If necessary, check with your instructor for alternate instructions.

No student data files are required to complete this chapter.

 SNAP If you are a SNAP user, go to your SNAP Assignments page to complete the Precheck, Tutorials, and Recheck.

Learning Objectives

2.1 Describe the Internet, World Wide Web, web browser, web page, hyperlink, and web address

2.2 Navigate the web using Microsoft Edge, use tabs to view multiple websites, and add a page to the Favorites list

2.3 Navigate the web using Chrome, use tabs to view multiple websites, bookmark a page, and use the Find bar

2.4 Navigate the web using Firefox, use tabs to view multiple websites, bookmark a page, use the Find bar, and locate information using the Search bar

2.5 Use a search engine website to find information, refine a search using advanced search tools, and print a web page

2.6 Download a picture from a web page

2.1 Introduction to the Internet and the World Wide Web

Skills

Define Internet

Define World Wide Web

Explain web browser, web page, and hyperlink

Describe web address

The **Internet** or **net** is a global network that links other networks such as individuals, businesses, schools, government departments, nonprofit organizations, and research institutions. The worldwide system of interconnected networks using standardized communication protocols to transmit data is known as the Internet. High-speed communications and networking equipment are used on the Internet to transmit data from one computer to another. For example, a request sent from your computer to display a web page showing flight times from an airline schedule, would travel through several other networks, such as telephone, cable, or satellite company networks, to reach the airline web server.

The World Wide Web

The collection of electronic documents circulated on the Internet in the form of web pages make up the **World Wide Web** or **web**. A **web page** is a document that contains text and multimedia content, such as images, video, sound, and animation (Figure 2.1, page 41). Web pages also contain hyperlinks. A **hyperlink**, also called a **link**, is text, a picture, an icon, or another object on a web page that moves you to another related page when clicked. Web pages are stored in a format that is read and interpreted for display within a **web browser**, which is a software program used to view web pages. A website is a collection of related web pages for one organization or individual. For example, the collection of web pages linked to the main page for your school make up your school website. All the web pages and resources such as photos, videos, sounds, and animations that make the website work are stored on a computer called a *web server*. Web servers are connected to the Internet continuously.

You connect your PC or mobile device to the Internet by subscribing to Internet service through an **Internet service provider (ISP)**, a company that provides access to Internet infrastructure for a fee. The ISP will provide you with the equipment needed to connect to its network, as well as instructions for installing and setting up the equipment to work with your computer and other mobile devices. Once your account and equipment are set up, you can start browsing the web using a browser, such as Microsoft Edge, Chrome, or Firefox.

Many people connect wirelessly to the Internet from multiple devices.

Web Addresses

Each web page has a unique text-based **web address** that allows you to navigate to the page. A web address is also called a **uniform resource locator (URL)**. One way to navigate the web is to type a web address into the Address bar of a web browser. For example, to view the main web page for the publisher of this textbook, you would use the web address *http://paradigm.emcp.com*. If *http* is left out when typing a web address in a browser, it is assumed. For this reason, many people often go to a page by typing only the portion of the address after *http://.* The parts of the web address (URL) https://www.nasa.gov/topics/journeytomars/index.html shown in Figure 2.1 are explained in Table 2.1.

Many times when you go to the web to look for information, you do not know the web addresses for the pages you need. In this case, search tools help you find web pages. You will learn to find web pages using search tools in Topic 2.5.

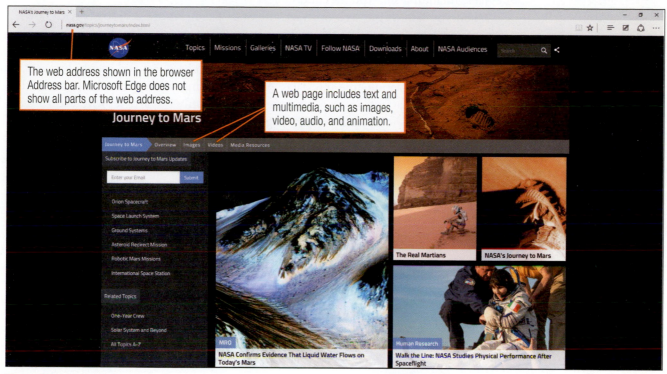

Figure 2.1

A web page from NASA that includes multimedia content is shown in a Microsoft Edge browser window.

Table 2.1

The Parts of the Web Address (URL) https://www.nasa.gov/topics/journeytomars/index.html

Part of URL	What It Means	Examples
https://	Hypertext Transfer Protocol Secure. A set of rules for transmitting data in encrypted format over the Internet. Https is used for exchanging data in online transactions. Other websites are starting to use the protocol to provide authentication of the website and to protect data. Other protocols used are http to transfer data without encryption and ftp to transfer a file.	http ftp https
www.nasa.gov	Domain name. A text-based address for the web server on the Internet that stores the web page. The last three letters (gov) are the extension and tell you the type of organization. Some domain names end with a two-character country code, such as eBay.ca for the Canadian auction site eBay. Using *www* in a domain name is becoming less common. In the *Examples* column are four domain names to show a sampling of a variety of extensions. More extensions are used, such as .net, .biz, .legal, and .realtor.	google.com (Google) harvard.edu (Harvard University) navy.mil (US Navy) wikipedia.org (Wikipedia online encyclopedia)
/topics/journeytomars/	Folder path to the web page. The forward slash (/) and text after the slash are the folder names on the web server where the page is stored.	path folder names will vary
index.html	Web page file name. The file name extension indicates the language used to create the web page. Html stands for *Hypertext Markup Language*, which uses tags to describe content and is widely used.	file names will vary

2.2 Navigating the Web Using Microsoft Edge

Microsoft Edge is the web browser included with Windows 10. The new browser replaces the Internet Explorer web browser included with earlier Windows versions. Similar to the Windows apps you learned about in Chapter 1, where an app looks and feels the same on all devices, Microsoft Edge is the browser for all Windows 10 devices, including smartphones and tablets.

When you launch the web browser, the Start page (Figure 2.2) loads with news and information feeds unless the browser has been customized to display another page. For example, many schools will customize browsers to show the student portal when started. Microsoft Edge was designed to be a fast and secure browser, with a clean, distraction-free interface. A new Reading view option strips away all the sidebars and navigation on a web page to show only the article. Microsoft Edge also provides tools to annotate a web page with markup, highlights, handwriting, or typed notes, saving the page as a web note for reading in the Reading list, a OneNote notebook, or shared via email. See Beyond Basics for information on creating a web note.

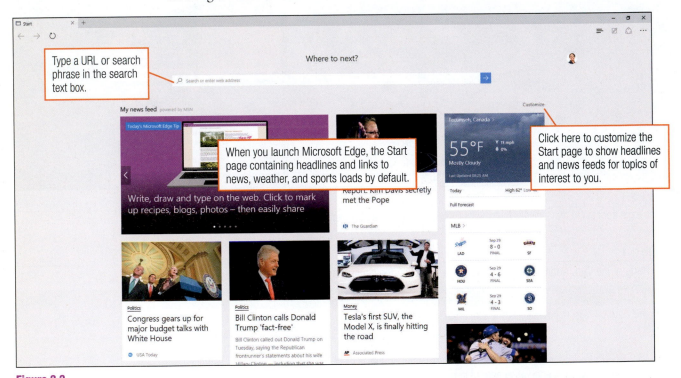

Type a URL or search phrase in the search text box.

Where to next?

When you launch Microsoft Edge, the Start page containing headlines and links to news, weather, and sports loads by default.

Click here to customize the Start page to show headlines and news feeds for topics of interest to you.

Figure 2.2

The Microsoft Edge Start page is shown.

Starting Microsoft Edge and Displaying a Web Page

1. Click the Microsoft Edge icon on the taskbar.

   The Microsoft Edge browser window opens to the Start page shown in Figure 2.2 unless the browser has been customized to display another page.

2. With the insertion point positioned in the search text box (contains the dimmed text *Search or enter web address*), type nps.gov/grca and then press Enter. If a message box appears asking you to sign up for national park news, close the message using the × at the top right corner. Review the layout and tools in the Microsoft Edge window shown in Figure 2.3 (page 43).

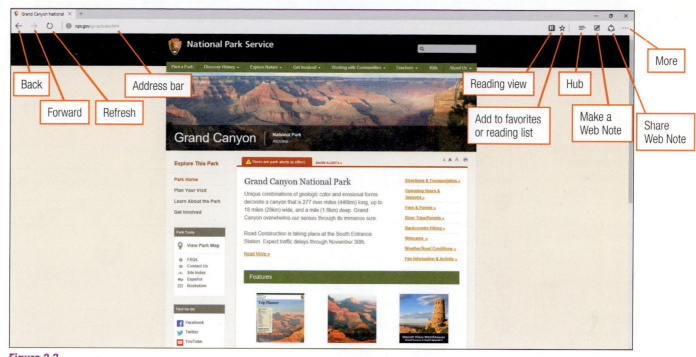

Figure 2.3

The National Park Service Grand Canyon web page is shown in a Microsoft Edge browser window.

3 Click <u>About Us</u> near the top right of the web page.

4 Click <u>Overview</u>.

Navigating web pages involves clicking hyperlinks to move from one page to another at a website or to a web page stored at another website. Click **Back** to return to the previous page viewed or click **Forward** to move forward one web page viewed. To display a different web page, click in the **Address bar** to select the current entry, or drag to select the current text in the Address bar, type the web address, and then press Enter.

5 Click in the Address bar or drag to select the existing web address, type loc.gov, and then press Enter.

The Library of Congress home page appears. The starting page for a website is called the **home page**.

6 Click <u>Digital Preservation</u> in the *Resources & Programs* section at the right side of the Library of Congress page.

7 Click Back to return to the Library of Congress home page (move back one web page).

8 Click Forward to return to the Digital Preservation page (move forward one web page).

App Tip

As you begin typing a URL, Microsoft Edge displays matches below the Address bar that begin with similar text. Click the page if it appears in the list before you finish typing.

9 Click <u>View the tips</u> in the *Personal Digital Archiving: Preserving Your Digital Memories* section.

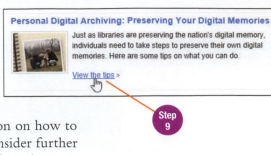

Step 9

The Personal Archiving page at the Library of Congress contains links to articles with useful information on how to preserve your personal memories. Consider further exploring the links at this web page if you have an interest in this topic.

10 Click Reading view on the toolbar.

Step 10

All the sidebars and navigational items disappear from the page leaving only the text for distraction-free reading.

11 Scroll down to view the page in Reading view.

12 Click Reading view to turn off the view.

13 Click in the Address bar or drag to select the existing web address, type www.nasa.gov/topics/history, and then press Enter.

Step 13

14 Scroll down and click a link to a story that interests you. Read the story using Reading view.

15 Turn off Reading view and then click Back to return to the NASA History page.

Displaying Multiple Web Pages and Adding a Page to Favorites

Open multiple web pages within the same Microsoft Edge window by displaying each page in its own tab. Click **New tab** (displays as a plus symbol) to open a new page in the window. Switch between web pages by clicking the tab for the page you want to view. This is called **tabbed browsing**. The web address for a page you visit frequently can be saved to the **Favorites** list by displaying the page and then using the star on the toolbar to add the page to Favorites.

16 Click New tab (plus symbol) next to the tab for the NASA History page at the top of the window.

Step 16

17 Type www.flickr.com/commons in the search text box and then press Enter.

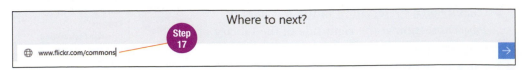

Step 17

The Commons at flickr.com contains public photography archives from around the world. Images in the commons have no known copyright restrictions. The next time you need a picture or photo for a project, consider sourcing an image from this website.

18 Click the NASA History page tab to change the web page displayed in the window.

19 Click Close tab (displays × on tab) on the NASA History tab.

20 With The Commons page at flickr.com displayed in the window, click the star at the right end of the Address bar on the toolbar.

21 Click the Add button to accept the default option to save the page in Favorites. The page will display in the Favorites list with the name Flickr: The Commons.

The star turns gold for a page that is added to the Favorites list.

22 Click Hub to open the Hub panel at the right side of the window.

Open the **Hub** to view the Favorites list, Reading list, History list, or Downloads list.

23 If necessary, click the star in the Hub panel to view the Favorites list if another list is currently active.

24 Click Hub to close the Hub panel.

25 Close the Microsoft Edge window.

Your Favorites list may vary with more links.

Beyond Basics   **Annotating a Web Page in Microsoft Edge**

Add a note, highlight text on a page, or draw markings using the tools from Make a Web Note. Click Make a Web Note to display the Web Note toolbar on the current page. The tab control and Web Note toolbar display purple. Use the Pen, Highlighter, Eraser, Typed note, and Clip tools to annotate the web page. When finished, save the annotated page to a OneNote notebook using the Save Web Note button or click the Share Web Note button to send the Web Note in an email message or via OneNote. Click Exit to remove the Web Note tools when finished.

Make a Web Note

Web Note Tools

2.3 Navigating the Web Using Chrome

Skills

Navigate the web using Chrome

View multiple websites within the same window

Bookmark a page

Use the Find bar

Google **Chrome** is a free web browser for Windows-compatible PCs, Macs, or Linux PCs. Chrome was reported in 2015 by StatCounter (a company that measures web traffic) to be the browser of choice for approximately 50 percent of desktop and tablet users, and 30 percent of smartphone users worldwide. The browser has gained a significant following since its release in 2008 due to its fast page loading and searching directly from the Address bar using the Google search engine. Complete the steps in this topic if you have the Chrome web browser installed on your device; otherwise, skip to the next topic.

Starting Chrome and Displaying a Web Page

Oops! !

Insertion point not positioned in the blank Address bar? On some tablets and PCs, Chrome starts with the insertion point in the search text box. Click to select the current entry in the web address first, before typing the address at Step 2.

1 Click the Chrome icon on the taskbar or double-click the icon on the desktop, and then review the layout and tools in the Chrome window shown in Figure 2.4.

If Chrome is not currently set as the default browser, you may be prompted to set it as the default browser in a message bar below the Address bar.

2 With the insertion point positioned in the Address bar, type techmeme.com and then press Enter.

Techmeme is a technology news site that posts each day the top news headlines from the industry.

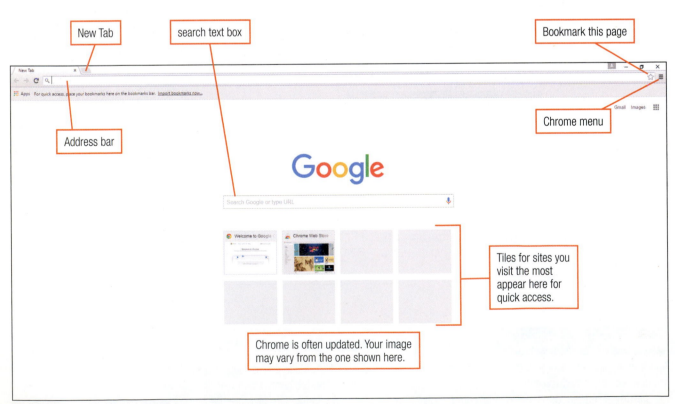

New Tab   search text box   Bookmark this page

Address bar

Chrome menu

Tiles for sites you visit the most appear here for quick access.

Chrome is often updated. Your image may vary from the one shown here.

Figure 2.4
The Google search engine is shown displayed in the Chrome browser.

3 Click a link to a technology story that interests you on the Techmeme page.

Techmeme is a content aggregator—a company that searches, collects, and organizes content in one place. Clicking a link to a story on the Techmeme page takes you to another website.

4 Click Back to return to the Techmeme page.

Step 4

Your web address will vary depending on the story you clicked at Step 3.

Displaying Multiple Web Pages and Bookmarking Pages

Similar to Microsoft Edge, Chrome uses tabbed browsing to display more than one web page within the same browser window. Open a new tab and navigate to a web address to browse a new site without closing the existing web page. Chrome displays a page name at the top of the tab along with a close control. Web pages you visit frequently can be bookmarked by clicking the white star at the end of the Address bar (displays the ScreenTip *Bookmark this page*).

5 With techmeme.com the active web page, click the white star at the end of the Address bar.

The white star changes to gold, and the Bookmark dialog box appears.

6 Click Done to close the Bookmark dialog box.

White star changes to a gold star when page has been bookmarked.

Bookmark added!

Name: Techmeme

Folder: Bookmarks bar

Remove   Edit...   Done

Step 6

Good to Know 🎓

If you sign in to Chrome using your Google account, your bookmarks and other browser preferences are saved and can be accessed on other PCs or mobile devices.

Good to Know 🎓

A button or icon with three bars is called a *hamburger button* because it resembles a hamburger patty between a top and bottom bun. The button is commonly used in apps to indicate a menu.

The **Bookmarks bar** displays below the Address bar and is turned on by default when Chrome is installed. To turn the Bookmarks bar off or on, display the menu system by clicking the **Chrome menu** (displays three bars, often called a *hamburger button*) near the top right of the window.

7 Click the Chrome menu (displays as three bars with the ScreenTip *Customize and control Google Chrome*) and then point to *Bookmarks*. A check mark next to *Show bookmarks bar* (shown below) means that the bar is turned on. In that case, click in a blank area away from the menu to close the menu; otherwise, click *Show bookmarks bar* to turn the bar on.

App Tip ▶

By default, bookmarks are added to the bookmarks bar, which is docked below the Address bar.

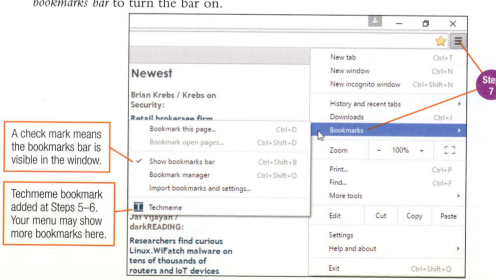

Newest

Brian Krebs / Krebs on Security:

Retail brokerage firm

Bookmark this page...   Ctrl+D
Bookmark open pages...   Ctrl+Shift+D

✓ Show bookmarks bar   Ctrl+Shift+B
Bookmark manager   Ctrl+Shift+O
Import bookmarks and settings...

Techmeme

Jai Vijayan /
darkREADING:

Researchers find curious Linux.WiFatch malware on tens of thousands of routers and IoT devices

New tab   Ctrl+T
New window   Ctrl+N
New incognito window   Ctrl+Shift+N

History and recent tabs   ▶
Downloads   Ctrl+J
Bookmarks   ▶

Zoom   —   100%   +   ⌐⌐

Print...   Ctrl+P
Find...   Ctrl+F
More tools   ▶

Edit   Cut   Copy   Paste

Settings
Help and about   ▶

Exit   Ctrl+Shift+Q

Step 7

A check mark means the bookmarks bar is visible in the window.

Techmeme bookmark added at Steps 5–6. Your menu may show more bookmarks here.

8 Click New tab next to the Techmeme tab to open a new page in the window, type www.youtube.com, and then press Enter.

Step 8

9 Click the white star and then click Done to add youtube.com to the Bookmarks bar.

The Bookmarks bar with the two bookmarks added in this topic appears below the Address bar. Your bar may show more bookmarks.

Searching the Web from the Address Bar

Like Microsoft Edge, Chrome allows you to type a search phrase in the Address bar. Search results from Google (or other default search engine) appear in the window.

10 Click New tab next to the YouTube tab, type seven wonders of the world, and then press Enter.

Notice that as you type, searches that match your entry appear in a drop-down list below the Address bar. If one of the searches is what you are looking for, click the entry in the list.

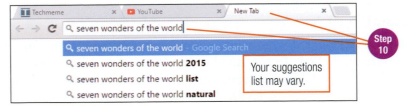

Step 10

Your suggestions list may vary.

11 Click Images for seven wonders of the world in the Google search results and then scroll down to view the pictures.

12 Click in the Address bar to select the current web address, type great wall of china, and then press Enter.

13 Click Great Wall of China - Wikipedia, the free encyclopedia, then scroll down to review the *Wikipedia* page about the Great Wall of China.

14 Scroll back up to the top of the *Wikipedia* page.

Finding Text on a Page

Use the **Find bar** to locate a specific word or phrase on a web page. The Find bar is turned on from the Chrome menu. Using the Find bar is helpful when viewing a long story or article and you need to locate a reference you know exists in the text.

15 Click the Chrome menu and then click *Find*.

The Find bar opens at the right side of the window below the bookmarks bar with the insertion point positioned in the Find bar text box.

16 Type first emperor in the Find bar text box.

The currently selected match is shaded with an orange background on the page, and the number of matches found on the web page displays in the Find bar next to the find text. Additional matches below the current selection are shaded yellow.

Step 15

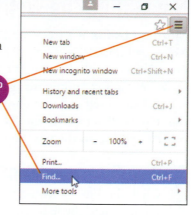

17 Click Next (down-pointing arrow) on the Find bar to scroll the page to the next occurrence of *first emperor*.

Step 16   **Step 17**

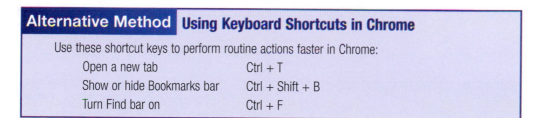

18 Continue clicking Next until you have seen all occurrences.

19 Close the Chrome window.

Alternative Method | **Using Keyboard Shortcuts in Chrome**

Use these shortcut keys to perform routine actions faster in Chrome:

Open a new tab	Ctrl + T
Show or hide Bookmarks bar	Ctrl + Shift + B
Turn Find bar on	Ctrl + F

Beyond Basics | **Incognito Browsing in Chrome**

Click the Chrome menu and then click *New incognito window* to open a new tab for private browsing. Web pages you visit while using an incognito tab will not appear in the browser history or search history, and any cookies that are created are automatically deleted after all incognito windows are closed. Be aware that browsing using incognito tabs does not prevent an employer or ISP from tracking the websites you visit because they use special software to monitor usage.

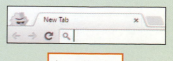

Incognito tab

2.4 Navigating the Web Using Mozilla Firefox

Skills

Navigate the web
using Firefox

View multiple websites
within the same window

Bookmark a page

Use the Find bar

Use the Search bar

Mozilla **Firefox** is a free web browser that runs on Windows-compatible PCs, Macs, or Linux PCs. The software program is published by the nonprofit Mozilla Foundation. At the time of writing, Firefox was the third most popular web browser. Firefox enthusiasts prefer the nonprofit open source software environment, which lends itself to more customization options in the Firefox browser. Complete the steps in this topic if you have the Firefox web browser installed on your device; otherwise, skip to the next topic.

Starting Firefox and Displaying a Web Page

① Click the Firefox icon on the taskbar or double-click the icon on the desktop. Click Not now if prompted to set Firefox as the default browser. Review the layout and tools in the Firefox window shown in Figure 2.5.

② At the Firefox Start page, click in the Address bar that displays the dimmed text *Search or enter address*, type commons.wikimedia.org, and then press Enter.

Wikimedia Commons is a media file repository maintained by volunteers. Go to this site to find images, sound, and video clips that are free to use or copy by following the terms specified by the author, which often require only that you credit the source.

Check This Out ✓

http://CA2.Paradigm
College.net/Firefox

Go here to download and install the Firefox web browser on your PC or mobile device.

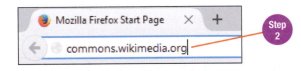

Open a new tab

Reload current page

Search bar

Bookmark this page

Show your bookmarks

Open menu

At the Start page, you can type a search request in either the Address bar or Search bar.

Firefox is often updated. Your image may vary from the one shown here.

Figure 2.5
The Firefox window is shown.

③ Click in the *Search Wikimedia Commons* text box near the top right of the Wikimedia Commons page that displays the dimmed word *Search*, type clownfish, and then press Enter or click Go (magnifying glass).

④ Scroll down and view the pictures of clownfish.

Displaying Multiple Web Pages and Bookmarking Pages

Similar to Microsoft Edge and Chrome, Firefox provides for tabbed browsing. Click **Open a new tab** (displays as a plus symbol), type a web address in the Address bar, and then press Enter to display a new page in its own tab. Web pages you visit frequently can be saved by clicking the star on the toolbar next to the Search bar (displays the ScreenTip *Bookmark this page*). Click **Show your bookmarks** located on the toolbar next to the star to display the bookmarks menu, which is used to navigate to a bookmarked page or turn on the Bookmarks toolbar.

⑤ Click the star on the toolbar (displays the ScreenTip *Bookmark this page*).

   The star turns blue for a bookmarked page.

⑥ Click Open a new tab (displays as a plus symbol) next to the Wikimedia tab. If a message box opens on the new page with information about updates made to New Tab, click the Got it! button.

⑦ If necessary, click in the address bar on the new page. Type wikitravel.org and then press Enter.

   Wikitravel is a worldwide travel guide written and updated by travelers.

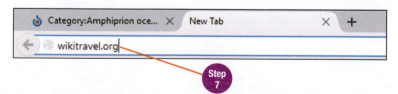

⑧ Click the star to bookmark the page.

9 Click Show your bookmarks next to the star on the toolbar, click *Recently Bookmarked*, and then click *Category:Amphiprion ocellaris - Wikimedia Commons* at the side menu.

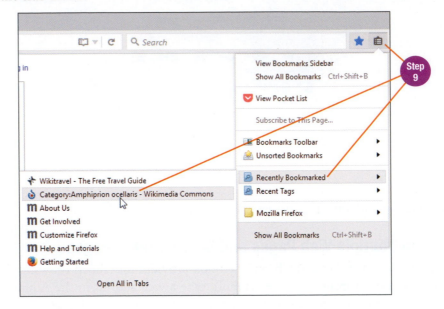

10 Click Close Tab (displays as × on the tab) in the second tab titled Category:Amphiprion ocellaris – Wikimedia Commons to close the page.

Finding Text on a Page

Open the Find bar in Firefox to search for all occurrences of a word or phrase on a web page. The Find bar is useful when you need to quickly locate a specific reference or other text on a page.

11 Click Open menu (displays three bars) at the end of the toolbar and then click *Find*.

The Find bar opens at the bottom left of the Firefox window with an insertion point positioned in the Find bar text box. As you begin typing an entry in the text box, Firefox immediately begins highlighting matches for the search text on the current web page. If no matches are found, Firefox shades the Find box pink. Use the next and previous navigation buttons to move to each occurrence of the search word or phrase.

12 With the insertion point positioned in the Find bar text box (contains dimmed text *Find in page*), type aquarium.

The first matched occurrence of the search text is shaded green on the page.

13 Click Next to move to the next occurrence of *aquarium* on the page.

Matches are shaded green.

The number of matches found and position of images on the page may vary as more images become available.

Step 12

Step 13

The number of matches found and the occurrence you are viewing is shown here.

14 Continue clicking Next until you return to the first occurrence on the page.

15 Press Esc or click Close find bar (displays with an ×) at the right end of the Find bar.

Searching the Web Using the Search Bar

Search for information using a search engine, such as Google, Yahoo, or Bing, by typing a search word or phrase in the **Search bar** next to the Address bar in the Firefox window.

16 Click in the Search bar, type tropical fish, and then press Enter or click Search (right-pointing arrow).

Step 16

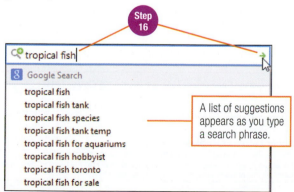

A list of suggestions appears as you type a search phrase.

17 Click a link to a web page that interests you from the search results list.

18 Close the Firefox window.

2.5 Searching for Information and Printing Web Pages

Skills

Use search tools to find information on the web

Narrow search results by applying advanced search options

Print a web page, including selected pages only

A **search engine** is a company that indexes web pages by keywords and provides a search tool with which you can search their indexes to find web pages. To create indexes, search engines use a program called a **spider** or **crawler** that reads web pages and other information supplied by the website owner to generate index entries. Search engines also provide advanced tools, which you can use to narrow search results.

Using a Search Engine to Locate Information on the Web

Several search engines are available, with Google, Bing, and Yahoo! the leading companies; however, consider searching using other search engines, such as Dogpile, Ask, and DuckDuckGo, because you get different results from each company. Spider and crawler program capabilities and timing for indexing create differences in search results among companies. Depending on the information you are seeking, performing a search in more than one search engine is a good idea.

1. Click the Microsoft Edge icon on the taskbar.

By default, Microsoft Edge performs searches using Bing, the search engine from Microsoft.

2. With the insertion point positioned in the search text box (displays dimmed text *Search or enter web address*) on the Start page, type cover letter examples and then press Enter or click Go (right-pointing arrow).

Notice that Bing provides search suggestions that match the characters you type in a drop-down list as soon as you begin typing. Click a search suggestion in the list if you see a close match.

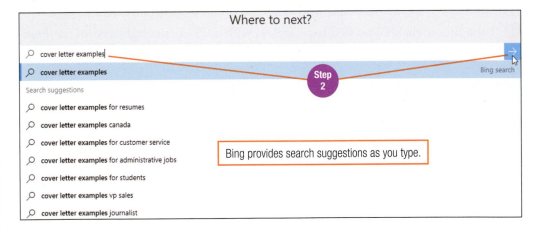

Bing provides search suggestions as you type.

Check This Out ✓

http://CA2.Paradigm College.net/Dogpile

Go here to perform a search using a metasearch search engine. Metasearch search engines send your search phrase to other search engines and show you one list of search results from the wider group.

3. Scroll down the search results page and click a link to a page that interests you.

Your results will vary. Choose a page that interests you.

④ Read a few paragraphs about cover letters at the page you selected.

⑤ Click Back to return to the search results list.

⑥ Click in the Address bar or drag to select the current web address in the Address bar, type yahoo.com, and then press Enter.

⑦ With the insertion point positioned in the search text box, type cover letter examples and then press Enter, or click the Search Web button.

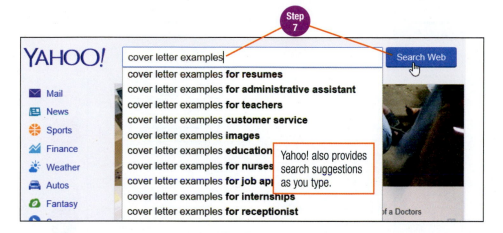

App Tip

Click a category above a search box or along the top of a search engine web page to restrict search results to a specific type of content such as Images, Videos, Maps, or News.

⑧ Scroll down the search results page. Notice that some links are to the same web pages that you saw in the Bing search results; however, the same page may be in a different order in the list or you may notice new links not shown by Bing.

Using Advanced Search Options

In both Bing and Yahoo!, the search results for the cover letter examples resulted in millions of links in the results pages. Search engines provide tools to help you narrow the search results. Each search engine provides different tools. Explore the options at the search engine you prefer or look for a help link that provides information on how to use advanced search tools.

⑨ Click in the Address bar or drag to select the current web address, type google.com, and then press Enter.

⑩ With the insertion point positioned in the search text box, type cover letter examples, and then press Enter.

⑪ Click the Options button (displays as gear icon near top right of the page), and then click Advanced search at the drop-down list.

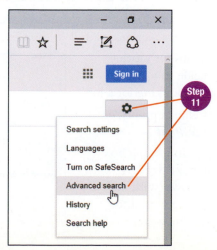

12 Click in the *site* or *domain* text box, and then type edu. You may need to scroll down the screen to locate the option.

An edu domain is restricted to US accredited postsecondary institutions.

13 Click the Advanced Search button near the bottom of the Advanced Search page.

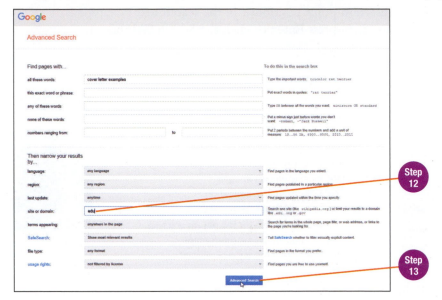

14 Click Search tools above the search results, click Any time, and then click Past month.

The search results list is now significantly reduced from the millions in the prior results list.

15 Scroll down the search results page and then click a link to a page that interests you. Notice that all URLs are .edu domains and all dates are within the past month.

Printing a Web Page

Printing a web page can sometimes be frustrating, because web pages are designed for optimal screen viewing (not printing). Many times you may print a web page only to discard a second or third page that you did not need, or the content printed did not fit the width of the paper and the printout was unusable.

16 Click in the Address bar or select the current web address, type www.usf.edu /career-services/students/cover-letter-dos-and-donts.aspx, and then press Enter.

This page contains several tips for writing cover letters. Assume you decide to print the page for later use when you are looking for a job.

17 Click More (displays as ellipsis points, or three dots) and then click *Print*.

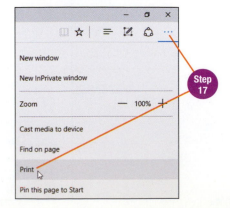

The page displays in the Print dialog box shown in Figure 2.6 on the next page. A preview of the first page is shown with a right-pointing arrow above the preview used to navigate and view the remaining pages in the printout.

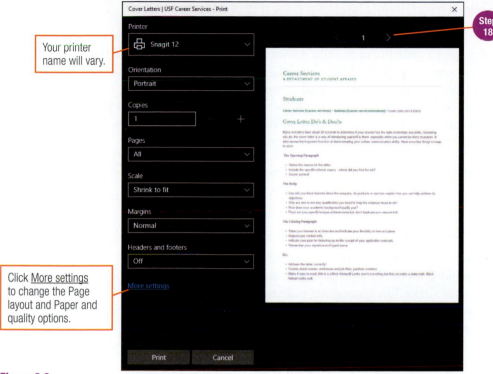

Your printer name will vary.

Click More settings to change the Page layout and Paper and quality options.

Figure 2.6

The Print dialog box in Microsoft Edge.

Option boxes are available to change the destination printer, page orientation, number of copies, pages to print, scaling, margins, and to add a header and/or footer.

18 Click the Next page arrow (right-pointing arrow) above the preview page to view the second page that will be printed.

19 Click the Next page arrow again to view the third page that will be printed.

The web page needs three pages to print, with the last page being unnecessary as it contains links only.

20 Click the *Pages* option box and then click *Page range* at the drop-down list.

21 Click in the *Range* text box and then type 1-2.

22 Click the Print button or click the Cancel button if you are not connected to a printer to close the Print dialog box.

Only pages 1 and 2 of the web page print. Avoid wasting paper whenever possible when printing web pages by taking the time to preview, ensuring you use only the pages that you need.

23 Close the Microsoft Edge window. Click Close all if prompted to close all tabs.

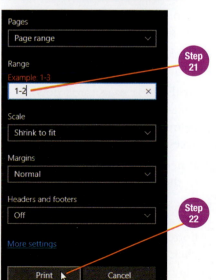

Good to Know

Some web pages provide a link to view the page in a printable format. Look for a View Printable hyperlink, which reformats the content for a letter-size page that prints only the article or story.

2.6 Downloading Content from a Web Page

Copying an image, audio clip, video clip, or music file (such as MP3) from a web page to your PC or mobile device is referred to as **downloading** content. Most content is protected by copyright law from being used by someone without permission. Before you download content, check the website for restrictions against copying information. Look for a contact link and request permission from the website owner to use the content if no restrictions are shown. Always cite the original source of any content you copy from a web page.

Saving a Picture from a Web Page

Saving content generally involves selecting an object and displaying a context menu from which you can select to copy or save the object.

1 Click the Microsoft Edge icon on the taskbar.

Assume you want to find an image from the Grand Canyon for a project. You decide to use the flickr The Commons page to find a picture in the public domain that can be used without copyright restrictions.

2 Click Hub.

3 Click the star if the Favorites list is not currently displayed in the Hub.

4 Click Flickr: The Commons in the Favorites list (you added this link to the Favorites list in Topic 2.2).

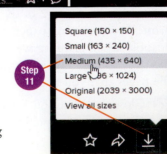

5 If necessary, scroll down to the section titled *A Commons Sampler*.

6 Click in the *Search The Commons* text box, type grand canyon, and then press Enter, or click the SEARCH button.

7 Scroll down and view the images in the search results page.

8 Click a picture that you like and want to save.

9 At the next page, scroll down if necessary and read the description below the photograph. Note the access and use restrictions, if any.

10 If necessary, scroll back up the page to the picture preview.

11 Click the Download this photo icon (displays as a down-pointing arrow above a bar) at the bottom right of the picture preview area and then click <u>Medium</u> at the pop-up list. Note that the dimensions for each size option will vary depending on the picture that is currently selected.

Your list may vary. Choose any picture you want to save.

12 Close the message bar that displays when the picture has finished downloading. A message box appears if you do not have a Yahoo account connected with your Microsoft account. Click the close button at the top right of the message box and proceed to Step 13 if you do not wish to create a new Yahoo account by clicking the Sign up with Yahoo button.

13 Click in the Address bar to select the current web address, type facebook.com/GrandCanyonNationalPark, and then press Enter. You may be prompted to sign in to Facebook to view the page. If necessary, skip to Step 18 if you cannot view the Facebook page.

14 Scroll down the timeline to a picture that you would like to save, right-click the picture, and then click *Open in new tab*.

15 If necessary, click the new tab to view the photo on its own page.

16 Right-click the photo and then click *Save picture as*.

The timeline pictures will vary. Choose any picture you want to save.

Step 14

Step 16

17 At the Save As dialog box, click Downloads in the Navigation pane, select the current text in the *File name* text box, type GrandCanyon, and then click Save.

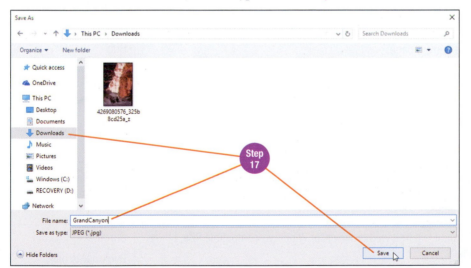

Step 17

18 Close the Microsoft Edge window. Click Close all when prompted to close all tabs.

19 Click the File Explorer icon on the taskbar.

20 Click Downloads in the Navigation pane.

You will see the two downloaded pictures in the Content pane.

21 Double-click each picture you downloaded to view the picture in the Photos app or other picture viewing window and then close the window.

22 Close File Explorer.

Topics Review

Topic	Key Concepts	Key Terms
2.1 Introduction to the Internet and the World Wide Web	The Internet is the information infrastructure that represents the global network that links together individuals and organizations. All the web pages circulated on the Internet form the World Wide Web. Documents stored on a web server that contain text and multimedia elements are called *web pages*. A web page can contain hyperlinks, also called *links*, that move from one page to another when clicked. A web browser is a software program used to view web pages, such as Microsoft Edge, Chrome, and Firefox. An Internet service provider is a company that provides Internet service to individuals and businesses for a fee. A web address is a unique text-based address, also called a *uniform resource locator (URL)*, that identifies each web page on the Internet. A web address is made up of parts that define the protocol, domain, folder, and file name for the web page.	Internet (net) World Wide Web (Web) web page hyperlink (link) web browser Internet service provider (ISP) web address uniform resource locator (URL)
2.2 Navigating the Web Using Microsoft Edge	Microsoft Edge is the new web browser included with Windows 10. Start Microsoft Edge by clicking the Microsoft Edge icon on the taskbar. To display a web page, type the URL or enter a search phrase in the search text box and then press Enter. Click Back to move back one page; click Forward to move forward one page. Click in the Address bar and type a new URL or search phrase to display a different web page in the window. The first page that displays for a website is called the *home page*. Click New tab to open a new page in the window and click Close tab to remove the current page. Tabbed browsing means you have multiple web pages open in a window, with each web page on a separate tab. Display a web page, click the star on the toolbar, and then click the Add button to save a page to the Favorites list. Open the Hub to show Favorites, Reading list, History, and Downloads in a panel at the right side of the window. Click Make a Web Note on the current page to display the Web Note tools for annotating a web page that can be saved to a OneNote notebook or shared via email.	Microsoft Edge Back Forward Address bar home page New tab tabbed browsing Favorites Hub
2.3 Navigating the Web Using Chrome	Google Chrome is a free web browser for Windows-compatible PCs, Linux PCs, or Macs. Start Chrome by clicking the Chrome icon on the taskbar or by double-clicking the icon on the desktop. Display a web page by typing a URL or search phrase in the Address bar. Click the white star to bookmark the current web page. Use the Chrome menu to access the Bookmarks menu and turn on or turn off the display of the Bookmarks bar that docks below the Address bar. Open the Find bar from the Chrome menu to locate all occurrences of a word or phrase on the current web page.	Chrome Bookmarks bar Chrome menu Find bar

continued...

Topic	Key Concepts	Key Terms
2.4 Navigating the Web Using Firefox	Mozilla Firefox is a free web browser for Windows-compatible PCs, Linux PCs, or Macs. Start Firefox by clicking the Firefox icon on the taskbar or by double-clicking the icon on the desktop. Display a web page by typing a URL or search phrase in the Address bar and then pressing Enter. Bookmark the current web page by clicking the star. A bookmarked page displays a blue star. Click Show your bookmarks to show the bookmarks menu system. Click Open menu, then click *Find* to open the Find bar and locate all occurrences of a word or phrase on the current web page. Type a search word or phrase in the Search bar to search for information using the default search engine.	Firefox Open a new tab Show your bookmarks Search bar
2.5 Searching for Information and Printing Web Pages	A company that indexes web pages by keywords using programs called *spiders* or *crawlers* and provides a search tool to find the pages is called a *search engine*. Use more than one search engine because you will get different results from each due to differences in spider and crawling programs and timing differentials. To find information using a search engine, launch a web browser, type the URL for the desired search engine, type a search word or phrase in the search text box, and then press Enter. Each search engine provides tools for advanced searching, which narrows the search results list. For example, at Google, the Advanced Search page provides options to include or exclude words in the search and restrict search results to types of domains, file formats, country, or language. Preview and print a web page by accessing the Print option from a menu in the browser (in Microsoft Edge, go to the More menu, then click *Print*).	search engine spider crawler
2.6 Downloading Content from a Web Page	Copying an image, audio clip, video clip, or music file from a web page to your PC or mobile device is called *downloading*. Some websites, such as flickr.com, provide a download tool that facilitates downloading an image on their website to the Downloads folder. If no download tool is on the web page, saving content generally involves right-clicking the object and selecting a save option at the shortcut menu.	downloading

Recheck
Recheck your understanding of the topics covered in this chapter.

Workbook
Chapter review and assessment resources are available in the *Workbook* ebook.

Chapter 3

Exploring Microsoft Office 2016 Essentials

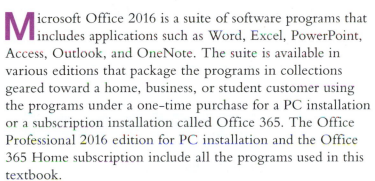

Precheck

Check your understanding of the topics covered in this chapter.

Microsoft Office 2016 is a suite of software programs that includes applications such as Word, Excel, PowerPoint, Access, Outlook, and OneNote. The suite is available in various editions that package the programs in collections geared toward a home, business, or student customer using the programs under a one-time purchase for a PC installation or a subscription installation called Office 365. The Office Professional 2016 edition for PC installation and the Office 365 Home subscription include all the programs used in this textbook.

One reason the Microsoft Office suite is popular is because several features or elements are common to all the programs. Once you learn your way around one of the applications in the suite, another application looks and operates similarly, making the learning process faster and easier.

In this chapter, you will navigate the Microsoft Office 2016 interface and perform file-related tasks or routines common to all the applications. You will customize the Quick Access Toolbar and choose options using multiple methods. You will also save and open files to and from OneDrive, an online file storage option.

SNAP If you are a SNAP user, go to your SNAP Assignments page to complete the Precheck, Tutorials, and Recheck.

Data Files

Before beginning this chapter, be sure you have copied the student data files for this course to your storage medium. Steps on downloading and extracting the data files are provided in Chapter 1, Topic 1.8, on pages 22–23.

Learning Objectives

3.1 Start an Office program, identify common features, and explore the ribbon interface

3.2 Open, save, print, export, close, and start new documents in the backstage area

3.3 Customize and use the Quick Access Toolbar

3.4 Select text and objects, perform commands using the ribbon and Mini toolbar, and select options in a dialog box

3.5 Copy and paste using buttons in the clipboard and format an object using a task pane

3.6 Use the Tell Me feature to find options and help resources

3.7 Save files to and open files from OneDrive, navigate longer documents, and use Undo

3.8 Change the Zoom setting and screen resolution

Skills

Start an Office program

Switch between programs

Start a new presentation

Explore the ribbon interface

Opening a Blank Document

Opening a Document from the Recent Options List

3.1 Starting and Switching Programs, Starting a New Presentation, and Exploring the Ribbon Interface

All the programs in the Microsoft Office suite begin at a Start screen and share some common features and elements. An application can be started using an icon on the desktop, from a tile or program name on the Start menu, or by using the search text box on the taskbar.

Microsoft Office 2016 Editions

The Microsoft Office 2016 suite is packaged in various collections of programs as one-time purchased software. Table 3.1 lists three of these editions. The suite is also available through a monthly or annual subscription as Office 365. **Office 365** includes Word, Excel, PowerPoint, OneNote, Outlook, Publisher, and Access, along with Skype calling minutes, additional OneDrive storage, and technical support from Microsoft. Other advantages to an Office 365 subscription is the suite is accessible on up to five devices, which can include Apple and Android devices, and updates to the software are automatically pushed to each installation.

Microsoft also offers a special edition of Office 365 for students called Office 365 University, which is available only to verified students and faculty of postsecondary institutions. Other editions for installation on a Mac computer and collections of software and services geared toward business environments are available.

Table 3.1

Microsoft Office 2016 One-Time Purchase Editions

Edition	What It Includes
Office Home and Student 2016	Word, Excel, PowerPoint, and OneNote
Office Home and Business 2016	Word, Excel, PowerPoint, OneNote, and Outlook
Office Professional 2016	Word, Excel, PowerPoint, Access, Publisher, OneNote, and Outlook

Microsoft Office 2016 System Requirements

Table 3.2 on the next page provides the standard system requirements for installing Microsoft Office 2016. Generally, if the PC can successfully run Windows 7, Windows 8.1, or Windows 10, then Office 2016 will also work on the same hardware, provided there is enough free disk space for the program files.

Starting a Program in the Microsoft Office 2016 Suite

To start a program in the Microsoft Office 2016 suite, double-click the program icon on the desktop or click the program tile in the right pane on the Start menu. If a tile for the application is not in the right pane on the Start menu, click *All apps* in the left pane, scroll down, and then click the program name in the left pane.

A quick way to find and start an application is to type the first few letters of the program name in the search text box on the taskbar and then click the name in the search results list or press Enter if you have typed the entire program name. For example, type Word 2016 in the search text box and then press Enter to start Microsoft Word 2016.

Table 3.2

Microsoft Office 2016 System Requirements

Hardware Component	Requirement
Processor	1 gigahertz (GHz) or faster processor
Memory	2 gigabytes (GB) of RAM Mac computer requires 4 GB RAM
Disk space	3.0 GB available free space Mac computer requires 6.0 GB available disk space
Operating system	Windows 7, Windows 8.1, or Windows 10 Mac computer requires Mac OS X 10.10 or newer
Display setting	1280 x 800 screen resolution or higher
Browser	Current versions of Microsoft Edge, Chrome, Firefox, Safari, Internet Explorer

1 Click in the search text box on the taskbar, type *word 2016*, and then press Enter.

Word 2016 is the application within the Microsoft Office suite used to create, edit, and format text-based documents. When you start Word, the Start screen appears. The **Start screen** for Microsoft Office applications shows the *Recent* option list used to open a document worked on recently.

2 Click the Maximize button ☐ near the top right of the window if Word does not fill the screen. Skip this step if you see the Restore button ❐ since the window is already maximized.

3 Compare your screen with the Word Start screen shown in Figure 3.1.

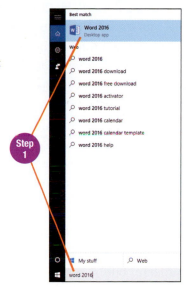

Step 1

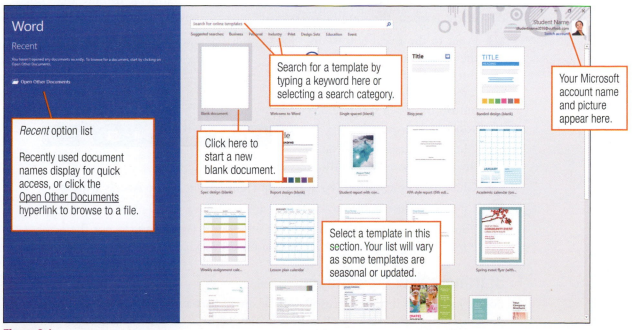

Figure 3.1

Word 2016 opens with the Word Start screen shown above.

④ Click the Start button and then click the Excel 2016 tile in the right pane on the Start menu. If an Excel 2016 tile is not available, click *All apps* in the left pane, scroll down the program list, and then click *Excel 2016.* Click the Maximize button if the Excel window does not fill the screen.

Use **Excel 2016** when your focus is on entering, calculating, formatting, and analyzing numerical data.

⑤ Compare the Excel Start screen with the Word Start screen shown in Figure 3.1 on the previous page.

Each Office application offers templates in the template gallery on the Start screen. A template is a preformatted document, worksheet, presentation, or database. Click a thumbnail to preview a larger picture of the template. The first thumbnail is used to start a new blank document, workbook, presentation, or database.

⑥ Start PowerPoint 2016 by typing PowerPoint 2016 in the search text box and pressing Enter or by using the Start menu. If necessary, maximize the window.

Create slides for an oral or kiosk-style presentation that includes text, images, sound, video, or other multimedia in **PowerPoint 2016**.

⑦ Compare the PowerPoint Start screen with the Word Start screen shown in Figure 3.1.

⑧ Start Access 2016 using any method described in Step 1 or Step 4. If necessary, maximize the window and then compare the Access Start screen with the Word Start screen shown in Figure 3.1.

Access 2016 is a database program in which you organize, store, and manage related data, such as information about customers or products.

Switching between Office Programs

As learned in Chapter 1, you switch to another open program by clicking the program button on the taskbar. Point to a button on the taskbar to see a thumbnail appear above the button with a preview of the open document. If more than one document is open, the thumbnails make it easy to switch to the correct window.

⑨ Point to the Excel button on the taskbar to preview the thumbnail of the Excel Start screen and then click the button to switch to the Excel window.

Step 9

⑩ Click the Access button on the taskbar to switch to the Access window.

⑪ Click the Close button (✕) at the top right corner of the Access window.

⑫ Click the Close button to close Excel.

⑬ You should now see the PowerPoint window. If PowerPoint is not the active window, click the PowerPoint button on the taskbar.

Starting a New Presentation

For any program in the Microsoft Office suite, start a new document, workbook, presentation, or database by clicking the new blank document, workbook, presentation, or database thumbnail in the Templates gallery on the Start screen.

14 Click *Blank Presentation* in the Templates and themes gallery on the PowerPoint Start screen.

Exploring the Ribbon Interface

The ribbon interface appears along the top of each Office application. Buttons within the ribbon are used to access commands and features within the program. The ribbon is split into individual tabs, with each tab divided into groups of related buttons, as shown in Figure 3.2. Word, Excel, PowerPoint, and Access all have the File and Home tabs as the first two tabs in the ribbon. The File tab is used to perform document-level routines, such as saving, printing, and exporting. This tab is explored in the next topic. The Home tab always contains the most frequently used features in each application, such as formatting and editing buttons.

This is the active tab. Click a tab name to change the active tab.

A new Draw tab was added to Word, Excel, and PowerPoint for touchscreen users.

Your background design may vary.

Ribbon Display Options

The Font Group organizes related buttons into one part of the ribbon.

Figure 3.2
The PowerPoint ribbon appears along the top of the application window, as it does for all Office applications. Shown here is the PowerPoint ribbon for a touchscreen device.

The **Ribbon Display Options** button near the top right of the window is used to change the ribbon display from *Show Tabs and Commands* to *Show Tabs*, which hides the command buttons until you click a tab, or *Auto-hide Ribbon*, which displays the ribbon only when you click along the top of the window.

15 Click the Insert tab to view the groups and buttons on the Insert tab ribbon for PowerPoint.

16 Click the Word button on the taskbar and then click *Blank document* on the Start screen to start a new Word document.

17 Click the Insert tab to view the groups and buttons on the Word Insert tab ribbon.

18 Click the Design tab in Word and then review the buttons on the Word Design tab ribbon.

19 Switch to PowerPoint and then click the Design tab.

20 Spend a few moments exploring other tabs and then close PowerPoint.

21 Spend a few moments exploring other tabs in Word and then close Word.

Oops!
Ribbon is showing more or fewer buttons, or the buttons look different than shown in Figure 3.2? The screen resolution for your PC or mobile device affects the ribbon display. In Topic 3.8, you will learn how to change the display. For now, the instructions will not ask you to click anything you cannot see or identify by the button label or icon.

3.2 Using the Backstage Area to Manage Documents

The File tab opens the **backstage area** in the Microsoft Office applications. The backstage area is where you find file management options, such as *Open, Save, Save As, Print, Share, Export,* and *Close.* You also go to the backstage area within an application to start a new document. Other tasks performed in the backstage area include displaying information about a document, protecting a document, and managing the document properties and versions.

Each application in the Microsoft Office suite provides options that can be personalized at the backstage view in the Options dialog box. Your Microsoft account and/or connected services and the background or theme for all the Office applications are managed in the Account backstage area.

Note: If necessary, insert your USB flash drive into an empty USB port before starting this topic. Close the File Explorer window if it opens after you insert the USB flash drive.

1. Start Word 2016.

2. At the Word Start screen, click Open Other Documents in the left pane.

3. Click the *Browse* option.

4. If necessary, scroll down the list of places in the Navigation pane in the Open dialog box until you can see the entry for your USB flash drive.

5. Click the entry in the Navigation pane for your USB flash drive.

6. Double-click *Student_Data_Files* in the Content pane.

7. Double-click *Ch3.*

The Word document **Cottage_rental_listing** is displayed in the Content pane. Within an Office application, the Open dialog box by default shows only files created in the active application.

8. Double-click *Cottage_rental_listing.*

In the next steps, you will use the Save As command to save a copy of the document in another folder.

9. Click the File tab to display the backstage area.

When a document is open, the backstage area displays the *Info* option with document properties for the active file and buttons to protect, inspect, and manage versions of the document.

Step 2

Open Other Documents

Step 3

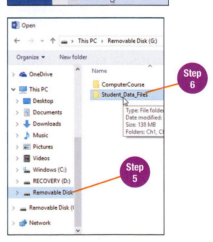

Step 6

Step 5

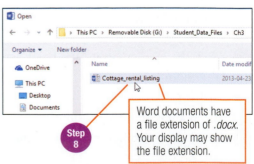

Step 8

Word documents have a file extension of *.docx.* Your display may show the file extension.

10 Click *Save As*.

The Save As command is used to save a copy of a document in another location or to save a copy of the document in the same location but with a different file name.

11 With *This PC* already selected in the Save As backstage area, click the <u>More options</u> hyperlink below the location, file name text box, and file type option box that displays at the top of the right panel.

Step 10

Step 11

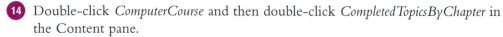

12 Click the Up arrow button at the left of the Address bar.

13 Click the Up arrow button a second time.

Step 12

14 Double-click *ComputerCourse* and then double-click *CompletedTopicsByChapter* in the Content pane.

15 Click the New folder button on the Command bar, type **Ch3**, and then press Enter.

16 Double-click *Ch3*.

Step 15

17 Click in the *File name* text box or drag to select the current file name, type CottageListing-YourName, and then press Enter or click the Save button.

Step 17

18 Click in the document next to *List Date:* below the address *3587 Bluewater Road* (in the second column of the table) and then type the current date.

19 Click next to *Assigned Agent:* below the date (in the second column of the table) and then type your name.

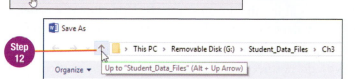

COTTAGE LISTING	
Owner:	Sandstone Cottage Rentals Inc.
	P. O. Box 364
	St. Paul, MN 55102
	info@sandstone.emcp.net
Cottage:	Cottage Number 13-598
Address:	3587 Bluewater Road
List Date:	October 8, 2018
Assigned Agent:	Student Name

Step 18

Step 19

Your date will vary.

Quick Steps

Open a Document
1. At Word Start screen, click the hyperlink <u>Open Other Documents</u>.
2. Click *Browse*.
3. Navigate to drive and/or folder and then double-click file name.

Use Save As to Save a Copy of a Document
1. Click File tab.
2. Click *Save As*.
3. Click <u>More options</u> hyperlink in right panel.
4. Navigate to desired drive and/or folder.
5. If necessary, change file name.
6. Click Save.

App Tip

You do not need to open the Save As dialog box if you want to save a copy of the document in another folder on the same drive or change the file name. Use the Up arrow button and/or the file name text box at the top of the right panel at the Save As backstage area to make the desired changes.

Oops!

Typing mistake? Press Backspace to delete what you have typed if you make an error and then retype the text. You can also drag across text and press Delete.

Printing a Document

Display the Print backstage area when you want to preview and print a document. Before printing, review the document in the **Print Preview** panel of the backstage area shown in Figure 3.3. The bottom of the Print Preview panel shows the number of pages needed to print the document, and navigation buttons are included to move to the next page and previous page in a multipage document.

In the Print panel, choose the printer on which to print the document in the *Printer* section, and modify the print settings and page layout options in the *Settings* section. When you are ready to print, click the Print button.

20 Click the File tab and then click *Print*.

21 Examine the document in the Print Preview panel, check the name of the default printer, and review the default options in the *Settings* section.

When print settings are changed, the options are stored with the document, so you do not need to change them again the next time you want to print.

22 Click the Print button to print the document.

The document is sent to the printer, and the backstage area closes.

Figure 3.3

Print backstage area

Exporting a Document as a PDF File

Many people exchange documents in PDF format via email or websites. The advantage of a PDF file is that the document looks and prints as it would in the application in which it was created but without having to open or install the source program. Export a Word document as a PDF if you need to send a document to someone who does not have Word installed on his or her computer.

A **PDF document** is a document saved in portable document format, an open standard for exchanging electronic documents developed by Adobe Systems. A PDF document can be viewed in any web browser.

23 Click the File tab and then click *Export*.

24 With *Create PDF/XPS Document* selected in the Export backstage area, click the Create PDF/XPS button.

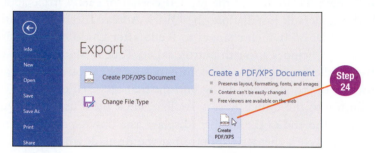

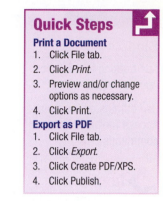

Quick Steps

Print a Document
1. Click File tab.
2. Click *Print.*
3. Preview and/or change options as necessary.
4. Click Print.

Export as PDF
1. Click File tab.
2. Click *Export.*
3. Click Create PDF/XPS.
4. Click Publish.

25 Click the Publish button at the Publish as PDF or XPS dialog box.

By default, the PDF is created in the same drive and folder in which the Word document resides and with the same file name but with the file extension *.pdf.* Because PDF files have a different file extension, the same name can be used for both the Word document and the PDF document.

26 The published PDF document opens in a reader app, such as Windows Reader or Adobe Reader, or in a browser window. If no program is associated with PDF documents, a message box similar to the one shown at the right opens with the prompt *How do you want to open this file?* In that case, click OK to open the PDF document in the selected option *Microsoft Edge.*

27 Close the window in which the PDF document opened to return to Word.

28 Click the File tab and then click *Close.*

29 Click the Save button when prompted at the message box asking if you want to save your changes. Leave Word open for the next topic.

Close a document when you are finished editing, saving, printing, and publishing. A blank window displays when no documents are open.

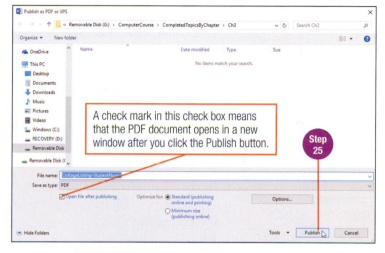

A check mark in this check box means that the PDF document opens in a new window after you click the Publish button.

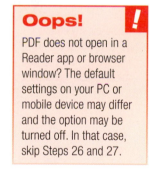

Oops!

PDF does not open in a Reader app or browser window? The default settings on your PC or mobile device may differ and the option may be turned off. In that case, skip Steps 26 and 27.

View
Model Answer
Compare your completed file with the model answer.

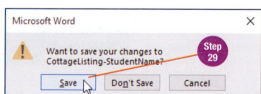

Beyond Basics   **Pinning Documents and Folders to *Recent* option lists**

At the Open backstage area, you can pin a frequently used file to the *Recent* option list so that the file name always appears in the *Recent* option list. Point to the file name in the *Recent* option list and then click the pushpin icon that displays at the right to pin the item to the list.

3.3 Customizing and Using the Quick Access Toolbar

The **Quick Access Toolbar** is at the top left corner of each Office application window. With the default installation, the Quick Access Toolbar has buttons to Save, Undo, and Repeat (which changes to Redo after Undo has been used). A touchscreen device includes a fourth button called Touch/Mouse Mode that is used to optimize the spacing between commands for the mouse or for touch. Most people customize the toolbar by adding buttons that are used often.

To add a button to the Quick Access Toolbar, click the Customize Quick Access Toolbar button at the end of the toolbar (displays as a down-pointing arrow with a bar above) and then click the desired button at the drop-down list.

Note: Skip this topic if the Quick Access Toolbar on the computer you are using already displays the New, Open, Quick Print, and Print Preview and Print buttons in Word, PowerPoint, and Excel. Skip any steps in which a drop-down list option already displays with a check mark, which means the button is already on the toolbar.

1. At a blank Word screen, click the Customize Quick Access Toolbar button.

2. Click *New* at the drop-down list.

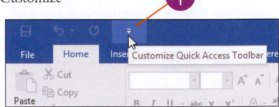

Options displayed with a check mark are already added to the toolbar. Buttons added to the Quick Access Toolbar are placed at the end of the toolbar.

3. Click the Customize Quick Access Toolbar button.

4. Click *Open* at the drop-down list.

5. Click the Customize Quick Access Toolbar button.

6. Click *Quick Print* at the drop-down list.

7. Click the Customize Quick Access Toolbar button.

8. Click *Print Preview and Print* at the drop-down list.

Touch-enabled devices also show the Touch/Mouse Mode button.

Toolbar after buttons added in Steps 1–8

9. Click the Open button on the Quick Access Toolbar.

The Open backstage area opens with the *Recent* option list.

10 Click the file named *CottageListing-StudentName* in the *Recent* options list.

Step 10

Quick Steps

Add a Button to the Quick Access Toolbar
1. Click Customize Quick Access Toolbar button.
2. Click desired button at drop-down list.

11 Click the Print Preview and Print button on the Quick Access Toolbar.

The Print backstage area opens.

Step 11

12 Click the Back button (left-pointing arrow inside circle) to return to the document without printing.

13 Start PowerPoint 2016.

14 Click Blank Presentation at the PowerPoint Start screen.

Step 12

App Tip

The Quick Print button added to the Quick Access Toolbar automatically sends the current document to the printer using the active printer and print settings.

Notice the customized Word Quick Access Toolbar does not carry over to other Microsoft applications.

15 Customize the Quick Access Toolbar in PowerPoint by adding the New, Open, Quick Print, and Print Preview and Print buttons.

Step 15

16 Close PowerPoint.

17 Start Excel 2016 and then click Blank Workbook at the Excel Start screen.

18 Customize the Excel Quick Access Toolbar to add the New, Open, Quick Print, and Print Preview and Print buttons.

Step 18

19 Close Excel.

20 At the Word document, click the File tab and then click *Close*.

21 Click the New button on the Quick Access Toolbar.

A new blank document window opens. Leave this document open for the next topic.

Step 21

New Blank Document (Ctrl+N)
Create a document.

App Tip

For Office applications, the keyboard command Ctrl + F4 closes the current document and Alt + F4 closes the program.

Beyond Basics   **Moving the Quick Access Toolbar below the Ribbon**

You may prefer to have the Quick Access Toolbar display below the ribbon so that the buttons are closer to the area in which you are working. To do this, click the Customize Quick Access Toolbar button and then click *Show Below the Ribbon* at the drop-down list.

Quick Access Toolbar below the ribbon

3.4 Selecting Text or Objects, Using the Ribbon and Mini Toolbar, and Selecting Options in Dialog Boxes

Skills

Select text and objects

Perform commands using the ribbon and Mini toolbar

Display a task pane and dialog box

Choose options in a dialog box

Tutorials

Selecting, Replacing, and Deleting Text

Applying Font Formatting Using the Mini Toolbar

Applying Font Formatting Using the Font Dialog Box

Creating a document, worksheet, presentation, or database involves working with the ribbon to select options or perform commands. Some options involve using a button, list box, or gallery, and some commands cause a task pane or dialog box to open in which you select options.

In many instances, before you choose an option from the ribbon, you first select text or an object as the target for the action. Select text by clicking within a word, paragraph, cell, or placeholder or by dragging across the text you want to select. Select an object, such as a picture or other graphic, by clicking the object.

Selected objects display with a series of selection handles. A **selection handle** is a circle at the middle and/or corners of an object, or at the beginning and end of text on a touch-enabled device. Selection handles are used to manipulate the object or to define the selection area on a touch-enabled device. Table 3.3 provides instructions for selecting text using a mouse or touch and for selecting an object.

Table 3.3

Selecting Text and Objects Using the Mouse and Touch

Selecting Text Using a Mouse	Selecting Text Using Touch	Selecting Objects
Point at the beginning of the text or cell to be selected. The pointer displays as I, called an *I-beam* in Word and PowerPoint, or as ✛, called a *cell pointer* in Excel. Hold down the left mouse button and drag to the end of the text or cells to be selected. Release the mouse button. Summer vacation destinations I In some cases, a Mini toolbar displays when you release the mouse after selecting text. Summer vacation destinations I Mini toolbar	Tap at the beginning of the text to be selected. A selection handle appears below the text (displays as an empty circle). Summer vacation destinations ◯ — selection handle Slide the selection handle right to move the insertion point to edit text. Alternatively, double-tap at the beginning of the text to select the first word and show a second selection handle, then slide the right selection handle to extend the selection. Summer vacation destinations ◯ ◯ After releasing your finger, you can slide the left or right selection handle to redefine the area if necessary. To display the Mini toolbar, press and hold inside the selected text area. The toolbar appears when you release your finger and displays already optimized for touch. Summer vacation destinations Mini toolbar	Click anywhere over the object. Selection handles appear at the ends, corners, and middle (depending on width and height) of each side of the object. A Layout Options button also appears next to a selected object with options for aligning and moving the object with surrounding text. Layout Options button for selected object

1 At a blank Word screen, type Summer vacation destinations and then press Enter.

2 Type Explore the beaches of Florida and experience the Florida sunset with friends or family. and then press Enter.

3 Select the title *Summer vacation destinations* to display the Mini toolbar.

The **Mini toolbar** appears near text after text is selected or with the shortcut menu when you right-click a selection. The toolbar contains frequently used formatting commands for quicker access within the work area than using the ribbon.

If necessary, refer to the instructions in Table 3.3 (page 74) for selecting text using the mouse or touch.

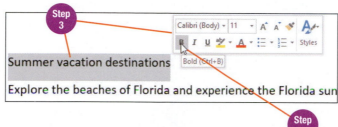

4 Click the Bold button on the Mini toolbar.

5 Click in the blank line below the sentence that begins with *Explore* to deselect the title.

6 Click the Insert tab on the ribbon.

7 Click the Pictures button in the Illustrations group.

This opens the Insert Picture dialog box.

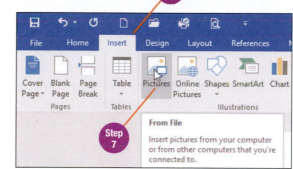

8 If necessary, scroll down the list of places in the Navigation pane. Click the entry in the Navigation pane for your USB flash drive.

9 Double-click *Student_Data_Files*, double-click *Ch3,* and then double-click *FloridaSunset*.

The FloridaSunset picture is inserted in the document and is automatically selected.

10 Drag the selection handle at the bottom right corner of the image until the picture is resized to the approximate height and width shown at the right.

The mouse pointer changes shape to a double-headed diagonal arrow () when you point

at the bottom right corner of the image. Drag downward and to the right when you see this icon. The pointer changes shape to a crosshairs—a large, thin, black cross (✛)—while you drag the mouse. When you release the mouse, the selection handles reappear.

Quick Steps

Select Text or an Object
Click or drag in word, paragraph, cell, or placeholder, or click the object.

Format Text Using the Mini Toolbar
1. Select text.
2. Click desired button on Mini toolbar.

App Tip

All the buttons available on the Mini toolbar are also available on the ribbon.

Oops!

No selection handles? If you click away from the object, the selection handles disappear. Click the picture to redisplay the selection handles.

11 Click the Save button on the Quick Access Toolbar.

12 Click *This PC* and then click the More options hyperlink in the right panel. At the Save As dialog box, navigate to the Ch3 folder within CompletedTopicsByChapter and ComputerCourse on your USB flash drive.

13 Type VacationDestinations-YourName and then press Enter or click the Save button.

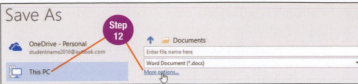

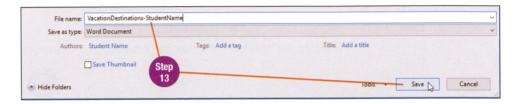

Working with Objects, Contextual Tabs, and Dialog Boxes

An **object** is a picture, shape, chart, or other item that can be manipulated separately from text or other objects around it. When an object is selected, a contextual ribbon tab appears. More than one contextual tab may appear. A **contextual tab** contains commands or options that are related to the type of object that is currently selected.

Some buttons in the ribbon display a drop-down gallery. A **gallery** displays visual representations of options for the selected item in a drop-down list or grid. Pointing to an option in a gallery displays a **live preview** of the text or object if the option is applied. Live previews let you see how formatting will look before applying an option.

14 With the image still selected, click the Picture Tools Format tab if the tab is not the active tab, and then click the Corrections button in the Adjust group.

15 Point to the last option in the Corrections gallery. Notice the picture brightens significantly when you point to the option.

16 Click the last option in the Corrections gallery to apply the *Brightness: +40% Contrast: +40%* correction.

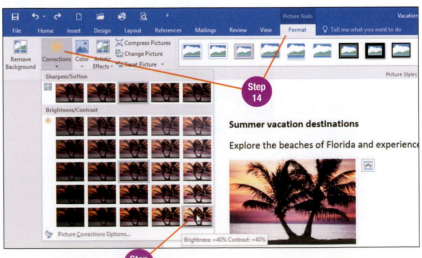

Some ribbon groups have a small button at the bottom right corner of the group that is a diagonal downward-pointing arrow (⌐). This button is called a **dialog box launcher**. Clicking the button causes a task pane or a dialog box to appear. A **task pane** appears at the left or right side of the window, whereas a **dialog box** opens in a separate window above (or, from the viewer's perspective, in front of) the document. Task panes and dialog boxes contain more options as buttons, lists, sliders, check boxes, text boxes, and option buttons.

17 Click the dialog box launcher at the bottom right corner of the Picture Styles group.

The Format Picture task pane opens at the right. You will work in a task pane in the next topic.

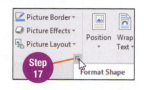

18 Click the Close button (✕) at the top right corner of the Format Picture task pane to close the pane.

19 Click the dialog box launcher at the bottom right corner of the Size group.

The Layout dialog box opens with the Size tab active.

20 Select the current value in the *Absolute* text box in the *Width* section and then type 3.

21 Click the Text Wrapping tab and then click the *Square* option in the *Wrapping style* section.

22 Click OK.

23 With the picture still selected, position the pointer on top of the picture and then drag the picture up to the top of the document, releasing it when the green horizontal and vertical alignment guides show the picture aligned at the top and left margins.

An **alignment guide** is a colored horizontal or vertical line that appears when you are moving an object to help you align and place the object within the document boundaries or in relation to surrounding text or other nearby objects.

24 Click the Save button on the Quick Access Toolbar. Leave the document open for the next topic.

Because the document has already been saved once, the Save button saves the changes using the existing file name and location.

Alignment guides show the picture is aligned at the top left margin.

3.5 Using the Office Clipboard and Formatting Using a Task Pane

The Clipboard group is standardized across Microsoft Office programs. The buttons in the Clipboard group are Cut, Copy, Paste, and Format Painter. You used Cut, Copy, and Paste in Chapter 1 when you learned how to move and copy files and folders. Cut, Copy, and Paste are also used to move or copy text or objects.

Format Painter is used to copy formatting options from selected text or an object to other text or another object.

1 With the **VacationDestinations-YourName** document still open, start PowerPoint 2016, and then click *Blank Presentation*.

2 Click anywhere in *Click to add title* on the blank slide and then type Florida Sunset.

3 Click the Word button on the taskbar.

4 Select the sentence that begins with *Explore* below the title and then click the Copy button in the Clipboard group on the Home tab.

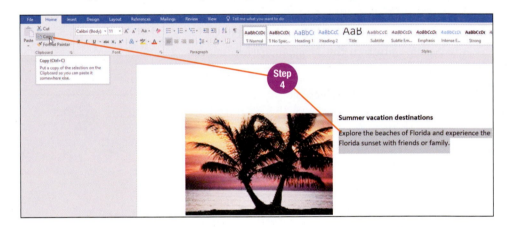

5 Click the PowerPoint button on the taskbar to switch to PowerPoint and then click anywhere in *Click to add subtitle* on the slide.

6 Click the top part of the Paste button in the Clipboard group. (Do *not* click the down-pointing arrow on the button.)

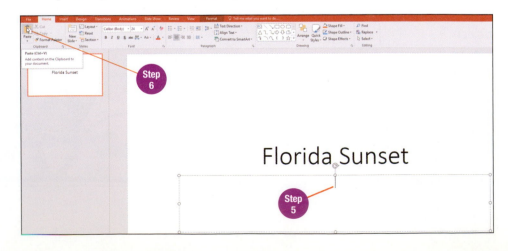

The selected text that was copied from Word is pasted on the slide in PowerPoint. Notice also the Paste Options button that appears below the pasted text.

7 Click the Paste Options button to display the Paste Options gallery.

Paste Options vary depending on the pasted text or object. Buttons in the gallery allow you to change the appearance or behavior of the pasted text or object in the destination location.

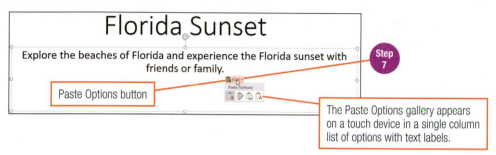

8 Click the Word button on the taskbar to switch to Word.

9 Click to select the picture, click the Copy button, click the PowerPoint button on the taskbar, and then click the Paste button. (Remember not to click the down-pointing arrow on the Paste button.)

The pasted picture is dropped onto the slide overlapping the text.

10 Click the Paste Options button. Notice that the Paste Options for a picture are different than the Paste Options for text.

11 Click in white space away from the picture to remove the Paste Options gallery.

12 If necessary, click to select the picture object.

13 Drag the selected picture below the text. Release the picture when the orange guide shows the picture is aligned with the middle of the text placeholders.

14 Click the Save button on the Quick Access Toolbar.

Oops!

A Design Ideas task pane opened after you pasted the picture on the slide? Click No Thanks at the bottom of the task pane to close it. PowerPoint Designer is a new service that opens when you add a photo to a title slide.

App Tip

Paste Options are helpful if different formatting exists between the source or destination documents, and you want to control which formatting is applied to the pasted text or object.

15 At the Save As backstage area, click *This PC* and then click *Browse*.

16 Navigate to the Ch3 subfolder in the CompletedTopicsByChapter folder on your USB flash drive.

17 Select the current text in the *File name* text box, type FloridaSunset-YourName, and then press Enter or click Save.

Using Format Painter to Copy Formatting Options

Sometimes instead of copying text or an object, you want to copy formatting options. Format Painter copies to the clipboard the formatting attributes for selected text or an object.

Click the Format Painter button to do a one-time copy of formatting options or double-click the button if you want to paste the formatting options multiple times. Double-clicking the Format Painter button turns the feature on until you click the button again to turn the feature off. A button that operates as on or off is called a **toggle button**.

18 Select the first occurrence of the word *Florida* in the subtitle on the slide.

19 Click the Font Color button arrow (down-pointing arrow at the right of the Font Color button) in the Font group on the Home tab.

20 Click the *Purple* square (the last square in the *Standard Colors* section).

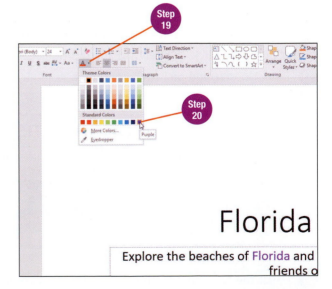

21 With the text still selected, click the Italic button in the Font group on the Home tab.

22 With the text still selected, click the Format Painter button (displays as a paint brush) in the Clipboard group.

23 Drag the mouse pointer with the paintbrush icon across the second occurrence of the word *Florida* in the subtitle.

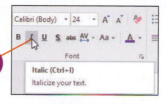

Step 21

Step 22

Explore the beaches of *Florida* and experience the Florida sunset with friends or family.

Step 23

24 Click in white space away from the selected text to deselect the text.

25 Click the Save button on the Quick Access Toolbar.

26 Click to select the picture and display the contextual tab.

27 Click the Picture Tools Format tab.

28 Click the *Drop Shadow Rectangle* option in the Picture Styles gallery (fourth picture style option).

Step 27

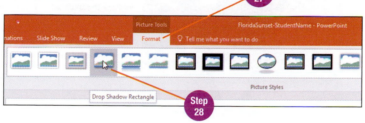

Step 28

Drop Shadow Rectangle

29 With the picture still selected, click the Home tab, click the Format Painter button, and then click the title text *Florida Sunset* at the top of the slide.

30 With the Florida Sunset placeholder selected, click the dialog box launcher in the Drawing group on the Home tab to open the Format Shape task pane at the right side of the window.

31 Click the Size & Properties tab (last option) in the Format Shape task pane.

32 Click Text Box to reveal the options below Text Box in the task pane.

33 Click the *Vertical alignment* option box arrow (down-pointing arrow at the right of *Bottom*) and then click *Middle*.

34 Close the Format Shape task pane.

35 Save and then close PowerPoint. Leave the Word document open for the next topic.

Step 34

Step 31

Step 32

Step 33

Florida Sunset

title aligned in the middle of the placeholder vertically at Steps 31–34

Explore the beaches of *Florida* and experience the *Florida* sunset with friends or family.

3.6 Finding Help or Options Using the Tell Me Feature

Skills

Use Tell Me to locate an option

Use Tell Me to look up a definition

Use Tell Me to find help information

 **Tutorials**

Using the Tell Me Feature

Using the Help Feature

A text box at the right of the last tab on the ribbon containing the text *Tell me what you want to do* is the new **Tell Me** feature in Microsoft Office 2016. Use the text box to type an option and quickly access the command or feature directly from the *Tell Me* text box instead of navigating the ribbon tabs. You can also use the *Tell Me* text box to type a word or phrase and search resources in a Help window to learn how to use an option or feature. The **Smart Lookup** option on the Tell Me drop-down list opens a task pane at the right side of the window that lets you see a definition of the term or explore web resources related to the term.

1. With the **VacationDestinations–YourName** document still open, click to place the insertion point after the period in the sentence below the document title, press Enter, and then type Visit http://ParadigmCollege.net/FloridaTours for information on our Florida vacation packages..

2. Press Enter and then type Our exclusive Got2GoSunset package is our most popular booking..

3. Click to place the insertion point after the last "t" in Got2GoSunset.

   Assume that Got2GoSunset is a trademarked name and you want to insert the trademark sign but do not know where the symbol feature is located.

4. Click in the *Tell Me* text box that contains the text *Tell me what you want to do* at the right of the View tab on the ribbon and then type insert trademark symbol.

5. Click *Insert a Symbol* in the drop-down list and then click the trademark symbol ™ in the Symbol palette that appears.

Check This Out ✓

http://CA2.Paradigm College.net/Office Support

Go here to find help at the Office website. Use the search text box to type a word or feature you need help with, or navigate the support website using the links below the search text box. Microsoft provides tutorials and video training by product at the Office Training Center accessed at this page.

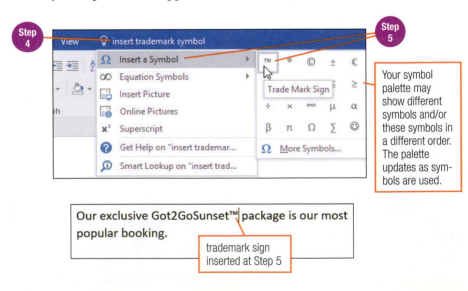

Your symbol palette may show different symbols and/or these symbols in a different order. The palette updates as symbols are used.

Our exclusive Got2GoSunset™ package is our most popular booking.

trademark sign inserted at Step 5

6 Click in the *Tell Me* text box and then type watermark.

7 Click *Smart Lookup on "watermark"*.

The Smart Lookup task pane opens at the right side of the window. In the Explore tab of the task pane are web links to resources related to watermark. In the Define tab, you can view a definition of watermark.

8 Click the Got it button in the Smart Lookup task pane if a message displays about data sent to Bing regarding the highlighted term and surrounding content. Skip this step if no message about privacy displays in the task pane.

9 Click the Define tab in the Smart Lookup task pane and then read the definition for watermark in the pane.

10 Close the Smart Lookup task pane.

11 Click in the *Tell Me* text box, type watermark, and then click *Get Help on "watermark"*.

A Word 2016 Help window opens with links to help articles about watermarks.

12 Scroll down if necessary and then click Insert a Watermark in Word 2016 for Windows in the Help window. Read the next page that provides information and the steps to add a watermark.

13 Click Back to return to the previous page.

14 Close the Help window.

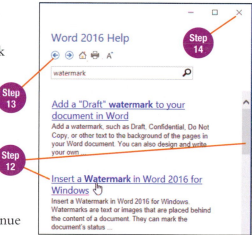

Inside the Help window, you can continue searching for other help information by typing a keyword or phrase in the search text box just below the toolbar and pressing Enter.

15 Click in the *Tell Me* text box, type watermark, click *Watermark* at the drop-down list, and then click *SAMPLE 1* in the watermark gallery.

16 Scroll down to view the SAMPLE watermark added to the document background.

17 Click the Save button on the Quick Access Toolbar and then close the Word window. Click No if prompted to save the copied item.

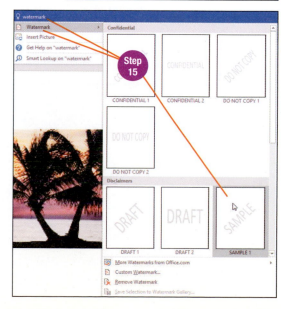

3.7 Using OneDrive for Storage, Scrolling in Documents, and Using Undo

Skills

Save and open files to and from OneDrive

Navigate in longer documents

Use Undo

Tutorials
Scrolling
Using Undo and Redo

OneDrive is secure online storage available to individuals signed in with a Microsoft account (often referred to as *cloud storage*). You can save files to and open files from OneDrive, giving you the ability to access the files from any Internet-connected device. The default storage location at the Save As backstage area when saving a new document, workbook, or presentation is OneDrive for Office programs.

When working with longer documents, you will need to scroll the display or change the zoom setting to see more or less text in the window. You will learn how to use the Zoom feature in the next topic. Undo reverses an action and is used often to restore a document when you make a mistake.

Note: To complete this topic, you need to be signed in with a Microsoft account. If you do not have a Microsoft account, skip the OneDrive section in Steps 2 to 19 and proceed to Step 20 after opening the presentation.

1. Start PowerPoint 2016 and then open the presentation **SpeechTechniques** from the Ch3 subfolder within the Student_Data_Files folder on your USB flash drive.

2. Click the File tab and then click *Save As*.

3. Click the *OneDrive - Personal* option at the Save As backstage area.

4. Click the *Documents* folder in the right panel.

App Tip

You can access the file you will save to OneDrive from any device, even if the device does not have a local copy of PowerPoint. You will learn about the Office Online applications in Chapter 15.

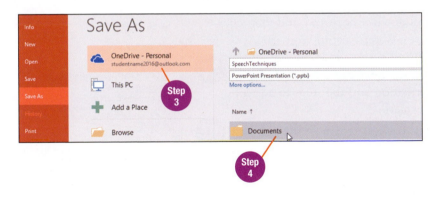

5. Select the text in the *File name* text box at the top of the right panel, type SpeechTechniques–YourName and then click the Save button.

App Tip

The Save button on the Quick Access Toolbar displays with two rounded arrows in a circle when you are working with a document saved to OneDrive.

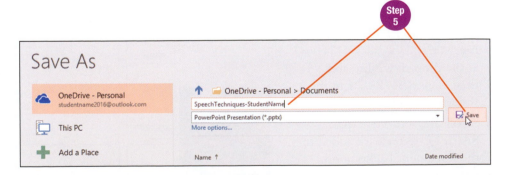

A progress message displays in the Status bar as the file is uploaded.

If you decide to use OneDrive for saving work, you should use folders to organize files. In the next steps, you will save the presentation to OneDrive a second time by creating a folder as you save the file.

6   Click the File tab and then click the *Save As* option.

7   With your OneDrive account already selected at the Save As backstage area, click the *Browse* option.

8   At the Save As dialog box, click the New folder button on the Command bar.

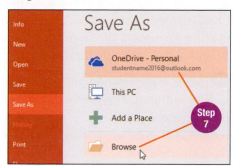

Notice that the file saved at Step 5 appears in the Content pane.

9   Type CompletedTopicsByChapter and then press Enter.

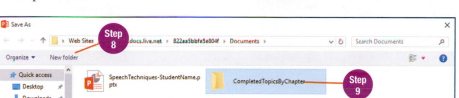

10   Double-click *CompletedTopicsByChapter*.

11   Click the New folder button on the Command bar, type Ch3, and then press Enter.

12   Double-click *Ch3*.

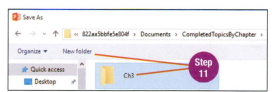

13   Click the Save button to save the presentation using the same name as before.

You now have two copies of the presentation saved on OneDrive.

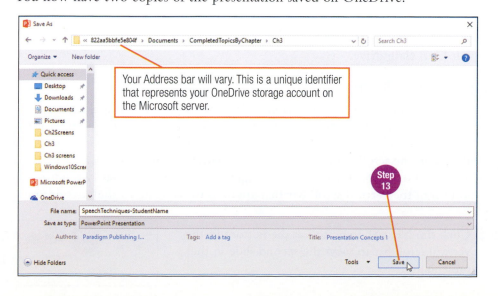

Your Address bar will vary. This is a unique identifier that represents your OneDrive storage account on the Microsoft server.

14 Click the File tab and then click *Close*.

Depending on the speed of your Internet connection, you may notice an Uploading to OneDrive message box while the file is uploaded to your OneDrive account.

15 Click the Open button on the Quick Access Toolbar.

Because you just closed the presentation, you will notice the presentation appears twice in the *Recent* list in the right panel. Notice that your OneDrive account is associated with each of these entries. You could reopen the presentation from OneDrive using the *Recent* list; however, in the next steps, you will use Browse to practice opening a file from OneDrive in case a file you need is not in the *Recent* panel for the Office application.

16 Click *Browse* at the Open backstage area.

You will notice the file you want to reopen is already in the Content pane because the Quick access feature shows recently opened presentations; however, you will practice navigating in case you ever need a file that has not been recently opened.

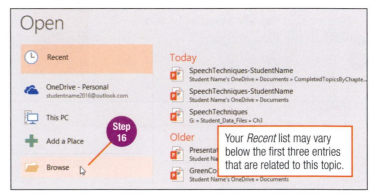

17 Click OneDrive in the Navigation pane.

18 In the Content pane, double-click the folders *Documents*, *CompletedTopicsByChapter*, and *Ch3*.

19 Double-click *SpeechTechniques-YourName* in the Content pane.

Navigate folders on OneDrive using the same methods as navigating folders on your USB flash drive.

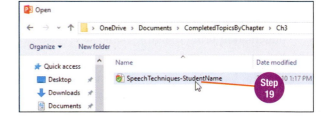

Using the Scroll Bars

Larger files with content beyond the current view display in the application window with a horizontal and/or vertical scroll bar. A **scroll bar** has arrow buttons at the top and bottom or left and right ends used to scroll up, down, left, or right. A **scroll box** in the scroll bar between the two arrow buttons is also used to scroll by dragging the box up or down, or left or right. PowerPoint also has two buttons at the bottom of the vertical scroll bar that are used to navigate to the next or previous slide.

20 Click the Next Slide button at the bottom of the vertical scroll bar to view the second slide in the presentation.

(21) Click the Next Slide button two more times to move to Slide 4.

(22) Click the up arrow button at the top of the vertical scroll bar repeatedly until you are returned to the first slide.

(23) Drag the scroll box at the top of the vertical scroll bar downward until you reach the end of the slides.

As you drag the scroll box downward, a ScreenTip displays the slide numbers and titles for the slides so you know when to release the mouse.

(24) Press Ctrl + Home.

Ctrl + Home is the universal keyboard shortcut for returning to the beginning of a file.

(25) Press Ctrl + End.

Ctrl + End is the universal keyboard shortcut for navigating to the end of a file.

Using Undo

The **Undo** command in all Office applications can be used to restore a document, presentation, worksheet, or Access object to its state before the last action that was performed. If you make a change to a file and do not like the results, immediately click the Undo button on the Quick Access Toolbar. Some actions, such as Save, cannot be reversed with Undo.

(26) Navigate to the first slide in the presentation.

(27) Select the title text *Speech Techniques*.

(28) Click the Bold button on the Mini toolbar or in the Font group on the Home tab.

(29) With the text still selected, click the Underline button on the Mini toolbar or in the Font group on the Home tab.

(30) Click on any part of the slide away from the selected text to deselect the title text.

(31) Click the Undo button on the Quick Access Toolbar. Do *not* click the down-pointing arrow on the button.

The underline is removed from the title text and the text is selected.

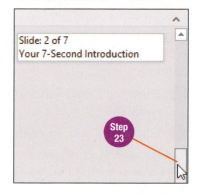

(32) Click the Undo button a second time to remove the bold formatting.

(33) Click on the slide away from the selected text to deselect the text and then click the Save button on the Quick Access Toolbar.

(34) Close the presentation and then close PowerPoint.

Quick Steps

Scroll within a File
Click arrow button for desired direction to scroll in horizontal or vertical scroll bar, or drag scroll box in horizontal or vertical scroll bar.

Use Undo
Click Undo button on the Quick Access Toolbar.

Oops!

Using Touch Keyboard? Skip Steps 24 and 25 because the Touch Keyboard does not have Home and End keys.

App Tip

Undo stores multiple actions. To undo an action that was not the last action performed, click the Undo button arrow and then click the desired action in the drop-down list.

App Tip

Use the Redo button on the Quick Access Toolbar if you change your mind after using Undo and want the action reapplied to the document.

View

Model Answer
Compare your completed file with the model answer.

3.8 Changing the Zoom Option and Screen Resolution

Skills

Change the Zoom setting

View the screen resolution and change the setting if possible to match textbook illustrations

Tutorial
Changing the Zoom

Word, Excel, and PowerPoint display a **Zoom slider** bar near the bottom right corner of the window. Using the slider, you can zoom out or zoom in to view more or less of a document, worksheet, slide, or presentation. At each end of the slider bar is a **Zoom In** (plus symbol) and a **Zoom Out** (minus symbol) control that increases or decreases the magnification by 10 percent at each click.

1. Start Excel 2016 and then open the workbook named **CutRateRentals** from the Ch3 folder in Student_Data_Files.

Notice the current Zoom setting near the bottom right corner of the Excel window, which is 100 percent if the setting is at the default option.

2. Click the Zoom In control (displays as a plus symbol).

3. Click the Zoom In control two more times.

The worksheet is now much larger in the display area with the magnification increased 30 percent.

4. Click the Zoom Out control (displays as a minus symbol).

5. Drag the Zoom slider left or right and watch magnification of the worksheet decrease or increase as you move the slider.

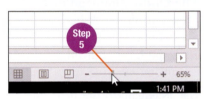

6. Drag the Zoom slider to the middle of the slider bar to change the zoom to 100 percent.

7. Click *100%* at the right of the Zoom In control.

This opens the Zoom dialog box, where you can choose a predefined magnification, type a custom percentage value, or choose the *Fit selection* option to fit a group of selected cells to the window.

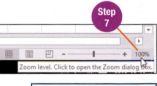

8. Click *75%* and then press Enter or click OK.

9. Return the zoom magnification to 100 percent by dragging the Zoom slider to the middle of the slider bar.

Zoom In, Zoom Out, the Zoom slider, and the Zoom dialog box function the same in Word and PowerPoint.

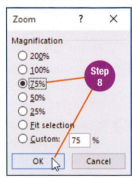

Oops! !

Having trouble using the Zoom slider on a touch device? Tap the controls or the percentage number to open the Zoom dialog box (Steps 7–8), or use the Zoom buttons on the View tab on the ribbon.

App Tip

The View tab in Word, PowerPoint, and Excel contain a Zoom group with buttons to change zoom magnification. Access does not include the Zoom feature.

Viewing and Changing Screen Resolution

The ribbon on your computer may show fewer or more buttons than the illustrations in this textbook, or some buttons may show icons only (no labels). The appearance of the ribbon is affected by the screen resolution (Figure 3.4 on page 89).

Screen resolution refers to the number of picture elements, called *pixels*, that form the image on the display. A pixel is a square with color values. Millions of pixels are used to render display images. Resolution is expressed as the number of horizontal pixels by the number of vertical pixels.

Note: Check with your instructor before proceeding onward. Some schools do not allow the display properties to be changed. If necessary, perform Steps 10–18 on your home PC.

10. Minimize the Excel window to display the desktop.

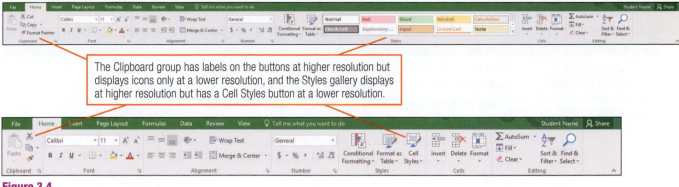

Figure 3.4

Shown is the Excel Home tab at 1920 x 1080 (top) and at 1280 x 1024 (bottom).

⑪ Right-click a blank area on the desktop and then click *Display settings*.

⑫ Click *Advanced display settings* at the bottom of the Customize your display panel in the Settings app with *Display* selected in the left pane. You may need to scroll down to the bottom of the right panel to find the option.

⑬ If the current setting for *Resolution* is *1920 x 1080*, skip to Step 16; otherwise, click the *Resolution* option box and then click *1920 x 1080* if the option is available; otherwise, use the resolution option that displays with *(Recommended)*.

⑭ Click the Apply button.

⑮ Click the Keep changes button.

⑯ Close the Settings app window.

⑰ Click the Excel button on the taskbar to view the Excel window at the new screen resolution.

⑱ If you changed your screen resolution, examine the screen to see if you like the new setting. If the interface is not of good quality, repeat Steps 10 to 16 to restore the resolution option to its original setting. The setting that displays with *(Recommended)* is usually the best choice for your device.

You do not need to change the screen resolution to the same setting used for the images in this textbook. Just be aware that some illustrations may not match exactly what is on your display if your resolution is at a different setting.

⑲ Close the **CutRateRentals** worksheet and then close Excel.

Quick Steps

Change Zoom Magnification
Click Zoom In or Zoom Out button or drag Zoom slider to desired setting
OR
1. Click zoom percentage.
2. Select desired zoom option at Zoom dialog box.
3. Click OK.

App Tip

Screen resolution is an operating system setting. A higher resolution uses more pixels and means the image quality is sharper or clearer. It also means more content can be displayed in the viewing area.

Topics Review

Topic	Key Concepts	Key Terms
3.1 Starting and Switching Programs, Starting a New Presentation, and Exploring the Ribbon Interface	The Microsoft Office suite is sold in various one-time purchase editions or as a subscription plan called Office 365.	Office 365
	Word 2016 is the application used for text-based documents.	Word 2016
	An Office program starts with the Start screen, which shows a list of recently opened files and a templates gallery.	Start screen
	Excel 2016 is used when the focus is on working with numeric data.	Excel 2016
	Create slides for an oral or kiosk-style presentation using PowerPoint 2016.	PowerPoint 2016
	Access 2016 is a database program used to organize, store, and maintain related data, such as information about customers or products.	Access 2016
	All programs display the ribbon along the top of the window, which contains buttons divided into tabs and groups for commands and features within the program.	Ribbon Display Options
	The Ribbon Display Options button lets you control whether the ribbon shows tabs only, tabs with commands, or no ribbon.	
3.2 Using the Backstage Area to Manage Documents	The backstage area is where you perform document-level options, such as *Open*, *Save*, *Save As*, *Print*, *Share*, *Export*, and *Close*.	backstage area
	Use the <u>Open Other Documents</u> hyperlink at the Word Start screen to navigate to a document not in the *Recent* option list.	Print Preview
	Use the *Save As* option to save a copy of a document in another location or in the same location but with a different file name.	PDF document
	At the Print backstage area, preview a document in the Print Preview panel to see how the document will look with the current print settings before printing.	
	A PDF document is an open standard created by Adobe Systems for exchanging electronic documents.	
	Use the *Export* option at the backstage area to publish a Word document in PDF format.	
3.3 Customizing and Using the Quick Access Toolbar	The Quick Access Toolbar is at the top left of each Office application window.	Quick Access Toolbar
	Add to or remove buttons from the Quick Access Toolbar by clicking the Customize Quick Access Toolbar button and then clicking the desired option at the drop-down list.	
	The Quick Access Toolbar can be customized individually for each Office application.	

continued...

Topic	Key Concepts	Key Terms
3.4 Selecting Text or Objects, Using the Ribbon and Mini Toolbar, and Selecting Objects in Dialog Boxes	A selected object displays with circles around the perimeter called *selection handles*, which are used to resize or otherwise manipulate the object. The Mini toolbar contains the same buttons as the ribbon and appears near selected text or with the shortcut menu. Dragging a selection handle resizes a picture, shape, chart, or other item referred to as an *object*. A contextual tab appears when an object is selected with buttons related to the object. A gallery is a drop-down list or grid with visual representations of options. A live preview shows the text or object as it will look if the option on which the mouse is resting is applied. The dialog box launcher is located at the bottom right of a group on the ribbon and displays a task pane or dialog box when clicked. Task panes and dialog boxes provide additional options for the related ribbon group as buttons, lists, sliders, check boxes, text boxes, and option buttons. Colored horizontal and vertical alignment guides appear when moving an object to help you place and align the object with text, margins, or other nearby objects.	selection handle Mini toolbar object contextual tab gallery live preview dialog box launcher task pane dialog box alignment guide
3.5 Using the Office Clipboard and Formatting Using a Task Pane	The Clipboard group on the ribbon is standardized across all Office applications. Use Cut, Copy, and Paste buttons to move or copy text or objects. A Paste Options button appears when you paste text or an object with options for modifying the paste action. Use the Format Painter button to copy formatting options. A button, such as the Format Painter button, that operates in an on or off state is called a *toggle button*.	Format Painter toggle button
3.6 Finding Help or Options Using the Tell Me Feature	Click in the *Tell Me* text box that displays *Tell me what you want to do* and then type a term or option to locate the option directly from the Tell Me list or to look up help resources about the feature in a Word 2016 Help window. Click the *Smart Lookup* option from the Tell Me drop-down list to open a task pane with web links and definitions for the term.	Tell Me Smart Lookup
3.7 Using OneDrive for Storage, Scrolling in Documents, and Using Undo	OneDrive is cloud storage where you can save files that can be accessed from any other Internet-connected device by selecting your OneDrive account at the Open and Save As backstage areas. Horizontal and vertical scroll bars with arrow buttons and a scroll box are used to navigate larger documents. The Undo feature is used to reverse an action performed restoring a document to its previous state.	OneDrive scroll bar scroll box Undo

continued…

Topic	Key Concepts	Key Terms
3.8 Changing the Zoom Option and Screen Resolution	Use the Zoom In, Zoom Out, Zoom slider, and Zoom dialog box in Word, Excel, and PowerPoint to increase or decrease the magnification setting. Screen resolution refers to the number of horizontal and vertical pixels used to render an image on the display. The screen resolution setting for your PC or mobile device affects the display of the ribbon.	Zoom slider Zoom In Zoom Out screen resolution

 Recheck

Recheck your understanding of the topics covered in this chapter.

 **Workbook**

Chapter review and assessment resources are available in the *Workbook* ebook.

Chapter 4

Organizing and Managing Class Notes Using OneNote

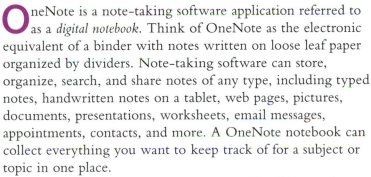

 Precheck

Check your understanding of the topics covered in this chapter.

OneNote is a note-taking software application referred to as a *digital notebook*. Think of OneNote as the electronic equivalent of a binder with notes written on loose leaf paper organized by dividers. Note-taking software can store, organize, search, and share notes of any type, including typed notes, handwritten notes on a tablet, web pages, pictures, documents, presentations, worksheets, email messages, appointments, contacts, and more. A OneNote notebook can collect everything you want to keep track of for a subject or topic in one place.

OneNote notebooks can be stored on OneDrive so that you can access the notes from any Internet-connected device. Another advantage to storing a notebook on OneDrive is that you can share it with others so that more than one person can edit a page. For group projects, OneNote is a useful tool for collaborating and sharing ideas, research, and content.

In this chapter, you will learn how to open an existing notebook; create a new OneNote notebook; add sections, pages, and content; tag and search notes; and share a notebook with others.

Learning Objectives

4.1 Open an existing notebook, add a note, apply color to note text, add a section, and add a page

4.2 Insert web content into a notebook

4.3 Insert a picture and document, and embed a copy of a presentation into a notebook

4.4 Tag a note, view tags, and jump to a tagged note

4.5 Search notes and close a notebook

4.6 Create a new notebook and share a notebook

 SNAP If you are a SNAP user, go to your SNAP Assignments page to complete the Precheck, Tutorials, and Recheck.

 Data Files

Before beginning this chapter, be sure you have copied the student data files for this course to your storage medium. Steps on downloading and extracting the data files are provided in Chapter 1, Topic 1.8, on pages 22–23.

4.1 Opening a Notebook and Adding Notes, Sections, and Pages

OneNote 2016 is the note-taking application within the Microsoft Office suite. A **OneNote notebook** is organized into sections, which are accessed by tabs across the top of the notebook. Think of sections as the dividers you would use in a binder to organize notes by subject, topic, or category. Within each section you add pages. Notes or other content are added to a page. You can add as many sections and pages as you like to organize a notebook. Notes or other content can be added anywhere on a page.

1 Click in the search text box on the taskbar, type OneNote 2016, and then press Enter. Windows 7 users may be prompted to sign in with their Microsoft account before OneNote starts for the first time.

The OneNote Start screen (Figure 4.1) appears with tips for using OneNote and links to how-to videos. By default a personal notebook for your Microsoft account is created and opened. The notebook is stored in OneDrive. You can put everything in one notebook or create separate notebooks for keeping notes organized. For example, you may want to create one notebook for school-related content and another for personal content.

Note: The Start screen shown in Figure 4.1 may not appear depending on the configuration for the computer you are using.

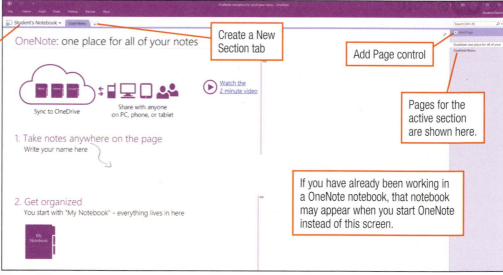

Figure 4.1

The OneNote Start screen is shown with your personal notebook open and the Quick Notes tab active.

2 Click Student's Notebook (where *Student's* is your first name), or other notebook name, and then click *Open Other Notebooks.*

This entry may vary and display *Quick Notes, My Notebook,* or *Personal (Web).*

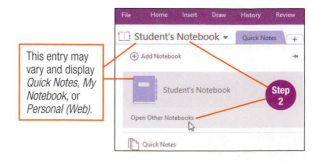

3 At the Open Notebook backstage area, click *Browse* in the *Open from other locations* section.

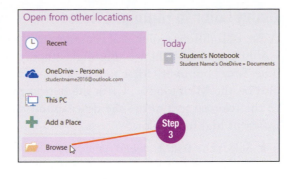

4 Navigate to the entry for your storage medium in the Navigation pane and then double-click the *Student_Data_Files* and *Ch4* folder names in the Content pane at the Open Notebook dialog box.

5 Double-click the *TechnologyCourse* folder name.

App Tip

You can also open a notebook at the Open backstage area by clicking the File tab and then clicking *Open*.

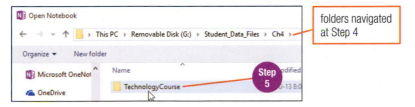

6 Double-click **Open Notebook**.

By default, OneNote creates a table of contents file named Open Notebook saved within a folder that is named the notebook name. The Open Notebook file is similar to a table of contents for a book in that it stores the name of each section added to the notebook. OneNote saves each section as a separate file within the notebook folder.

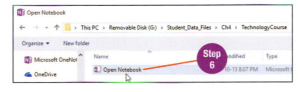

App Tip

OneNote differs from other Office programs in that the name that appears as the notebook name is the name of a *folder* (not an individual file). Each OneNote section is a *separate file saved within the notebook folder*.

7 Review the information on the Business Technology page in the *Course Information* section. If necessary, scroll down to view all information.

8 Click anywhere in the text *Prof. J. Wickham*.

Notice that a box surrounds the note text. The box is referred to as a **note container**. All OneNote content is placed inside a note container. Notice also that a selection handle appears at the left side of the note container when the pointer is resting on note text.

App Tip

To delete a note, select the note text using the selection handle or the gray bar at the top of the note container and then press the Delete key on the keyboard or choose Cut in the Clipboard group on the Home tab.

9 Click the selection handle at the left of the note container to select the text *Prof. J. Wickham*.

10 Click the Bold button on the Mini toolbar.

11 Click in any blank space at the right of the Prof. J. Wickham note container and then type Office hours every Tuesday from 12:00 to 1:00.

12 Drag the gray bar at the top of the note container to move the note below the *Prof. J. Wickham* note container in the approximate location shown at the right.

A four-headed white arrow pointer appears when you point to the top gray bar on the note container. Drag when you see this pointer.

OneNote keeps track of the author name for new notes. To show or hide author initials, click the History tab and then click Hide Authors. The Hide Authors button is a toggle that turns on or turns off the author initials.

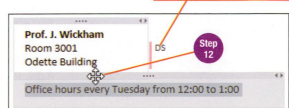

Using Color to Highlight Notes

A Text Highlight Color tool is used to apply color highlighting to notes just as you would use a highlighter to highlight important points while reading a textbook. The Text Highlight Color tool is in the Basic Text group on the Home tab and also on the Mini toolbar. Click the button to apply the default yellow highlighting to the selected text or use the down-pointing arrow on the button to choose a different highlight color.

13 Select the text *Examine various social media and communications applications* in the Learning Objectives note container and then click the Text Highlight Color button on the Mini toolbar to highlight the text using the default yellow color.

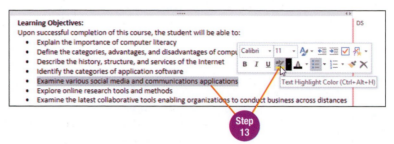

Step 13

Oops! !

Only ribbon tabs visible? Click the Home tab and then click the pushpin icon at the bottom right corner of the ribbon to keep the ribbon visible while you work.

14 Select the text *blogging, podcasting, VoIP, and Twitter* in the Course Description note container, click the Text Highlight Color button arrow on the Mini toolbar or in the Basic Text group on the Home tab, and then click the *Green* option (second color option in the first row).

Step 14

Adding Sections and Pages

A section is like a divider in a binder. Sections organize content by category, topic, or subject. Each section can have multiple pages. To create a new section in a notebook, click the **Create a New Section** tab (displays as a plus symbol), type a name for the section, and then press Enter. A new blank page opens for the section with an insertion point in the page title area. Type a title for the page and then press Enter.

Click the **Add Page** control (displays as a plus symbol inside a circle) at the top of the Pages pane to add a new page within the section. Type a title for the page and then press Enter.

15   Click the Create a New Section tab (displays as a plus symbol) next to the Course Information tab.

16   Type Web Pages and then press Enter.

17   With the insertion point positioned in the page title area, type Technology Web Pages and then press Enter.

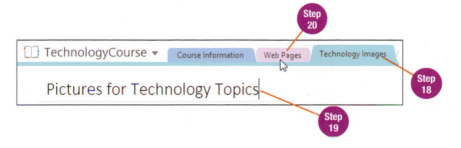

18   Click the Create a New Section tab, type Technology Images, and then press Enter.

19   Type Pictures for Technology Topics as the page title and then press Enter.

20   Click the Web Pages tab to display the section.

21   Click the Add Page control (displays as a plus symbol inside a circle) in the Pages pane.

22   Type Technology Web Links as the page title and then press Enter. Leave OneNote open for the next topic.

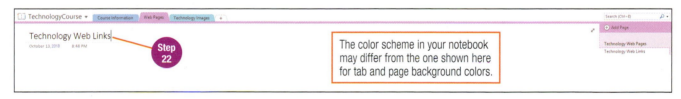

The color scheme in your notebook may differ from the one shown here for tab and page background colors.

Quick Steps

Open a Notebook
1. Click notebook name left of section tabs.
2. Click *Open Other Notebooks.*
3. Navigate to drive and/or notebook folder name.
4. Double-click *Open Notebook.*

Add a Note to a Page
1. Click at desired location on page.
2. Type note text.
3. Click in blank space outside note container.

Add a Section and Page to a Notebook
1. Click Create a New Section tab.
2. Type section title and press Enter.
3. Type page title and press Enter.

Add a Page to a Section
1. Make the desired section tab active.
2. Click Add Page control.
3. Type page title and press Enter.

App Tip

A Save button is not provided in OneNote; changes are saved automatically.

Beyond Basics   **Creating a Table in a Note and Calculating Expressions**

OneNote automatically formats text into a table when you type some text and then press the Tab key. Each time you press Tab, a new column is inserted into the table. Press Enter to add a new row. You can also perform a calculation in OneNote by typing an expression. When you press the spacebar after an equals sign, OneNote calculates the expression. For example, typing *15*12=* and then pressing the spacebar causes OneNote to calculate the result and show *15*12=180* in the note container.

4.2 Inserting Web Content into a Notebook

Web content can be inserted into a notebook as a link or as a screen clipping. The method you use will vary depending on the content and the frequency of updates to the content. For example, you may want to use a link if a web page updates frequently. Embed a screen clipping of a web page if you want to copy a portion of content from the web page and are not concerned with updates to the captured content.

You can also use the standard copy and paste tools to select and copy text from a web page and paste it into a notebook. When you paste text copied from a web page, OneNote automatically includes the source URL below the pasted text.

1 With the TechnologyCourse notebook open and an insertion point active on the Technology Web Links page, type https://twitter.com/Techmeme and then press Enter twice.

OneNote automatically formats web addresses as hyperlinks.

2 Type Techmeme provides daily summaries of leading technology stories and tweets headlines with links throughout the day. and then click in a blank area away from the note container.

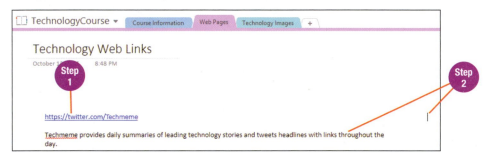

3 Click https://twitter.com/Techmeme to view the Twitter page in Microsoft Edge or other browser window. If a window opens with the message *How do you want to open this?*, click OK to accept Microsoft Edge.

4 Click in the Address bar to select the web address or drag to select the web address, type https://en.wikipedia.org/wiki/Tim_Berners-Lee, and press Enter.

To embed a copy of content from a web page into a notebook, use the *Take screen clipping* option from the **New quick note** button (OneNote button with scissors on the taskbar, or revealed by clicking Show hidden icons in the Notification area on the taskbar).

5 If necessary, click the Show hidden icons button in the Notification area on the taskbar to reveal the New quick note button.

6 Right-click the New quick note button.

7 Click *Take screen clipping* at the shortcut menu.

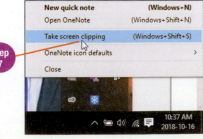

The screen dims and a crosshairs (✛) displays. A screen clipping is used to capture a portion of a web page that you want to save in a notebook by dragging the crosshairs across the content that you want to copy.

8 Drag the crosshairs from just above the *Wikipedia* page title to the bottom of the Sir Tim Berners-Lee box at the right side of the page as shown below.

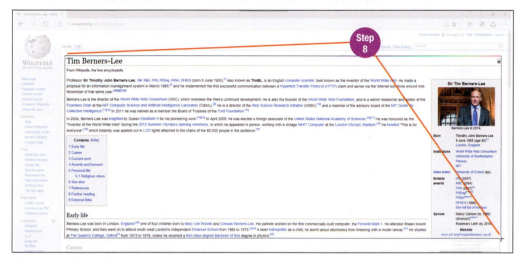

9 At the Select Location in OneNote dialog box, click the expand button (displays as a plus symbol) next to TechnologyCourse in the *All Notebooks* section.

10 Click the expand button next to Web Pages.

11 Click *Technology Web Pages* and then click the Send to Selected Location button.

12 Click in the Address bar or select the web address, type wikipedia.org/wiki/3D_printing, and then press Enter.

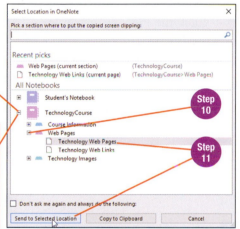

Expand button (plus symbol) changes to collapse button (minus symbol) when the list is expanded.

Assume you want to save the picture of the 3-D printer shown on the *Wikipedia* page about 3-D printing.

13 Repeat Step 5 to Step 7 to start the OneNote screen clipping tool.

14 Drag the crosshairs from the top left to the bottom right of the 3-D printer image at the right side of the *Wikipedia* page as shown at the right. Note that the image shown at the right may have been replaced with another picture.

15 At the Select Location in OneNote dialog box, click the expand button next to TechnologyCourse, click the expand button next to Technology Images, click *Pictures for Technology Topics*, and then click the Send to Selected Location button.

16 Close the browser window to return to the OneNote window and view the 3-D printer image embedded in a note container on the notebook page. Notice the date and time you took the screen clipping appears below the image. Leave OneNote open for the next topic.

Your image may vary.

A MakerBot 3d Printer

4.3 Inserting Files into a Notebook

Skills

Insert a picture

Insert a link to a document

Embed a copy of a presentation

Pictures, documents, presentations, workbooks, contacts, email messages, appointment details, and more can be added to a notebook as a link to the source content or as a copy of the content. Consider using a OneNote notebook as a repository to collect all the data related to a course, subject, or other topic. The advantage to assembling all the content in one place is that you no longer need to keep track of web links or web pages separately from documents and other notes for a subject.

Items are inserted into a OneNote page using buttons on the Insert tab. A document can be inserted as an icon that links to the source file, or the text and other content from the document can be embedded into the notebook. Once inserted, you can add annotations with your own notes.

1. With the TechnologyCourse notebook open and the Pictures for Technology Topics page active in the *Technology Images* section, click the Insert tab.

2. Click the Pictures button in the Images group.

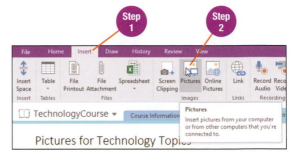

3. At the Insert Picture dialog box, navigate to the Ch4 folder in Student_Data_Files on your storage medium.

4. Double-click *AnalogCptr_1950s*.

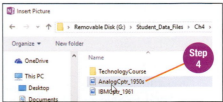

5. With an insertion point positioned in the note container below the image, type Analog computer from the 1950s.

6. Click the Create a New Section tab, type Documents as the section title, and then press Enter.

7. Type Course Documents as the page title and then press Enter.

App Tip

You can also drag and drop a picture onto a OneNote page from a File Explorer window.

Analog computer from the 1950s

Course Documents

8. Click the File Attachment button in the Files group on the Insert tab.

9 With the Ch4 folder in Student_Data_Files active, double-click the Word document ***Tech_Wk1_SocialMedia*** at the Choose a file or a set of files to insert dialog box.

10 Click Attach File at the Insert File dialog box.

11 With an insertion point positioned in the note container below the Word document icon and file name, type Week 1 Assignment.

The file is linked to the source document and can be opened from OneNote.

12 Double-click the document icon. If you are using a touchscreen, tap Open. Click OK at the Warning message that opening attachments could harm your computer or data.

13 Scroll down to review the Word document and then close Word.

14 Create a new section *Presentations* with a page title *Course Presentations*.

15 With the Course Presentations page active, click the File Printout button in the Files group on the Insert tab.

16 Double-click the PowerPoint presentation ***Tech_Wk1***.

Use File Printout to embed a copy of the content of the file onto the current page. This option allows you to add your own notes within the content. For example, embed a presentation in advance of a lecture so that you can type your own notes directly on or next to a slide as the teacher is teaching the class.

17 Scroll down and review the PowerPoint slides inserted on the OneNote page. Leave OneNote open for the next topic.

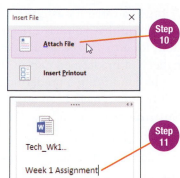

Step 10

Step 11

Step 12

Step 15

App Tip

Changes that occur in the file after it has been inserted into OneNote as a printout are not updated. Be aware that the note container may not contain the most up-to-date content.

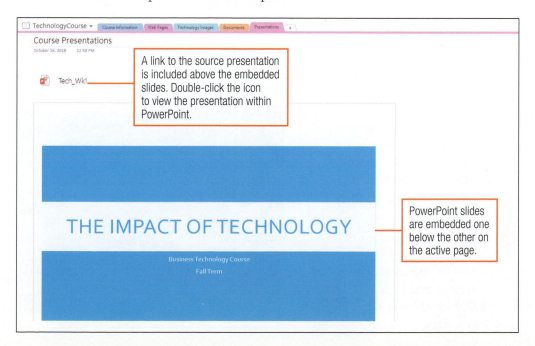

A link to the source presentation is included above the embedded slides. Double-click the icon to view the presentation within PowerPoint.

PowerPoint slides are embedded one below the other on the active page.

THE IMPACT OF TECHNOLOGY

Business Technology Course
Fall Term

4.4 Tagging Notes, Viewing Tags, and Jumping to a Tagged Note

Skills

Tag a note

View and jump to tagged notes

A **tag** is a category assigned to a note. The tag allows you to identify the note later as an item that you have flagged as important, as a question, as a definition, as an item for a to-do list, as an idea, or for some other purpose. OneNote includes a gallery of predefined tags in the Tags group on the Home tab. You can customize a tag by modifying a predefined OneNote tag or by creating a new tag of your own.

Once tags have been assigned to items in the notebook, you can display the **Tags Summary pane** and use the pane to navigate to a tagged item.

1   With the TechnologyCourse notebook open and the Course Presentations page active in the *Presentations* section, click the Web Pages section tab.

2   Click Technology Web Links in the Pages pane.

3   Click at the beginning of the note text *Techmeme provides daily summaries of leading technology stories…*, click the Home tab, and then click *Important* in the Tags gallery.

OneNote inserts a gold star (tag icon) for the Important tag next to the note text.

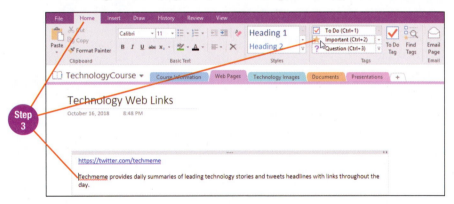

4   Click Technology Web Pages in the Pages pane.

5   Double-click after the bolded title *Tim Berners-Lee* to place an insertion point and then click *Important* in the Tags gallery.

The tag is inserted inside a new note container on the page.

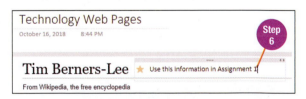

6   Type Use this information in Assignment 1. If necessary, drag the top gray bar to move the note container if it overlaps other text.

7   Click the Technology Images section tab.

8   Click at the beginning of the caption text below the picture of the analog computer and then click *? Question* in the Tags gallery.

Analog computer from the 1950s

A purple question mark appears next to the caption text.

9 Click the Presentations section tab.

10 Scroll down the page of embedded PowerPoint slides to the slide with the title *REFLECTION BLOG* and then double-click to place the insertion point at the right of the slide title.

11 Click the More button (displays as a down-pointing arrow below a bar) at the bottom of the Tags gallery to display more predefined tag options.

12 Click *Remember for blog* in the drop-down gallery.

13 Type Blog entry homework in the note container next to the tag icon. If necessary, move the note container if it overlaps existing text.

Once tags have been applied to notes, you can view all the tagged notes and jump to a specific item that you need to review. Click the **Find Tags** button in the Tags group on the Home tab to display in a Tags Summary pane at the right side of the OneNote window a list of tagged items grouped by tag category.

14 Click the Find Tags button in the Tags group.

The Tags Summary pane opens at the right side of the OneNote window.

15 Click Techmeme provides daily summa… in the Tags Summary pane. The hyperlinks may show more or less text in the Tags Summary pane on the computer you are using depending on the screen size and screen resolution.

OneNote jumps to the Technology Web Links page in the *Web Pages* section of the notebook and selects the entry in the note container.

16 Click each of the other tags in the Tags Summary pane to jump to each tagged item in the notebook.

17 Click the Close button at the top right of the Tags Summary pane. Leave OneNote open for the next topic.

Beyond Basics   Using the To-Do Tag

You can use the Tags feature to create a to-do list by tagging items in your notebook with the To Do tag. OneNote inserts a blank check box at the left of note text tagged with *To Do*. Click the check box when a task has been completed to mark the item finished.

4.5 Searching Notes and Closing a Notebook

Skills

Search notes

Close a notebook

An advantage to using an electronic notebook instead of a paper-based notebook is the ability to search all the pages in the notebook for a keyword or phrase and instantly locate each occurrence of the note text. The search feature in OneNote searches all the pages within all open notebooks. Type a search keyword or phrase in the search text box located above the Pages pane at the right side of the OneNote window. OneNote begins listing pages with matches in a drop-down list and highlights matches on each page as soon as you begin typing. Click a page in the search results to view the matches.

When you close OneNote, the active notebook is left open so that you are returned to the place you left off when you start OneNote again. You may instead choose to close a notebook when you are finished working. One reason to close a notebook is that when you want to search for a keyword in a specific notebook but it also exists in the current notebook, closing the current notebook will avoid pages showing up in search results that you are not interested in reviewing.

1 With the TechnologyCourse notebook open, click the Course Information section tab.

2 Click in the search text box above the Pages pane that contains the text *Search* and then type Tim Berners-Lee.

OneNote begins displaying matches as soon as you start typing.

3 Click the entry in the search results list to navigate to the page and review the highlighted text entries on the page.

Notice that OneNote is able to search content embedded from external sources.

4 Click *Tim Berners-Lee* or select the text in the search text box and then type analog.

5 Click *Pictures for Technology Topics* in the search results list to navigate to the page with the picture of the analog computer.

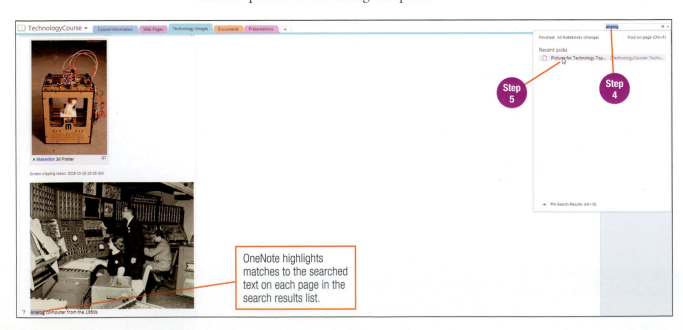

OneNote highlights matches to the searched text on each page in the search results list.

6 Click *analog* or select the text in the search text box and then type blog.

7 Click each entry in the search results list to review each item.

8 Click the Close button at the right of the search text box to close the search results list.

Your search results may vary.

Because OneNote automatically saves changes to notebooks as you work, you can leave all your notebooks open and be assured that changes are being saved. However, if you want to close a notebook, click the *Close* option in the Settings button drop-down list at the **Notebook Information backstage area**.

9 Click the File tab.

10 Click the Settings button next to TechnologyCourse at the Notebook Information backstage area.

11 Click *Close* at the Settings button drop-down list to close the notebook.

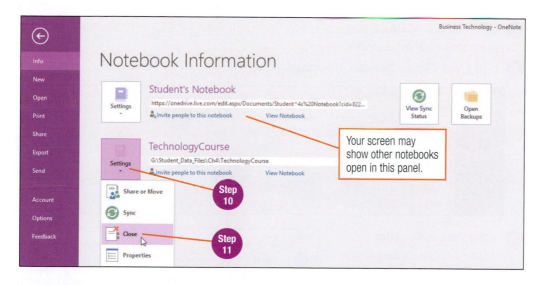

Your screen may show other notebooks open in this panel.

Alternative Method | **Searching Content on the Active Page Only**

To search for a keyword or phrase on the active page only, press Ctrl + F. OneNote displays *Find on page* in a yellow box next to the search text box. Type the search keyword or phrase in the search text box. OneNote displays an up and down arrow in the yellow box with the number of matches found. Click the arrows to navigate to the matches on the current page.

Beyond Basics | **Printing and Exporting Notebook Sections**

Print a notebook section by making the desired section active, clicking the File tab, and then clicking *Print.* Use Print Preview at the Print backstage area to view and modify print settings, such as the print range, paper size, and page orientation. Consider exporting a page, section, or notebook as a PDF file instead of printing. Display the Export backstage area, choose to export a page, a section, or the entire notebook, choose the export file format, and then click the Export button. OneNote displays the Save As dialog box in which you choose the drive, folder, and file name for the exported file.

4.6 Creating a New Notebook and Sharing a Notebook

Skills

Create a new notebook

Share a notebook using OneDrive

Some people may choose to organize all their notes for all purposes within one notebook (the default notebook file), using sections and pages to create an organizational structure. Others may choose to create separate notebooks in which to organize notes. For example, you may want to maintain separate notebooks for home, work, and school items. A new notebook is created at the **New Notebook backstage area** where you specify the notebook name and storage location.

A shared notebook is useful if you are working on a group project. With a shared notebook, members of the group can each post research, links, ideas, or other notes in one place. Share an existing notebook at the **Share Notebook backstage area**. A shared notebook has to be stored on OneDrive. One person creates the notebook and is referred to as the notebook *owner*. He or she shares the notebook by entering each person's email address at the Share Notebook backstage area. Each group member receives an invitation to view the notebook by email message.

Note: Check with your instructor for the name of the classmate with whom you will share the notebook created in this topic. At Step 13 you will need the Microsoft account email address for the classmate. If necessary, you can practice sharing the notebook with a friend or relative.

1 Click the File tab and then click *New.*

2 Click your OneDrive account name at the New Notebook backstage area if OneDrive is not automatically selected.

3 Click in the *Notebook Name* text box and then type MyElectives-xx. Substitute your first and last initials for *xx.*

4 Click the Create Notebook button.

App Tip

By default, each person with whom you share a notebook can make changes to the notebook. To share a notebook and only allow the other person to view the content, click the permissions option box arrow (displays *Can edit*) at the right of the email addresses text box and then click *Can view.*

5 Click Not now at the Microsoft OneNote message box asking if you would like to share the notebook with other people.

You will set up the sharing feature later in this topic. OneNote opens a new notebook with one section created titled *New Section 1.*

6 Right-click the New Section 1 section tab and then click *Rename.*

7 Type Child Lit and then press Enter.

Oops!

Don't remember how to embed files? Refer to Topic 4.3, Step 15.

8 Type Children's Literature as the page title and then press Enter.

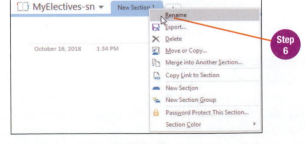

9 Embed a copy of the PowerPoint presentation **ChildLitPres** on the page.

10 Add a second section tab named *Film* with the page title *Film Genres* and then embed a copy of the Word document ***ApNowReflectionPaper***.

You decide to share the notebook with a classmate taking the same electives as you, so you can each add notes to the notebook.

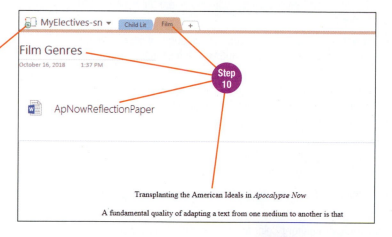

A notebook stored online displays this icon indicating the content is synced with OneDrive.

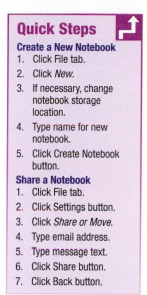
11 Click the File tab.

12 Click the Settings button next to MyElectives-xx and then click *Share or Move*.

13 At the Share Notebook backstage area, click in the *Type names or e-mail addresses* text box and then type the classmate's Microsoft account email address.

14 Click in the message box and then type Here is a notebook I created that we can use to share notes for our electives.

15 Click the Share button.

When sharing is completed, the student's name is shown in the *Shared with* section of the backstage area.

16 Click the Back button.

17 Close the MyElectives-xx notebook and then close OneNote.

Optional

18 Open a browser window, navigate to https://outlook.com, and sign in if you are not automatically signed in.

19 At the Inbox, click the message received from a classmate with subject text similar to *MyElectives-xx has been shared with you* to open the message.

20 Click the View in OneDrive button in the message to open the notebook in OneNote Online. Click Unblock if a message pops up informing you parts of the message have been blocked.

21 Add a note to one of the pages. (You determine the note text.)

22 Close the browser window.

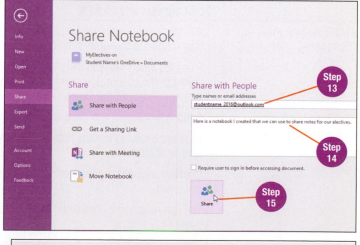

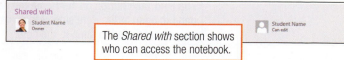

The *Shared with* section shows who can access the notebook.

Topics Review

Topic	Key Concepts	Key Terms
4.1 Opening a Notebook and Adding Notes, Sections, and Pages	OneNote 2016 is a note-taking software application that is the electronic equivalent of a binder with loose leaf notes separated by dividers.	OneNote 2016
	A OneNote notebook is organized into sections (dividers), which are tabs across the top of the window. Each section can have multiple pages.	OneNote notebook
	To open a notebook, click the active notebook name, click *Open Other Notebooks*, navigate to and then double-click the file *Open Notebook* stored within a folder named the same as the notebook name.	note container
	Each note on a page is stored within a note container, which is a box that surrounds the note text.	Create a New Section
	Drag the gray bar along the top of the note container to move a note.	Add Page
	Use the Text Highlight Color tool to add color to notes similarly to using a highlighter to emphasize text in a textbook.	
	Click the Create a New Section tab to type a new section heading, then type a page title to start a new section.	
	Pages are added to a section using the Add Page control at the top of the Pages pane.	
4.2 Inserting Web Content into a Notebook	A web address is automatically converted to a hyperlink in a note container.	New quick note
	A portion of a web page can be captured and inserted into a notebook. To do this, click the Show hidden icons button in the Notification area on the taskbar if the New quick note button is not already visible on the taskbar, right-click the New quick note button, and then click *Take screen clipping*. Drag the crosshairs over the portion of the web page you want to embed. Next, select the destination page in the Select Location in OneNote dialog box that appears and click OK.	
4.3 Inserting Files into a Notebook	A OneNote notebook can be used as a repository to collect all the files related to a course, subject, or other topic.	
	From the Insert tab, you can insert a picture, a file as an icon linked to the source document, or a file as a printout, which embeds a copy of the file contents into the notebook.	
4.4 Tagging Notes, Viewing Tags, and Jumping to a Tagged Note	A tag is a category assigned to a note.	tag
	Tags are useful to identify notes that you want to flag for later review or follow up.	Tags Summary pane
	OneNote includes a gallery of predefined tags, such as *Important* or *Definition*.	Find Tags
	You can modify the predefined tags or create a new tag of your own.	
	The Find Tags button on the Home tab causes the Tags Summary pane to open at the right side of the window with links to each tagged note used to jump to the note.	

continued...

Topic	Key Concepts	Key Terms
4.5 Searching Notes and Closing a Notebook	OneNote can search all pages in all open notebooks for a keyword or phrase typed in the search text box. OneNote begins highlighting matches to the search keyword or phrase and displays pages in the search results list as soon as you begin typing. Close a notebook using the Settings button at the Notebook Information backstage area.	Notebook Information backstage area
4.6 Creating a New Notebook and Sharing a Notebook	Create a new notebook at the New Notebook backstage area. A notebook saved to OneDrive can be shared with other people. More than one person can edit a page at the same time when a notebook is shared. Use the Settings button at the Notebook Information backstage area to access the Share Notebook backstage area. You can then share a notebook by typing the email address and a short message for each invitee.	New Notebook backstage area Share Notebook backstage area

Recheck

Recheck your understanding of the topics covered in this chapter.

Workbook

Chapter review and assessment resources are available in the *Workbook* ebook.

Chapter 5

Communicating and Scheduling Using Outlook

Precheck

Check your understanding of the topics covered in this chapter.

Microsoft Outlook is a software application often referred to as a **personal information management (PIM) program**. PIM programs organize items such as email messages, appointments or meetings, events, contacts, to-do lists, and notes. Reminders and flags help you remember and follow up on activities.

In the workplace, Outlook is often used with an Exchange server, which allows employees within the organization to easily share calendars, schedule meetings, and assign tasks with one another. Consider using Outlook on your home PC or mobile device to connect to the mail server supplied by your ISP and manage your messages. Outlook can also help you organize your time, activities, address book, and to-do list.

In this chapter, you will learn how to use the desktop application PIM program Microsoft Outlook 2016 for creating and sending email messages; scheduling appointments, events, and meetings; creating contacts; and maintaining a to-do list.

If you do not have access to the desktop software application, you can use Outlook.com to practice the skills taught in this chapter. However, be aware that the screens shown and the step-by-step instructions are for the desktop Outlook application. The web-based program Outlook.com uses a different interface. Where possible, instructions or marginal notes direct you to the method used in Outlook.com. In some cases, a feature available in the desktop application is not available in Outlook.com, and you will not be able to complete all the tasks in the topic.

Learning Objectives

5.1 Create, send, read, reply to, and forward email messages

5.2 Attach a file to a message, delete a message, and empty the Deleted Items folder

5.3 Preview, open, and save a file attachment

5.4 Schedule an appointment and an event

5.5 Schedule a recurring appointment and edit an appointment

5.6 Schedule and accept a meeting request

5.7 Add and edit contact information

5.8 Add, update, and delete a task

5.9 Search Outlook messages, appointments, contacts, or tasks

 **SNAP** If you are a SNAP user, go to your SNAP Assignments page to complete the Precheck, Tutorials, and Recheck.

 Data Files

Before beginning this chapter, be sure you have copied the student data files for this course to your storage medium. Steps on downloading and extracting the data files are provided in Chapter 1, Topic 1.8, on pages 22–23.

5.1 Using Outlook to Send Email

Skills

Create and send an email message

Reply to a message

Forward a message

Electronic mail (email) is communication between individuals by means of sending and receiving messages electronically. Email is the business standard for communication in the workplace. Individuals also use email to communicate with relatives and friends around the world. While text messaging is popular for brief messages between individuals, email is still used to send longer messages.

Setting Up Outlook

The screen that you see when you start Outlook for the first time depends on whether a prior version of Outlook existed on the computer you are using. **Outlook 2016** can transfer information from an older version of Outlook to a new data file or, if no prior data file exists, will present a Welcome to Outlook 2016 screen at start-up. Click Next at the welcome screen and then click Next at the second screen to accept the option to connect an email account. At the Add Account dialog box shown in Figure 5.1, enter your name, email address, and your password twice and then click Next. Outlook automatically configures the mail server settings displaying progress messages as each part is completed. Click Finish when completed to start Outlook.

In instances where Outlook cannot automatically set up your email account, additional information will be required such as the incoming and outgoing mail server address. Contact your ISP if necessary for this information.

Note: The instructions in this chapter assume Outlook has already been set up and that you are connected to the Internet with an always-on connection (high-speed Internet service) at school or at home. If necessary, connect to the Internet and sign in to your email account before starting the topic activities. Check with your instructor for assistance if you are not sure how to proceed.

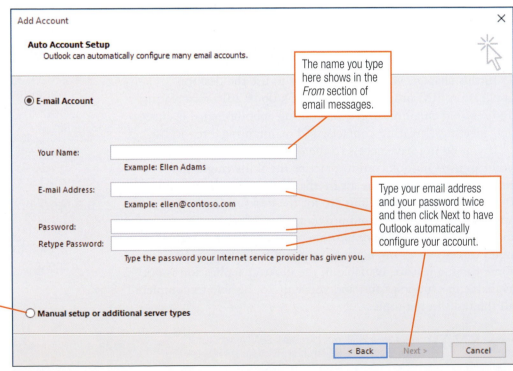

Figure 5.1

Outlook automatically sets up most email accounts once you enter your name, email address, and email password.

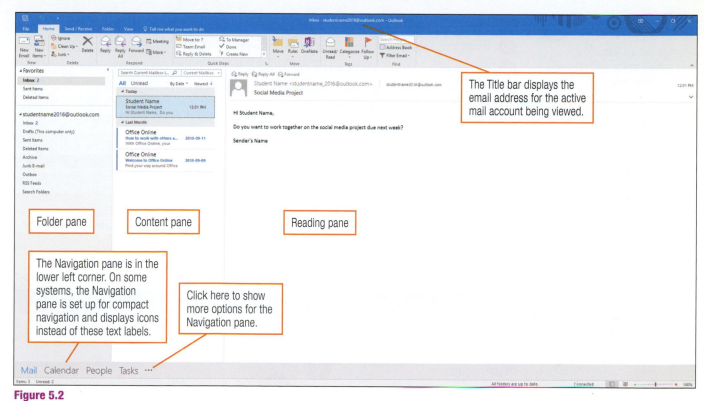

Figure 5.2

The Outlook window. Outlook checks for new messages automatically when the program is started.

Once Outlook has been set up with an email account, the Outlook window appears, similar to the one shown in Figure 5.2. By default, **Mail** is the active component, with **Inbox** the mail folder shown when Outlook starts. Messages in the Content pane are shown with the newest message received at the top of the message list. The left pane, called the Folder pane, is used to switch the display to another mail folder. At the bottom of the Folder pane is the Navigation pane, used to navigate to another Outlook component, such as Calendar. The right pane, called the Reading pane, displays the contents of the selected message.

Creating and Sending a Message

Click the **New Email** button in the New group on the Home tab to start a new email message. Type the recipient's email address in the *To* text box, type a brief description in the *Subject* text box, and then type your message text in the white message text box. Click the Send button when finished.

Note: Check with your instructor for instructions on whom you should exchange messages and meeting requests with for this chapter. Your instructor may designate an email partner for each person or allow you to choose your email partner. If necessary, send messages to yourself.

1. Click in the search text box on the taskbar, type Outlook 2016, and then press Enter to start Outlook 2016. If you are using Outlook.com, open a browser window, navigate to https://outlook.com, and sign in to your Microsoft account if you are not automatically signed in.

2. Click the New Email button in the New group on the Home tab or click the New button next to Outlook.com.

App Tip

Click the View tab, click the Folder Pane button in the Layout group, and choose *Normal*, *Minimized*, or *Off* to change the Folder pane. The Reading Pane button in the same group is used to display the Reading pane at the right or bottom of the screen or to turn off the Reading pane.

Oops!

A Welcome to Outlook 2016 message box appears? This screen appears the first time Outlook is started with no email account set up. Refer to the instructions on the previous page to configure your email account first before you can proceed to Step 2.

3. Type the email address for the recipient in the *To* text box.

4. Press the Tab key twice or click in the *Subject* text box.

5. Type Social Media Project and the press Enter.

6. With the insertion point positioned at the top of the Message text window, type the following text:

   Hi (type recipient's name), [press Enter twice]

   I think we should do our project on Pinterest.com. Pinterest is a virtual pinboard where people pin pictures of things they have seen on the Internet that they want to share with others. [press Enter twice]

   What do you think?

7. Press Enter twice at the end of the message text and then type your name as the sender.

8. Click the Send button.

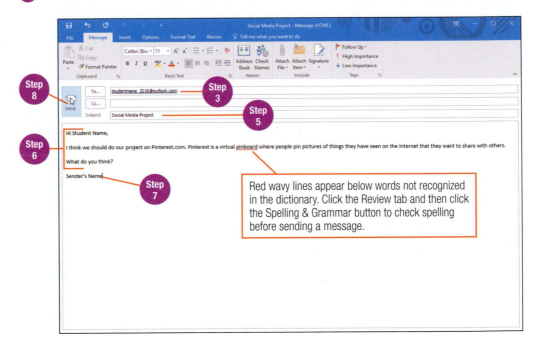

Red wavy lines appear below words not recognized in the dictionary. Click the Review tab and then click the Spelling & Grammar button to check spelling before sending a message.

Replying to a Message

New messages appear at the top of the message list in the Content pane with message headers that show the sender, subject, time, and first line of message. Click to select a message header and reply directly from the Reading pane using the **Reply** button. Replying from the Reading pane is called an **inline reply**. You can also double-click the message header in the Content pane to open the message in a Message window from which you can choose to Reply or Forward the message.

9. Click the Send / Receive tab and then click the Send/Receive All Folders button in the Send & Receive group to update the Content pane. Skip this step if you can already see the message sent to you by a classmate or yourself from Step 8.

10. Click to select the message header for the message received from Step 8 and then read the message text in the Reading pane.

⑪ Click the Reply button at the top of the message in the Reading pane.

⑫ Type the following reply message text and then click the Send button.

Step 10

Step 11

(Type the name of the person from whom you received the message), [press Enter twice]

I agree. I have a few pictures we can use to practice with if you want to set up a sample account at Pinterest.com. [press Enter twice]

(Type your name)

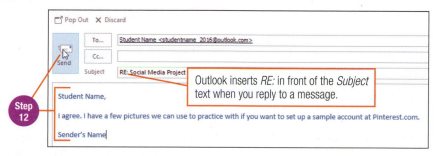

Outlook inserts *RE:* in front of the *Subject* text when you reply to a message.

Forwarding a Message

Forward a message if you want someone else to receive a copy of a message you have received. Choose the **Forward** button, type the email address for the person to whom you want to forward the message, and type a brief explanation if desired before sending the message.

Think carefully before forwarding a message to be certain that the original sender would not object to another person reading the message without his or her permission. If in doubt, do not forward the message.

⑬ With the message header for the message selected at Step 10 still active, click the Forward button in the Reading pane. In Outlook.com, click the arrow next to Reply and then click Forward.

Step 13

⑭ Type the email address for another classmate in the *To* text box.

⑮ Click in the message window above the original message text and then type the following text:

Hi (type recipient's name), [press Enter twice]

Do you want to join our group for the social media project? See message below discussing Pinterest.com. [press Enter twice]

(Type your name)

⑯ Click the Send button.

Step 16

Step 14

Step 15

Outlook inserts *FW:* in front of the *Subject* text when you forward a message.

Beyond Basics   Email Signatures

A signature is a closing automatically inserted at the bottom of each sent message. Signatures usually include contact information for the sender, such as name, title, department, company name, and contact telephone numbers. To create a signature, open a message window, click the Signature button on the Message tab in the Include group, and then click *Signatures*.

5.2 Attaching a File to a Message and Deleting a Message

Files are often exchanged between individuals via email. To attach a file to an email message, use the **Attach File** button in the Include group on the Message tab. The recipient of an email message with a file attachment can choose to open the file from the mail server or save it to a storage medium to open later.

Messages that are no longer needed should be deleted to keep your mail folders to a manageable size. You can delete a message in the Inbox folder if you replied to the message because you can view the original text with your reply from Sent Items. Consider setting aside a time each week to clean up your Inbox by deleting messages.

1 With Outlook open and Inbox the active folder, click the Home tab if Home is not the active tab.

2 Click the New Email button in the New group.

3 Type the email address for the recipient in the *To* text box.

Notice that as you begin typing an email address, Outlook provides email addresses that match what you are typing in a drop-down list. This feature is referred to as **AutoComplete**. Rather than type the entire email address, you can click the correct recipient in the AutoComplete list. People in the workplace who use Outlook connected to a special server called an Exchange server can send a message to someone within the organization by typing just the name of the recipient.

4 Press the Tab key twice or click in the *Subject* text box.

5 Type Picture for Pinterest and then press Enter.

6 With the insertion point positioned in the Message text window, type the following text:

Hi (type recipient's name), [press Enter twice]

Attached is a picture we can put on Pinterest.

7 Press Enter twice at the end of the message text and then type your name as the sender.

8 Click the Attach File button in the Include group on the Message tab in the message window.

9 Click *Browse This PC* at the drop-down list.

10 At the Insert File dialog box, navigate to the Student_Data_Files folder on your storage medium and then double-click *Ch5*.

11 Double-click the image file *MurresOnIce*.

Outlook adds the file to an *Attached* area below the *Subject* text box.

12 Click the Send button.

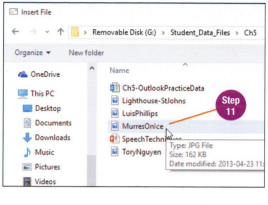

file attached at Step 11

Over time your mail folders (Inbox and Sent Items) become filled with messages that are no longer needed. To delete a message, select the message header in the Content pane and then click the Delete button in the Delete group on the Home tab. Deleted messages are moved to the **Deleted Items** folder where the message can be restored if needed. Periodically, empty the Deleted Items folder to permanently delete the messages.

13 Click the Send/Receive All Folders button on the Quick Access Toolbar (first button) to update the Inbox. Skip this step if you can already see the message sent to you by a classmate or yourself in Step 12.

14 Click *Sent Items* in the Folder pane.

15 If necessary, click to select the message header for the message with the picture attached that you sent to a classmate or yourself in this topic.

16 Click the Delete button in the Delete group on the Home tab.

17 Click *Deleted Items* in the Folder pane.

Notice the message you deleted appears in the Content pane.

18 Right-click *Deleted Items* in the Folder pane and then click *Empty Folder*.

19 Click *Yes* at the Microsoft Outlook message box asking if you want to continue to permanently delete everything in the Deleted Items folder.

20 Click *Inbox* in the Folder pane.

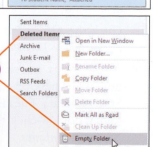

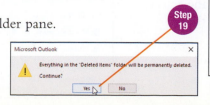

5.3 Previewing File Attachments and Using File Attachment Tools

When you receive an email message with a file attached, you can preview, open, print, save, remove, or copy the file attachment from the Reading pane or from a message window. Click the arrow next to the file name to select the desired action in the drop-down list or click the file name in the Reading pane to preview the attachment. During preview, the message text disappears, replaced with the contents of the attached file, and the Attachment Tools Attachments tab becomes active. Some files cannot be viewed within the Reading pane. In those instances, double-click the file name to open the file attached to the message.

1. With Outlook open and Inbox the active folder, click the message header for the message received in the previous topic with the file attachment. Skip this step if the message header is already selected.

2. Click the file name **MurresOnIce.jpg** in the Reading pane. Do *not* click the arrow at the right of the file name.

Outlook removes the message text and displays in the Reading pane the picture attached to the message. Notice also that the Attachment Tools Attachments tab is active on the ribbon.

App Tip

If you are using Outlook .com, clicking the file name in the message pane offers three options: View online, Save to OneDrive, and Download. When a picture is attached to an email the picture is automatically previewed when the message header is clicked and two options appear below the image: Download as zip and Save to OneDrive.

3. Click *Back to message* at the top of the Reading pane to restore the message text.

4. Click the New Email button and then type the email address for the recipient in the *To* text box.

5. Click in the *Subject* text box and then type Presentation for Business Class.

6. Type the following text in the Message text window:

   Hi (type recipient's name), [press Enter twice]

   Attached is the PowerPoint presentation for our group project. [press Enter twice]

   (Type your name)

7. Click the Attach File button in the Include group on the Message tab and then click *Browse This PC*.

8. At the Insert File dialog box, with the Ch5 folder in Student_Data_Files the active folder, double-click the file **SpeechTechniques**.

9. Click the Send button.

App Tip

If you are using Outlook .com, the file is uploaded to the server and may take a few seconds to complete the upload.

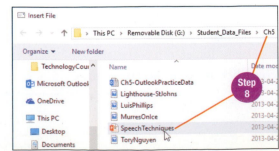

10 Click the Send/Receive All Folders button on the Quick Access Toolbar to update your Inbox folder. Skip this step if you can already see the message sent to you by a classmate or yourself in Step 9.

11 Click the message header for the message received with the subject *Presentation for Business Class.*

12 Click the arrow at the right of the file name **SpeechTechniques.pptx** in the Reading pane and then click *Preview.*

13 Scroll down the Reading pane to the last slide in the presentation.

14 Click the file name button in the Reading Pane, and then click the Open button in the Actions group on the Attachment Tools Attachments tab.

PowerPoint starts with the **SpeechTechniques** file open. Notice the Title bar and Message bar displayed below the ribbon tabs indicate the presentation is open in **Protected View**. Protected View allows you to read the file content in the source application; however, editing the file is not permitted until you click the Enable Editing button on the Message bar. The presentation remains in Read-Only mode with editing enabled until you save the file to another location because the file was opened from an email message.

15 Close PowerPoint to return to Outlook.

16 Click the file name button in the Reading Pane, and then click the Save As button in the Actions group on the Attachment Tools Attachments tab.

17 At the Save Attachment dialog box, navigate to the CompletedTopicsByChapter folder on your storage medium and then create a new folder *Ch5*.

18 Double-click the *Ch5* folder and then click the Save button.

19 Click the file name button in the Reading Pane, and then click the Show Message button in the Message group on the Attachment Tools Attachments tab to close the presentation preview in the Reading pane.

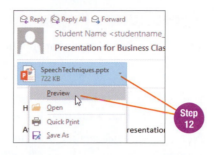

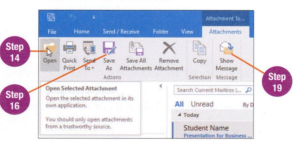

App Tip

Clicking the file name button in the Reading pane automatically previews the file. Click the arrow at the right of the file name button to access other attachment options.

Quick Steps

Preview a File Attached to a Message
Click file name button in Reading pane.
Open a File Attached to a Message
1. Click file name button in Reading pane.
2. Click Open button in Actions group on Attachment Tools Attachment tab.
Save a File Attached to a Message
1. Click file name button in Reading pane.
2. Click Save As button in Actions group on Attachment Tools Attachment tab.
3. Navigate to drive and/or folder.
4. Click Save.

Security Alert   Be Cautious Opening File Attachments!

Outlook blocks certain file types attached to messages that are known to be the target for viruses and are considered unsafe. However, even with the protection Outlook provides, you should exercise caution when opening a file received in an email message. Open files only from people you know and trust and always make sure you have real-time, up-to-date virus protection turned on. When in doubt, delete without opening any emailed files you were not expecting.

5.4 Scheduling Appointments and Events in Calendar

The **Calendar** component in Outlook is used to schedule appointments and events, such as meetings or conferences. An **appointment** is any activity where you want to track the day or time that the activity begins and ends in your schedule or when you want to be reminded to be somewhere. An appointment in Calendar can be a class, a meeting, a medical test, or a lunch date.

Note: In this topic and the next two topics, you will schedule appointments, an event, and a meeting in October 2018. Check with your instructor, if necessary, for alternate instructions that schedule these items in the current month or in October of the current year.

1. With Outlook open and Inbox the active folder, click Calendar in the Navigation pane.

Your Navigation pane may display icons instead of these text labels.

Outlook displays the current date, week, or month in the Content pane in Day, Week, or Month view. A **Date Navigator** displays at the top of the Folder pane with the current month and next month, along with directional arrow buttons to browse forward or back to upcoming or previous months. Open the **Go To Date** dialog box to navigate to a specific date that is not easily seen using the Date Navigator or the buttons in the Go To group on the Home tab.

2. If necessary, click the Day button in the Arrange group on the Home tab and then click the Go to Date launcher button (downward-pointing diagonal arrow) at the bottom right of the Go To group.

3. At the Go To Date dialog box, type 10/8/2018 and then press Enter or click OK.

4. Click next to 9:00 a.m. in the Appointment area, type Meet with program adviser, and then press Enter or click in another time slot outside the appointment box.

By default, Outlook schedules an appointment for a half hour.

5. Drag the bottom boundary of the appointment box to 10:00 a.m. If you are using a touchscreen, you may need to double-tap to open the appointment, tap the *End time* option box arrow, tap *10:00 AM (1 hour)*, and then tap Save & Close. Skip this step if you are using Outlook.com.

6. Click next to 11:00 a.m. in the Appointment area.

7. Click the New Appointment button in the New group on the Home tab.

Click the **New Appointment** button to open an Appointment window in which you can provide more details about the appointment or select more options to apply to the appointment.

8. Type Intern Interview in the *Subject* text box.

9. Press Tab or click in the *Location* text box and then type Room 3001. If you are using Outlook.com, click <u>View details</u> to open the Details panel in which you enter the location.

10. Click the *End time* option box arrow and then click *12:00 PM (1 hour)* at the drop-down list. Skip this step if you are using Outlook.com.

11. Click the Save & Close button in the Actions group on the Appointment tab in the Appointment window.

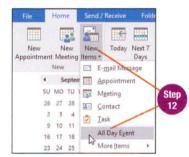

An **event** differs from an appointment in that it is an activity that lasts an entire day or longer. Examples of events include conferences, trade shows, or vacations. An event does not occupy a time slot in the Calendar. Event information appears in a banner along the top of the day in the Appointment area.

12. Click the New Items button in the New group on the Home tab and then click *All Day Event* at the drop-down list to open an Event window.

13. Type Career Fair in the *Subject* text box.

14. Press Tab or click in the *Location* text box and then type Student Center.

15. Click the Save & Close button in the Actions group on the Event tab in the Event window. Notice the event appears at the top of the Appointment area before the 8 a.m. time slot.

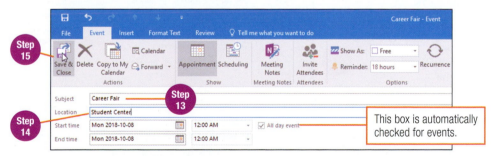

This box is automatically checked for events.

Quick Steps

Schedule an Appointment
1. Display the Calendar.
2. Navigate to appointment date.
3. Click next to appointment time.
4. Type appointment description.
5. Press Enter.
OR
1. Display the Calendar.
2. Navigate to appointment date.
3. Click next to appointment time.
4. Click New Appointment button.
5. Enter appointment details in Appointment window.
6. Click Save & Close.

Schedule an Event
1. Display the Calendar.
2. Navigate to event date.
3. Click New Items button and then click *All Day Event*.
4. Enter event description.
5. Enter event location.
6. Click Save & Close.

Oops!

Using Outlook.com? Click the down-pointing arrow on the New button and then click *Event* at the drop-down list.

App Tip

Change the End time date if the event lasts more than one day.

Beyond Basics   **Appointment Reminders and Tags**

By default, new appointments have a reminder set at 15 minutes. Turn off the reminder or change the reminder time by clicking to select an appointment in the Appointment area and then using the *Reminder* option box in the Options group on the Calendar Tools Appointment tab.

Assign one or more tags to a selected appointment with buttons in the Tags group on the Calendar Tools Appointment tab. Click the Private button for a personal appointment so that other people with access to your calendar cannot see the appointment details. Click the High Importance button to remind you of an important appointment.

5.5 Scheduling a Recurring Appointment and Editing an Appointment

Skills

Schedule a recurring appointment

Edit an appointment

An appointment that occurs on a regular basis at fixed intervals need only be entered once in Outlook and set up as a recurring appointment. To do this, open the Appointment Recurrence dialog box by clicking the **Recurrence** button in the Options group on the Appointment tab or the Calendar Tools Appointment tab. Enter the recurrence pattern for a repeating appointment, click OK, and then save and close the appointment.

1. With Outlook open and Calendar active for October 8, 2018, click the Forward button to display October 9, 2018 in the Appointment area.

2. Click next to 3:00 p.m. in the Appointment area, type Math Extra Help Sessions, and then press Enter.

3. With the Math Extra Help Sessions appointment box selected in the Appointment area, click the Recurrence button in the Options group on the Calendar Tools Appointment tab.

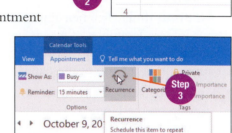

By default, Outlook sets the *Recurrence pattern* details for the appointment to recur *Weekly* at the same day and time as the appointment.

4. At the Appointment Recurrence dialog box, select *10* in the *End after* text box in the *Range of recurrence* section and then type 5.

5. Click OK.

A recurring icon displays at the right end of the appointment box in the Appointment area.

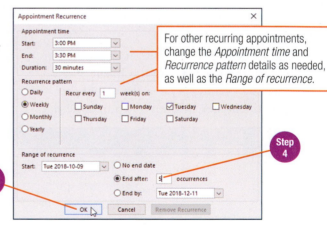

For other recurring appointments, change the *Appointment time* and *Recurrence pattern* details as needed, as well as the *Range of recurrence*.

6. Click *16* in the October 2018 calendar in the Date Navigator to move to the appointments for the following Tuesday and view the Math Extra Help Sessions appointment.

7. Click *23* in the October 2018 calendar in the Date Navigator to view the Math Extra Help Sessions appointment in the Appointment area.

8. Click the Go to Date launcher button in the Go To group, type 11/13/2018 in the Go To Date dialog box, and then press Enter or click OK.

Notice the Math Extra Help Sessions appointment does not appear in the Appointment area because the range of recurrence has ended.

Oops!

Calendar Tools Appointment tab not visible? Click the appointment box to select the appointment and display the contextual tab.

Oops!

Using Outlook.com? Set the recurrence option by displaying the appointment Details panel and then changing *How often* to *Weekly*.

App Tip

Consider entering your class schedule in the Outlook calendar for the current semester as recurring appointments.

Assign options or tags to an existing appointment by selecting the appointment and using the buttons in the Calendar Tools Appointment tab. Change the subject, location, day, or time of an appointment by double-clicking an appointment to open the Appointment window.

9 Display October 8, 2018 in the Calendar.

10 Click to select the appointment scheduled at 11:00 a.m.

A selected appointment box displays with a black outline. A pop-out opens at the left with the appointment details when you point at an appointment.

11 Click the Open button in the Actions group on the Calendar Tools Appointment tab. Click the <u>View details</u> hyperlink if you are using Outlook.com.

Assume the intern interview has been rescheduled to Tuesday.

12 Click the calendar icon at the right end of the *Start time* text box.

13 Click *9* in the drop-down calendar.

14 Click Save & Close in the Actions group on the Appointment tab.

15 Display October 9, 2018 in the Appointment area.

Notice the Intern Interview appointment appears next to 11:00 a.m.

Quick Steps

Schedule a Recurring Appointment
1. Display Calendar.
2. Navigate to appointment date.
3. Click next to appointment time.
4. Type appointment description.
5. Press Enter.
6. Click Recurrence button.
7. Enter recurrence pattern and/or range details.
8. Click OK.

Edit an Appointment
1. Select appointment.
2. Click Open button.
3. Change appointment details as needed.
4. Click Save & Close.

5.6 Scheduling a Meeting

Skills

Schedule a meeting

Respond to a meeting request

Schedule a **meeting** by selecting the day and time and then clicking the **New Meeting** button in the New group on the Home tab. At the Meeting window, enter the email addresses for the individuals to invite to the meeting in the *To* text box. Type the meeting subject, location, and other details as needed and then click the Send button. Meeting attendees receive a **meeting request** email message. Responses to the meeting request are sent back to the meeting organizer via buttons in the email message window or Reading pane.

1. With Outlook open and Calendar active with October 9, 2018 displayed in the Appointment area, click next to 1:00 p.m.

2. Click the New Meeting button in the New group on the Home tab.

3. Type the email address for the classmate with whom you have been exchanging email messages in the *To* text box.

Note: If you have been sending email messages to yourself in this chapter, send the meeting request message to a friend or relative, or use an email address for yourself that is different from your Microsoft account address because you cannot send a meeting request to yourself. You will not be able to complete Step 8 to Step 16 if you do not receive a meeting request message from someone else.

4. Press Tab or click in the *Subject* text box and then type Fundraising Planning Meeting.

5. Press Tab or click in the *Location* text box and then type Room 1010.

6. Click the *End time* option box arrow and then click *2:30 PM (1.5 hours)* at the drop-down list.

7. Click the Send button.

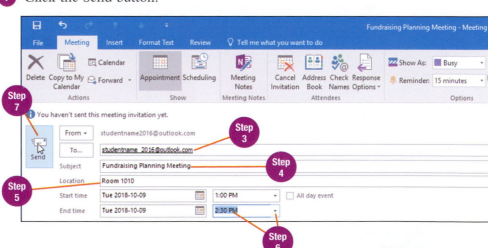

8. Click Mail in the Navigation pane.

9. Click the Send/Receive All Folders button on the Quick Access Toolbar to update your Inbox folder. Skip this step if you can already see the meeting request message sent to you by a classmate at Step 7.

Step 8

Oops!

Using Outlook.com? Create an appointment, display the Details panel, and click the Attendees tab (next to Details). Enter the email address for the meeting participant in the *Invite people* text box. Click Send invite to send the meeting request messages.

App Tip

Consider using the space in the message window below *End time* to type a meeting agenda or other explanatory text to inform attendees about the purpose of the meeting.

10 Click to select the message header for the meeting request to view the message details in the Reading pane.

Buttons along the top of the Reading pane or in the Respond group on the Meeting tab in a Message window for a meeting request are used to respond to the meeting organizer. Click the **Accept** button to send a reply that you will attend the meeting. Use the Tentative, Decline, or Propose New Time buttons if you are not required, or unable to attend a scheduled meeting. Outlook also provides a Calendar button at the top of the Reading pane, which is used to view your appointments for the meeting day and time to see if you are available before you send a response to the meeting organizer.

11 Double-click the message header to open the meeting request message with the Meeting tab active.

12 Click the Accept button in the Respond group on the Meeting tab.

13 Click *Send the Response Now* at the drop-down list.

Notice that the meeting request email message is deleted from your Inbox after you responded to the meeting invitation.

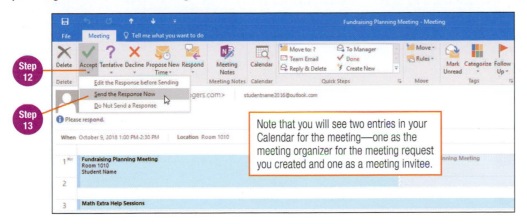

Note that you will see two entries in your Calendar for the meeting—one as the meeting organizer for the meeting request you created and one as a meeting invitee.

14 Click *Sent Items* in the Folder pane.

15 Click the message header for the message sent to the meeting organizer with your Accepted reply and then read the message you sent to the meeting organizer in the Reading pane.

16 Click *Inbox* in the Folder pane.

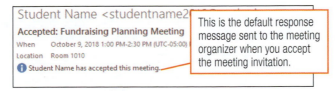

This is the default response message sent to the meeting organizer when you accept the meeting invitation.

Beyond Basics   Updating and Canceling a Meeting

If you need to reschedule a meeting, open the Meeting window; make the required changes to the day, time, or location; and then click the Send Update button. Outlook will send an email message to each attendee with the updated information. To delete a meeting, open the Meeting window and then click the Cancel Meeting button in the Actions group on the Meeting tab. Outlook sends an email message to each attendee informing them that the meeting is canceled and removes the meeting from each person's calendar.

5.7 Adding and Editing Contacts in People

Skills

Add a Contact

Edit a Contact

The **People** component in Outlook is used to store and organize contact information, such as email addresses, mailing addresses, telephone numbers, and other information about the people with whom you communicate. Think of People as an electronic address book. To add information about a person, click the **New Contact** button in the New group on the Home tab. Contacts display alphabetically in the People list in the Content pane. Click a contact name to display the information about the person in a **People card** in the Reading pane.

Oops!

Using Outlook.com? Click the Microsoft Apps button (it displays as a waffle icon) at the top left of the window and then click the People tile in the drop-down list. Click the New button to add a new contact. Click the expand buttons (plus symbol inside circle) to locate more options for entering data in the Add new contact panel.

App Tip

You can also add a picture using the Picture button in the Options group on the Contact tab.

1. With Outlook open and Inbox the active folder, click People in the Navigation pane.

2. Click the New Contact button in the New group.

3. Type Tory Nguyen in the *Full Name* text box in the Contact window.

4. Press Tab or click in the *Company* text box.

Notice the *File as* text box automatically updates when you move past the *Full Name* entry with the last name of the person followed by the first name. The *File as* entry is used to organize the People list alphabetically by last names.

5. Type NuWave Personnel in the *Company* text box.

6. Press Tab or click in the *Job title* text box and then type Recruitment Specialist.

7. Click in the *E-mail* text box and then type tory@emcp.net.

8. Click in the *Business* text box in the *Phone numbers* section and then type 8885559840.

9. Click in the *Mobile* text box in the *Phone numbers* section and then type 8885553256.

Notice that the phone numbers automatically format to show brackets around the area code, a space, and a hyphen in the number when you move past the field. In Outlook.com, the formatting of telephone numbers does not occur. Type the brackets and hyphens if desired.

10. Click the Add Contact Picture control that displays as a silhouette in a box between the name and business card sections of the Contact window.

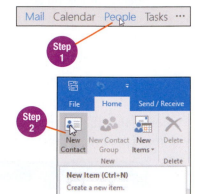

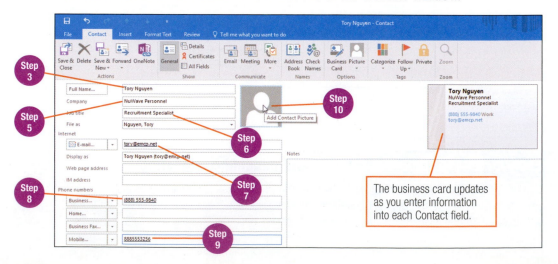

The business card updates as you enter information into each Contact field.

11 At the Add Contact Picture dialog box, navigate to the Ch5 folder in the Student_Data_Files folder and then double-click the image file *ToryNguyen*.

12 Click the Save & Close button in the Actions group on the Contact tab.

A selected person's information displays in the Reading pane in a People card with links to schedule a meeting or send an email to that person. Click *Edit* to open the People card fields for editing in the Reading pane, or double-click the name in the People list to add or modify information in a People card window.

13 Click Edit near the top right of the Reading pane with Tory Nguyen's information displayed.

14 Click at the end of the *Work* telephone number *(888) 555-9840,* press the spacebar, and then type extension 3115.

15 Click the Save button at the bottom right of the Reading pane.

16 Double-click in the white space below the last entry in the People list.

Double-clicking in blank space in the People list opens a new Contact window.

17 Enter the name of your instructor, the name of your school, and the email address of your instructor in the Contact window and then click Save & Close.

Alternative Method   **Editing Contact Information Using the Contact Window**

You can also edit a contact by opening the full Contact window and clicking the Outlook (Contacts) hyperlink below *View Source* in the Reading pane for the selected person. This causes the same Contact window to open that you used to add the person. Use this method if you need to change the picture of a contact or access the complete set of data or ribbon options.

Quick Steps

Add a Contact
1. Display People.
2. Click New Contact button.
3. Enter contact information as required.
4. Click Save & Close.

Edit a Contact
1. Click to select contact in People list.
2. Click Edit in Reading pane.
3. Edit information as required.
4. Click Save.

Oops!

No Reading pane? Your Contact view may be set to Business Card or Card. Click People in the Current View group on the Home tab. If necessary, click the View tab, Reading Pane button, then *Right*.

Oops!

A Check Full Name dialog box may appear after you enter a name. This happens when Outlook cannot determine the first and last name. Typing errors or hyphenated names can cause the dialog box to open. Enter or edit the text as needed and click OK when the dialog box appears.

5.8 Adding and Editing Tasks

Skills

Add a task

Edit a task

Delete a task

Mark a task complete

The **Tasks** component in Outlook maintains a to-do list. You can track information about each task, such as how much of the task is completed, how much time has been spent on the task, and the priority of the task. In **To-Do List** view, uncompleted tasks are shown grouped in descending order by the due date.

1. With Outlook open and People active, click *Tasks* in the Navigation pane and then click the *To-Do List* option in the Current View group on the Home tab if To-Do List is not the active view.

If Tasks is not visible, click here and then click *Tasks*.

Oops!

Using Outlook.com? Add a task by displaying the Calendar. Click the arrow on the New button and then click <u>Task</u>. Tasks display in a pane below the Calendar.

2. Click in the text box at the top of the To-Do List that displays the dimmed text *Type a new task* or *Click here to add a new Task*, type Do research on Pinterest, and then press Enter.

Outlook adds the task to the To-Do List under a flag with the heading *Today*.

3. Type Gather pictures to create Pinterest pinboard and then press Enter.

4. Type Create resume for Career Fair and then press Enter.

To add a task with more details and options, open a Task window by clicking the **New Task** button.

5. Click the New Task button in the New group on the Home tab.

6. Type Prepare study notes for exams in the *Subject* text box.

7. Click the *Priority* option box arrow and then click *High*.

8. Click the calendar icon at the right end of the *Due date* text box, navigate to the last month of the current semester, and then click the Monday that is one week before the last week of your semester.

9. Click the Save & Close button in the Actions group on the Task tab.

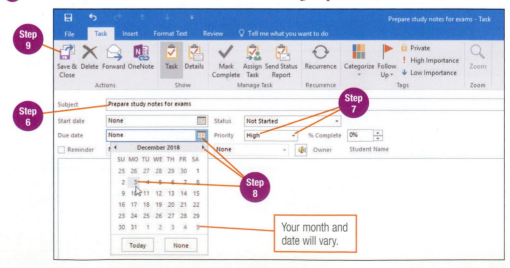

Your month and date will vary.

Editing or updating a task can include activities such as assigning or changing a due date, assigning a priority, entering the percentage of completion, or changing the status of a task. When a task is completed, use the Remove from List button in the Manage Task group on the Home tab or mark the task as complete in the Task window.

10 Click to select the task *Do research on Pinterest*.

11 Click the Remove from List button in the Manage Task group on the Home tab.

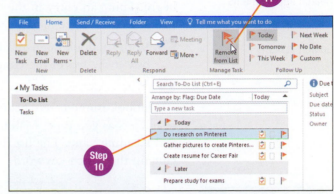

Notice the task is removed from the To-Do List. You can also use the Delete button in the Delete group on the Home tab to remove a task.

12 Double-click the task *Create resume for Career Fair* to open the Task window.

13 Click the Mark Complete button in the Manage Task group on the Task tab.

14 Double-click the task entry *Gather pictures to create Pinterest pinboard*.

15 Click the *Status* option box arrow and then click *Waiting on someone else*.

16 Click in the text box with white space below the *Reminder* options and then type Waiting for Leslie to send me pictures from her renovation clients..

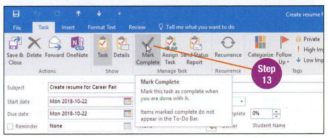

17 Click the Save & Close button. Notice the updated task details appear in the Reading pane for the selected task.

18 Click the Simple List option in the Current View group on the Home tab.

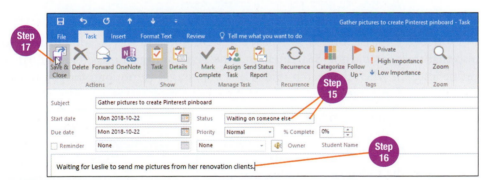

The list changes to include tasks marked completed. See Beyond Basics to read about the difference between Remove from List and Mark Complete options.

19 Display the Inbox folder in Mail.

Quick Steps

Add a Task
1. Display Tasks.
2. Click in *Type a new task* or *Click here to add a new Task* text box.
3. Type task description.
4. Press Enter.
OR
1. Display Tasks.
2. Click New Task button.
3. Enter task details.
4. Click Save & Close.

Beyond Basics   **Remove from List versus Mark Complete**

Remove from List deletes the task, while Mark Complete retains the task in the task list with a line drawn through the task and a gray check mark showing the task is finished. Mark Complete should be used if you need to retain task information for timekeeping or billing purposes. Display the complete Tasks list by clicking *Tasks* in the Folder pane.

5.9 Searching Outlook Items

Skills

Search messages

Search appointments

Search contacts

Search tasks

The search feature in Outlook, referred to as Instant Search, is used to locate a message, appointment, contact, or task. A search text box located between the ribbon and content is used to find items. Outlook begins a search as soon as you begin typing an entry in the search text box. Matched items are highlighted in the search results. Once located, you can open an item to view or edit the information.

1   With Outlook open and Inbox the active folder in Mail, click in the search text box at the top of the Content pane (contains the text *Search Current Mailbox*).

2   Type pinterest.

Outlook immediately begins matching messages in the Content pane with the characters as you type. Matched words are highlighted in both the Content pane and Reading pane. The list that remains is filtered to display all messages that contain the search keyword.

App Tip

The keyboard shortcut Ctrl + E opens the search text box.

App Tip

Instant Search shows the first 30 matches. Click the More hyperlink at the bottom of the search results list to view up to 250 matches.

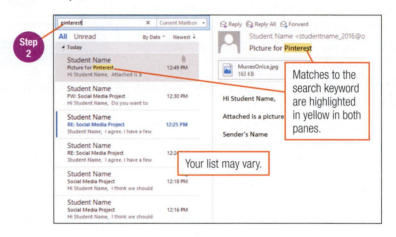

3   Click to view each message in the search results list.

Notice that for each message *Pinterest* is highlighted in the Reading pane. Notice also that the Search Tools Search tab is active. Use buttons on the tab to further refine a search when the search results list of messages is too long for you to easily find the right message.

4   Click the Close Search button in the search text box to close the search results list and return to the Inbox.

5   Display the Calendar.

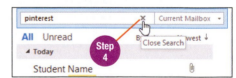

6   Click in the search text box at the top right of the Appointment area (contains the text *Search Calendar*) and then type career fair.

Outlook displays the appointment found with the search keywords in a filtered list.

7   Double-click the entry in the filtered list to view the details in the Event window and then close the window.

8   Click the Close Search button to restore the Calendar to the appointments for the current day.

9 Display People.

10 Click in the search text box at the top of the People list (contains the text *Search Contacts*) and then type nuwave.

Outlook displays the contact for Tory Nguyen who works at NuWave Personnel. You can use the search feature to find any Outlook item by any entry saved for the message, appointment, contact, or task. For example, you could find a contact by name, job title, company name, or even telephone number.

Quick Steps

Search for an Outlook Item
1. Display Mail, Calendar, People, or Tasks.
2. Click in search text box.
3. Type search keyword.

11 Click the Close Search button to restore the People list.

12 Display Tasks.

13 Click in the search text box at the top of the task list (contains the text *Search To-Do List* or *Search Tasks* depending on the active selection in the Folder pane.) and then type ex.

Outlook displays the task entry *Prepare study notes for exams*. Outlook can match items with only a partial entry for a word.

14 Click the Close Search button in the Close group on the Search Tools Search tab.

15 Display Mail with the Inbox folder.

16 Close Outlook.

Beyond Basics   **Other Search and Filter Techniques**

The Home tab for Mail, Calendar, People, and Tasks contains a search text box (contains the text Search People) in the Find group. Use this text box to find a contact from any Outlook area. You can click a person in the search results list to view the contact information in a People card.

In Mail, the Find group also contains a Filter Email button. Use this button to filter the message list by categories such as *Unread*, *Has Attachments*, *This Week*, *Flagged*, or *Important*.

Topics Review

Topic	Key Concepts	Key Terms
5.1 Using Outlook to Send Email	Microsoft Outlook 2016 is an application for organizing messages, appointments, contacts, and tasks referred to as a *personal information management (PIM) program.*	personal information management (PIM) program
	Electronic mail (email) is the exchange of messages between individuals electronically.	electronic mail (email)
	When Outlook is started, Mail is the active component within Outlook with the Inbox folder shown by default.	Outlook 2016
	The Inbox displays email messages with the newest message received at the top of the Content pane.	Mail
	Create and send a new email message using the New Email button in the New group on the Home tab.	Inbox
	Click the message header for a new message received to read the message contents in the Reading pane.	New Email
	Reply directly to a message by clicking the Reply button in the Reading pane (referred to as an *inline reply*).	Reply
	Send a copy of a message you have received to someone else using the Forward button.	inline reply
	A signature is a closing containing your name and other contact information that is inserted automatically at the end of each message.	Forward
5.2 Attaching a File to a Message and Deleting a Message	Attach a file to a message using the Attach File button in the Include group on the Message tab.	Attach File
	As you type an email address in the *To* text box, the AutoComplete feature shows email addresses in a drop-down list that match what you are typing. Click an entry in the list if the correct recipient appears before you finish typing.	AutoComplete
	Delete messages from mail folders that are no longer needed to keep folders to a manageable size.	Deleted Items
	Deleted messages are moved to the Deleted Items folder.	
	Empty the Deleted Items folder to permanently delete messages.	
5.3 Previewing File Attachments and Using File Attachment Tools	Preview a file attached to a message by clicking the file name in the Reading pane.	Protected View
	While a file is being previewed, message text is temporarily removed from the Reading pane.	
	Some files cannot be viewed in the Reading pane and must be viewed by double-clicking the file name to open the file.	
	Open or Save a file using buttons in the Actions group on the Attachment Tools Attachments tab or with options from the drop-down list accessed by clicking the arrow at the right of the file name in the Reading pane.	
	A file opened from a message is opened in Protected View in the source application, which allows you to read the contents. You cannot edit the file until you click the Enable Editing button on the Message bar.	
	Certain types of files known to contain viruses are automatically blocked by Outlook.	

continued...

Topic	Key Concepts	Key Terms
5.4 Scheduling Appointments and Events in Calendar	The Calendar component is used to schedule appointments and events. An appointment is any activity for which you want to record the occurrence by day and time. A Date Navigator at the top of the Folder pane displays the current month and next month with directional arrows to browse to the previous or next month. Use the Go To Date dialog box to navigate directly to a specific date you want to display in the Appointment area. Click next to the time in the Appointment area and then type a description to enter a new appointment. Click the New Appointment button in the New group on the Home tab to enter details for a new appointment in an Appointment window. An event is an appointment that lasts an entire day or longer. Click the New Items button in the New group on the Home tab and then click *All day event* to create an event in an Event window.	Calendar appointment Date Navigator Go To Date New Appointment event
5.5 Scheduling a Recurring Appointment and Editing an Appointment	An appointment that occurs at fixed intervals on a regular basis can be entered once, and Outlook schedules the remaining appointments automatically. Click the Recurrence button to set the recurrence pattern and range of recurrence details for a recurring appointment. Click to select an appointment in the Appointment area to assign options or tags to the appointment using buttons on the Calendar Tools Appointment tab. Open the Appointment window to make changes to the subject, location, day, or time of a scheduled appointment.	Recurrence
5.6 Scheduling a Meeting	A meeting is an appointment to which you invite people. Information about a meeting is sent to people via a meeting request, which is an email message sent to meeting participants. Click the New Meeting button in the New group on the Home tab to create a new meeting request by entering the email addresses for meeting attendees and the meeting particulars. A meeting attendee responds to a meeting request from the Reading pane or the message window by clicking a respond button, such as Accept. Meetings can be updated or canceled, and Outlook automatically informs all attendees via email messages.	meeting New Meeting meeting request Accept

continued…

Topic	Key Concepts	Key Terms
5.7 Adding and Editing Contacts in People	Use the People component to store and organize contact information for people with whom you communicate in a People card. Click the New Contact button or double-click in a blank area of the People list to open a Contact window and add information to a People card. Click the picture image control for a contact to select a picture of a contact in the Add Contact Picture dialog box. A contact's picture displays in the People card. The People card for a selected individual in the People list displays in the Reading pane. Click *Edit* in the Reading pane or double-click a person's name to edit the contact information in the People card.	People New Contact People card
5.8 Adding and Editing Tasks	Use Tasks in Outlook to maintain a To-Do List. Click in the *Type a new task* text box to add a task to the To-Do List or click the New Task button to enter a new task in a Task window. Open a Task window to add a due date or to add other task information, such as a priority or status. Select a task and use the Remove from List button when a task is completed. Open a task and use the Mark Complete button to indicate a task is completed; completed tasks are retained in the Task list but removed from the To-Do List.	Tasks To-Do List New Task
5.9 Searching Outlook Items	A search text box appears at the top of Mail, Calendar, People, and Tasks in which you can quickly search for an item by typing a keyword or phrase. Outlook begins to match items as soon as you begin typing a search keyword. Matches to the search keyword are highlighted in the filtered lists and in content in the Reading pane. A search text box also appears in the Find group on the Home tab in Mail, Calendar, People, and Tasks with which you can search for people and view the People card for an individual from any Outlook component.	

Recheck
Recheck your understanding of the topics covered in this chapter.

Workbook
Chapter review and assessment resources are available in the *Workbook* ebook.

Chapter 6

Creating, Editing, and Formatting Word Documents

Precheck
Check your understanding of the topics covered in this chapter.

Microsoft Word (referred to as *Word*) is a **word processing application** used to create documents that are mostly text for personal, business, or school purposes. Word documents can also include pictures, charts, tables, or other visual elements to make the document more interesting and easier to understand. Letters, essays, reports, invitations, recipes, agendas, contracts, and resumes are examples of documents that can be created using Word. Any text-based document you need to create can be generated using Word.

Word automatically corrects some errors as you type and indicates other potential spelling and grammar errors for you to consider. Other features provide tools to format and enhance a document. In this chapter, you will learn how to create, edit, and format documents. You will create new documents starting from a blank page and other documents by selecting from the template gallery in Word.

Note: If you are using a tablet, consider connecting it to a USB or wireless keyboard because parts of this chapter involve a fair amount of typing.

Learning Objectives

6.1 Create a new document and insert, delete, and edit text

6.2 Insert symbols and check spelling and grammar

6.3 Find and replace text

6.4 Move text and create bulleted and numbered lists

6.5 Format text using font options and change paragraph alignment

6.6 Indent paragraphs, change line spacing, and change spacing before and after paragraphs

6.7 Apply a style to text and change the style set

6.8 Create a new document from a template

SNAP If you are a SNAP user, go to your SNAP Assignments page to complete the Precheck, Tutorials, and Recheck.

Data Files
Before beginning this chapter, be sure you have copied the student data files for this course to your storage medium. Steps on downloading and extracting the data files are provided in Chapter 1, Topic 1.8, on pages 22–23.

6.1 Creating and Editing a New Document

Skills

Enter text

Describe AutoCorrect actions

Describe AutoFormat actions

Edit, insert, and delete text

Tutorials

Exploring the Word Screen

Entering Text

Undoing an AutoCorrect Correction

Adding and Deleting an AutoCorrect Entry

Saving with the Same Name

Saving with a New Name

Moving the Insertion Point and Inserting and Deleting Text

Recall from Chapter 3 that when Word starts, the Word Start screen appears, and you can choose to open an existing document, create a new blank document, or search for and select a template to create a new document. Creating a new document includes typing the document text, editing the text, and correcting errors. As you type, the AutoCorrect and AutoFormat features in Word help you fix common typing errors and apply formatting to characters or paragraphs.

1. Start Word 2016.

2. At the Word Start screen, click *Blank document* in the Templates gallery and compare your screen with the one shown in Figure 6.1.

If necessary, review the ribbon interface and Quick Access Toolbar described in Chapter 3. Table 6.1, p. 137, describes the elements shown in Figure 6.1.

3. Type Social Bookmarking and press Enter.

Notice that extra space is automatically added below the text before the next line.

4. Type the text on the next page allowing the lines to end automatically; press Enter only where indicated.

Word will move text to a new line automatically when you reach the end of the current line. This feature is called **wordwrap**. Word will also put a red wavy line below the word *bookmarklet*. Red wavy lines appear below words that are not found in the dictionary Word uses, indicating a possible spelling error. Correct typing mistakes as you go using the Backspace key to delete the character just typed and then retyping the correct character. You will learn other editing methods later in this topic.

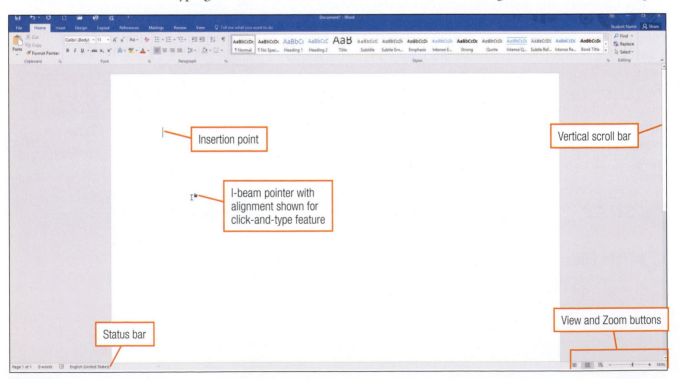

Figure 6.1
A new blank document screen is shown. The default settings in Word show a new document in Print Layout view and with rulers turned off; your display may vary if settings have been changed on the computer you are using. See Table 6.1, p. 137, for a description of screen elements.

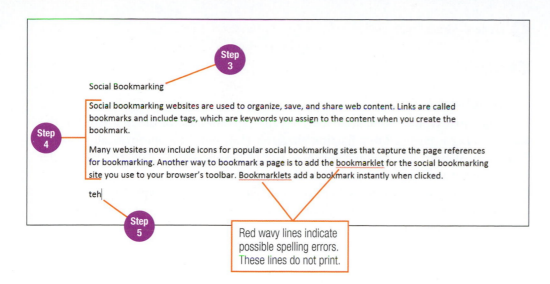

Red wavy lines indicate possible spelling errors. These lines do not print.

Social bookmarking websites are used to organize, save, and share web content. Links are called bookmarks and include tags, which are keywords you assign to the content when you create the bookmark. [press Enter]

Many websites now include icons for popular social bookmarking sites that capture the page references for bookmarking. Another way to bookmark a page is to add the bookmarklet for the social bookmarking site you use to your browser's toolbar. Bookmarklets add a bookmark instantly when clicked. [press Enter]

5 Type *teh* and press the spacebar. Notice that Word changes the text to *The*.

A feature called **AutoCorrect** changes commonly misspelled words as soon as you press the spacebar.

Table 6.1

Word Features

Feature	Description
Insertion point	Blinking vertical bar indicates where the next character typed will appear.
I-beam pointer	Pointer appearance for text entry or selection when you move the pointer using a mouse or trackpad. The I-beam pointer displays with a paragraph alignment option (left, center, or right) depending on the location of the I-beam within the current line. You can double-click and type text anywhere on a page, and the alignment option will be left-aligned, center-aligned, or right-aligned. This feature is called **click and type**.
Status bar	Displays page number with total number of pages and number of words in the current document. The right end of the Status bar has view and zoom options. The default view is Print Layout view, which shows how a page will look when printed with current print options.
Vertical scroll bar	Use the scroll bar to view parts of a document not shown in the current window.
View and Zoom buttons	By default, Word opens in Print Layout view. Other view buttons include Read Mode and Web Layout view. Read Mode maximizes reading space and removes editing tools, providing a more natural environment for reading. Zoom buttons, as you learned in Chapter 3, are used to enlarge or shrink the display.

App Tip

AutoCorrect also fixes common capitalization errors, such as two initial capitals, no capital at the beginning of a sentence, and no capital in the name of a day. AutoCorrect also turns off the Caps Lock key and corrects text when a new sentence is started with the key left on. Use Undo if AutoCorrect changes text that you don't want changed.

6 Type popular social bookmarking site Pinterest.com is used to pin pictures found on the Web to virtual pinboards. and press Enter.

7 Type a study by a marketing company found that 1/2 of frequent web surfers use a social bookmarking site, with Pinterest the 1st choice for most females. and press Enter.

Notice that Word automatically corrects the capitalization of the first word in the sentence, 1/2 is changed to a fraction character (½), and *1st* is automatically formatted as an ordinal, with the *st* shown as superscript text (superscript characters are smaller text placed at the top of the line). The **AutoFormat** feature automatically changes some fractions, ordinals, quotes, hyphens, and hyperlinks as you type. AutoFormat also converts straight apostrophes (') or quotation marks (") to smart quotes ('smart quotes'), also called *curly quotes* ("curly quotes").

8 Click the Save button on the Quick Access Toolbar.

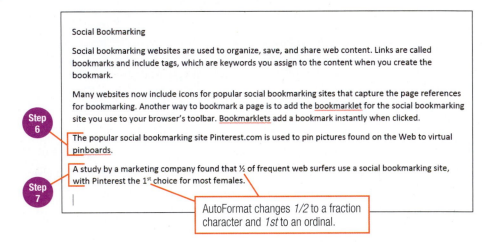

Social Bookmarking

Social bookmarking websites are used to organize, save, and share web content. Links are called bookmarks and include tags, which are keywords you assign to the content when you create the bookmark.

Many websites now include icons for popular social bookmarking sites that capture the page references for bookmarking. Another way to bookmark a page is to add the bookmarklet for the social bookmarking site you use to your browser's toolbar. Bookmarklets add a bookmark instantly when clicked.

Step 6 → The popular social bookmarking site Pinterest.com is used to pin pictures found on the Web to virtual pinboards.

Step 7 → A study by a marketing company found that ½ of frequent web surfers use a social bookmarking site, with Pinterest the 1st choice for most females.

AutoFormat changes *1/2* to a fraction character and *1st* to an ordinal.

Because this is the first time the document has been saved, the Save As backstage area appears.

9 Click the option at the Save As backstage area that represents the location for your storage medium and then click the Browse button. For example, click *This PC* if you are saving files to a USB flash drive and then click *Browse*.

10 At the Save As dialog box, navigate to the CompletedTopicsByChapter folder on your storage medium and then create a new folder, *Ch6*.

11 Double-click the *Ch6* folder name.

12 Select the current text in the *File name* text box, type SocialMediaProject-YourName, and then press Enter or click the Save button.

When creating a new document, changes often need to be made to the text after the text has been typed. You may need to correct typing errors, change a word or phrase to some other text, add new text, or remove text. Making changes to a document after the document has been typed is called **editing**. The first step to edit text is to move the insertion point to the location of a change.

13 Click to position the insertion point at the beginning of the last paragraph that begins with the text *A study by a marketing company* (insertion point will be blinking just left of *A*), type Experian Hitwise, and then press the spacebar.

Step 13 → Experian Hitwise A study by a marketing company found that ½ of frequent web surfers use a social bookmarking site, with Pinterest the 1st choice for most females.

Word automatically inserts the new text and moves existing text to the right.

14 With the insertion point still positioned at the left of *A* in *A study*, press the Delete key until you have removed *A study by*, type (, click to position the insertion point just after the *y* in *company*, and then type) conducted a survey of frequent web surfers and.

15 Position the insertion point just left of ½, press the Delete key until you have removed ½ *of frequent web surfers*, and then type one-half.

16 Position the insertion point at the left of *1st*, delete *1st*, and then type first.

17 Position the insertion point below the last paragraph and then type your first and last name.

18 Check your text with the document shown in Figure 6.2. If necessary, make further corrections by moving the insertion point and inserting and deleting characters as needed.

Oops!

Insertion point will not go below last paragraph? Click to place the insertion point at the end of the last sentence and then press Enter to move down to a blank line.

Social Bookmarking

Social bookmarking websites are used to organize, save, and share web content. Links are called bookmarks and include tags, which are keywords you assign to the content when you create the bookmark.

Many websites now include icons for popular social bookmarking sites that capture the page references for bookmarking. Another way to bookmark a page is to add the bookmarklet for the social bookmarking site you use to your browser's toolbar. Bookmarklets add a bookmark instantly when clicked.

The popular social bookmarking site Pinterest.com is used to pin pictures found on the Web to virtual pinboards.

Experian Hitwise (a marketing company) conducted a survey of frequent web surfers and found that one-half use a social bookmarking site, with Pinterest the first choice for most females.

Student Name

Figure 6.2

The document text for SocialMediaProject-StudentName is shown.

Quick Steps

Create a New Document
1. Start Word 2016.
2. Click *Blank document*.
3. Type text.

Save a New Document
1. Click Save button on the Quick Access Toolbar.
2. Navigate to drive and/or folder.
3. Enter file name.
4. Click Save.

Save a Document Using the Existing Name
Click Save button on the Quick Access Toolbar.

Edit a Document
1. Position insertion point at location of change.
2. Type new text or delete text as needed.
3. Save changes.

19 Click the Save button on the Quick Access Toolbar. Leave the document open for the next topic.

Because the document has already been assigned a file name at Step 12, the Save button saves the document changes using the same name.

Beyond Basics   **Line Breaks versus New Paragraphs**

You press Enter only at the end of a short line of text (such as the title) or at the end of a paragraph. Pressing Enter creates a hard return, and in Word creates a new paragraph. By default, line spacing in Word 2016 is set to 1.08, and 8 points of space is added after each paragraph. A point is a measurement system in which 1 point is approximately equal to the height of 1/72 inch. Think of 8 points as approximately .11 of an inch of space added after each hard return.

To end a short line of text and not add an extra 8-point space after the line, use the Line Break command Shift + Enter (hold down the Shift key while pressing Enter). A line break moves to the next line without creating a new paragraph. For example, use Shift + Enter when typing an address in a letter after the street number and street name to move to the next line.

Skills

Insert a symbol and special character

Check spelling and grammar

Tutorials

Inserting Symbols

Inserting Special Characters

Checking Spelling and Grammar

6.2 Inserting Symbols and Completing a Spelling and Grammar Check

In some documents, you need to insert a symbol or special character, such as a copyright symbol (©), registered trademark (™), or a fraction character for a fraction that AutoCorrect does not recognize, such as one-third (⅓). Symbols and special characters are inserted using the **Symbol gallery** or the Symbol dialog box.

The **Spelling & Grammar** button on the Review tab starts the Spelling and Grammar feature used to correct errors. The Spelling feature matches words in the document with words in a dictionary and flags a word not found as a possible error. The Spelling task pane opens at the right side of the window with suggestions for a word not found and with buttons to ignore, to change, or to add the word to the dictionary. The feature also checks for duplicate words and prompts you to remove the repeated word. Spelling and Grammar helps you correct many errors; however, you still need to proofread your documents. For example, the errors in the following sentence escaped detection: The plain fair was expensive!

1. With the **SocialMediaProject-YourName** document open, position the insertion point after *n* in *Experian* in the last paragraph.

Experian is a registered trademark, so you will add the registered trademark symbol after the name.

2. Click the Insert tab.

3. Click the Symbol button in the Symbols group.

4. Click *Trade Mark Sign*. Note that the symbol may appear in a different position than shown in the image at the right.

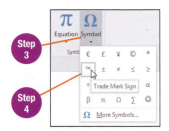

5. Position the insertion point after the period at the end of the paragraph that begins *Experian* and press the spacebar to insert a space.

6. Type The survey sample size of 1,000 interviews provides a standard error at 95% confidence of and then press the spacebar.

7. Click the Symbol button and then click *More Symbols*.

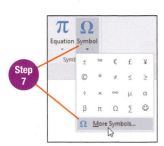

8. At the Symbol dialog box, with *Font* set to *(normal text)* and *Subset* set to *Letterlike Symbols*, scroll up the symbol list until the *Subset* changes to *Latin-1 Supplement*, click ±, and then click the Insert button.

Oops!

Different font and/or subset? Use the *Font* or *Subset* list arrow to change the option to *(normal text)* and *Letterlike Symbols* if the Symbol dialog box has different settings.

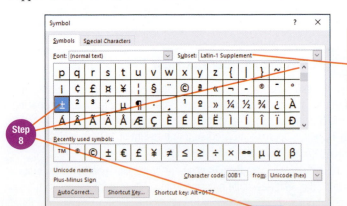

Scroll up the dialog box at Step 8 until the *Subset* changes to *Latin-1 Supplement* and you can see the plus-minus symbol.

9 Click the Close button to return to the document and then type 3%..

You may have noticed the Plus-Minus Sign symbol in the Symbol gallery. In Steps 7 to 9, you practiced using the Symbol dialog box so that you will know how to find a symbol that is not shown in the Symbol gallery.

10 Click the Review tab and then click the Spelling & Grammar button in the Proofing group.

11 When the word *bookmarklet* is selected, click the Ignore All button in the Spelling task pane.

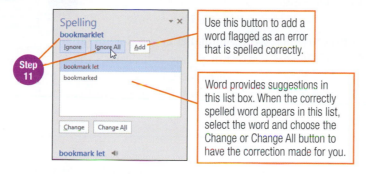

Use this button to add a word flagged as an error that is spelled correctly.

Word provides suggestions in this list box. When the correctly spelled word appears in this list, select the word and choose the Change or Change All button to have the correction made for you.

12 When the word *bookmarklets* is selected, click the Ignore All button in the Spelling task pane.

13 Choose Ignore when *pinboards* is selected.

14 Choose Ignore when *Hitwise* is selected.

15 Click OK at the message that the spelling and grammar check is complete.

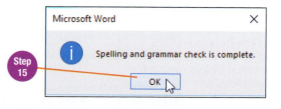

16 Save the document using the same name. Leave the document open for the next topic.

Alternative Method   **Inserting a Symbol Using a Keyboard Shortcut**

Type (c) to have AutoCorrect insert the copyright symbol ©.

Type (r) to have AutoCorrect insert the registered symbol ®.

Type (tm) to have AutoCorrect insert the registered trademark symbol ™.

To view the complete list of AutoCorrect entries, click the File tab, then click *Options* at the backstage area. At the Word Options dialog box, click *Proofing* in the left pane and then click the AutoCorrect Options button.

6.3 Finding and Replacing Text

Skills

Use Find Command

Use Replace Command

Tutorials

Finding Text

Finding and Replacing
Text

Navigating Using the
Navigation Pane

Using the Thesaurus

The **Find** feature moves the insertion point to each occurrence of a word or phrase. Find is helpful if you think you have overused a particular term and want to review how many times it appears in a document, or if you want to move the insertion point to a specific location in the document very quickly and a unique word or phrase exists near the location. **Replace** looks for each occurrence of a word or phrase and automatically changes the text to another word or phrase that you specify. Use Replace to make a global change throughout the document, such as changing a person's name in a will or legal contract.

1. With the **SocialMediaProject-YourName** document open, position the insertion point at the beginning of the document.

2. Click the Home tab.

3. Click the Find button in the Editing group.

This opens the Navigation pane at the left side of the Word document window.

4. Type social bookmarking in the search text box.

When you finish typing, Word highlights in the document all the occurrences of the search word or phrase and displays the search results below the search text box. The total number of occurrences appears at the top of the Results list. Each entry in the search results list is a link that moves to the search word location in the document when clicked (Figure 6.3).

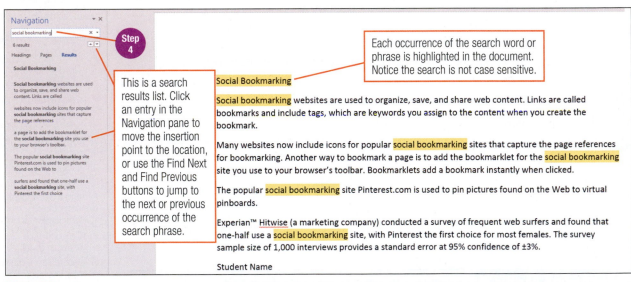

Figure 6.3

Search results for *social bookmarking* are highlighted in the SocialMediaProject-YourName document.

App Tip

You can find text using a partial word search. For example, entering *exp* would find *Experian*, *expert*, and *experience*. Use partial word searches if you are not sure of correct spelling.

5. Click each entry one at a time in the Results list in the Navigation pane.

Notice that each occurrence of the search word is selected as you move to the phrase location.

6. Click the Close button at the top right of the Navigation pane to close the pane.

7 Position the insertion point at the beginning of the document.

Find and Replace searches begin from the location of the insertion point in the document.

8 Click the Replace button in the Editing group on the Home tab.

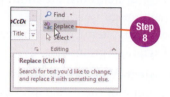

9 At the Find and Replace dialog box, with *social bookmarking* selected in the *Find what* text box, type bookmarklet.

10 Press Tab or click in the *Replace with* text box and then type bookmark button.

11 Click the Replace All button.

12 Click OK at the message that 2 replacements were made.

13 Click the Close button to close the Find and Replace dialog box.

Notice that Word matches the correct case of a word when the word is replaced at the beginning of a sentence, as seen in the last sentence of the paragraph that begins with *Many websites now include.*

14 Save the document using the same name. Leave the document open for the next topic.

App Tip

Ctrl + Home moves the insertion point to the beginning of a document.

App Tip

Click Find Next to replace text as you go through the document stopping at each occurrence if you do not want to globally replace the text.

Oops!

No replacements made? Click OK and check your spelling in the *Find what* text box. Correct the text and try Replace All again. Still have 0 replacements? Cancel the Replace command and check the spelling of *bookmarklet* within your document. Correct the text in the document and repeat Steps 8 to 12.

Quick Steps

Find Text
1. Position insertion point at beginning of document.
2. Click Find button.
3. Type search text.
4. Review Results list.

Replace Text
1. Position insertion point at beginning of document.
2. Click Replace button.
3. Type *Find what* text.
4. Press Tab or click in *Replace with* text box.
5. Type replacement text.
6. Click Replace All.
7. Click OK.
8. Click Close button.

Beyond Basics   **Using the Thesaurus to Replace a Word**

Sometimes when you find yourself overusing a word, you are stuck for an alternative word to use in its place in one or two occurrences. Consider using the Thesaurus to help you find a word with a similar meaning. Thesaurus is located in the Proofing group on the Review tab. Position the insertion point anywhere within a word you want to change, click the Thesaurus button, point to a word you want to change the occurrence to in the results list in the Thesaurus task pane, click the down-pointing arrow that appears, and then click *Insert*.

6.4 Moving Text and Inserting Bullets and Numbering

Skills

Move text

Create a bulleted list

Create a numbered list

Tutorials

Cutting, Copying, and
Pasting Text

Creating Bulleted Lists

Creating Numbered
Lists

The Cut button in the Clipboard group is used to move a selection of text from one location in the document to another location. Bulleted and numbered lists set apart information that is structured in short phrases or sentences. A bulleted list is text set apart from a paragraph with a list of items that are in no particular sequence. The **Bullets button** in the Paragraph group is used to create this type of list. A numbered list is used for a sequential list of tasks, items, or other text and is created using the **Numbering button** in the Paragraph group.

1. With the **SocialMediaProject–YourName** document open, position the insertion point at the beginning of the paragraph that begins with *The popular social bookmarking site.*

2. Click the Show/Hide button in the Paragraph group on the Home tab.

   Show/Hide turns on the display of hidden formatting symbols. For example, each time you press Enter, a paragraph symbol (¶) is inserted in the document. Revealing these symbols is helpful when you are preparing to move or copy text because you often want to make sure you move or copy the paragraph symbol with the paragraph.

3. Select the paragraph *The popular social bookmarking site Pinterest.com is used to pin pictures found on the Web to virtual pinboards.* Make sure to include the paragraph formatting symbol at the end of the text in the selection as shown below.

4. Click the Cut button in the Clipboard group on the Home tab.

   The text is removed from the current location and placed in the Clipboard.

5. Position the insertion point at the beginning of the paragraph that begins with *Many websites now include.*

6. Click the top of the Paste button in the Clipboard group (do *not* click the down-pointing arrow on the button).

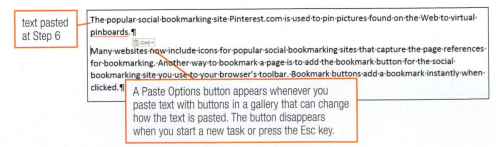

text pasted at Step 6

A Paste Options button appears whenever you paste text with buttons in a gallery that can change how the text is pasted. The button disappears when you start a new task or press the Esc key.

7. Click the Show/Hide button to turn off the display of hidden formatting symbols.

8 Position the insertion point after the period that ends the sentence *The popular social networking site Pinterest.com is used to pin pictures found on the Web to virtual pinboards*, press the spacebar, type Other social bookmarking sites include:, and then press Enter.

9 Click the left part of the Bullets button in the Paragraph group on the Home tab (do *not* click the down-pointing arrow on the button).

App Tip

The Bullets button arrow is used to choose a different bullet character from the Bullet library.

This action indents and then inserts the default bullet character, which is a solid round bullet.

10 Type StumbleUpon.com and press Enter.

11 Type Delicious.com and press Enter.

12 Type Digg.com and press Enter.

13 Type Newsvine.com.

14 Position the insertion point after the period that ends the sentence *Bookmark buttons add a bookmark instantly when clicked*, press the spacebar, type To add a bookmark button:, and then press Enter.

15 Click the left part of the Numbering button in the Paragraph group on the Home tab (do *not* click the down-pointing arrow on the button).

This indents and inserts 1.

16 Type Display the browser's Favorites toolbar or Bookmarks bar. and then press Enter.

17 Type Right-click the bookmark button and then choose Add to favorites. and then press Enter.

18 Type Choose Add button at dialog box that appears..

App Tip

The Numbering button arrow is used to choose a different number format from the Numbering library.

Step 9

Steps 10-13
The popular social bookmarking site Pinterest.com is used pinboards. Other social bookmarking sites include:
- StumbleUpon.com
- Delicious.com
- Digg.com
- Newsvine.com
Many websites now include icons for popular social bookm

Steps 16-18
Many websites now include icons for popular social bookmarking sites that capture the page references for bookmarking. Another way to bookmark a page is to add the bookmark button for the social bookmarking site you use to your browser's toolbar. Bookmark buttons add a bookmark instantly when clicked. To add a bookmark button:
1. Display the browser's Favorites toolbar or Bookmarks bar.
2. Right-click the bookmark button and then choose Add to favorites.
3. Choose Add button at dialog box that appears.

19 Save the document using the same name. Leave the document open for the next topic.

Quick Steps

Move Text
1. Select text.
2. Click Cut button.
3. Position insertion point.
4. Click Paste button.

Create a Bulleted List
1. Click Bullets button.
2. Type first list item.
3. Press Enter.
4. Type second list item.
5. Press Enter.
6. Continue typing until finished.

Create a Numbered List
1. Click Numbering button.
2. Type first numbered item.
3. Press Enter.
4. Type second numbered item.
5. Press Enter.
6. Continue typing until finished.

Alternative Method   Automatically Creating a Bulleted or Numbered List

The AutoFormat as You Type feature creates automatic bulleted and numbered lists when you do the following:

Type *, >, or –, press the spacebar, type text, and then press Enter (bulleted list).

Type 1., press the spacebar, type text, and press Enter (numbered list).

Immediately use Undo if an automatic list appears and you do not want to create a list.

6.5 Formatting Text with Font and Paragraph Alignment Options

Skills

Change font and font options

Change paragraph alignment

Tutorials

Applying Font Formatting Using the Font Group

Changing Paragraph Alignment

App Tip

Click the Font dialog box launcher to open the Font dialog box for more text effects options, such as double strikethrough and small caps.

Oops!

Mini toolbar disappeared? This happens when you move the mouse away from the toolbar after selecting text. Use the *Font* option box arrow in the Font group on the Home tab instead.

App Tip

Use the Font Color button arrow to choose a color other than red. Once the color is changed, the new color can be applied to the next selection without using the button arrow. The color on the button resets to red after Word is closed.

The process of altering the appearance of the text is referred to as **formatting**. Changing the appearance of characters is called **character formatting**. Changing the appearance of a paragraph is called **paragraph formatting**. In some cases, the first step in formatting is to select the characters or paragraphs to be changed.

Some people format text as they type the document. In that case, you can change the character or paragraph option before typing the text. Depending on the formatting option in use, you may have to turn off the option after typing text or change to another format option. For example, if you apply bold to text as you type, you turn on the bold feature, type the text, then turn off the bold feature.

Applying Character Formatting

The Font group on the Home tab contains the buttons used to change character formatting. A **font** includes the design and shape of the letters, numbers, and special characters of a particular *typeface*. A large collection of fonts is available from simple to artistic character design. The font size is set in points, with one point equal to approximately 1/72 of an inch in height. The default font in a new document is 11-point Calibri.

The Font group includes buttons to increase or decrease the font size; change the case; change the font style (bold, italic, or underline) or font color; highlight text; add the effects strikethrough, subscript, or superscript; or add an artistic look using the Text Effects and Typography button (outline, shadow, glow, and reflection accents).

1 With the **SocialMediaProject–YourName** document open, select the title text at the top of the document *Social Bookmarking*.

2 Click the *Font* option box arrow on the Mini toolbar.

3 Scroll down and then click *Century Gothic*.

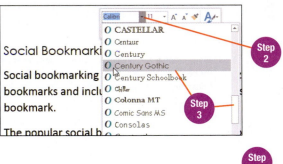

4 With the title text still selected, click twice the Increase Font Size button on the Mini toolbar or in the Font group on the Home tab.

The first time you increase the font size, the size changes to 12 points. The second time, the size changes to 14 points and continues to increase 2 point sizes each time until you reach 28. After 28 points, the size changes to 36, 48, and then 72.

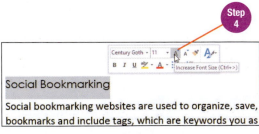

5 With the title text still selected, click the Bold button on the Mini toolbar or in the Font group on the Home tab.

6 With the title text still selected, click the Font Color button (do *not* click the Font Color button arrow) on the Mini toolbar or in the Font group on the Home tab.

The default Font Color is Red.

7 Click in the document away from the selected title to deselect the text.

Applying Paragraph Formatting

The Paragraph group on the Home tab contains the buttons used to change paragraph formatting. The bottom row of buttons in the group contains the buttons for changing the alignment of paragraphs from the default **Align Left** to **Center**, **Align Right**, or **Justify**. Justified text adds space within a line so that the text is distributed evenly between the left and right margins. You will explore other buttons in the Paragraph group in the next topic.

8 Click to place the insertion point anywhere within the title text *Social Bookmarking*.

To format a single paragraph, you do not need to select the paragraph text because paragraph formatting applies to all text within the paragraph to the point where a hard return was inserted.

9 Click the Center button in the Paragraph group on the Home tab.

10 Click to place the insertion point anywhere within the first paragraph of text and then click the Justify button in the Paragraph group on the Home tab.

Justified text spreads the lines out so that the text ends evenly at the right margin.

11 With the insertion point still positioned in the first paragraph, click the Align Left button in the Paragraph group on the Home tab.

12 Save the document using the same name. Leave the document open for the next topic.

Select more than one paragraph first if you want to apply the same alignment option to multiple paragraphs.

Step 9 — Center (Ctrl+E)
Center your content on the page.
Center alignment gives documents a formal appearance and is often used for cover pages, quotes, and sometimes headings.

Step 8 — **Social Bookmarking**
Social bookmarking websites are used

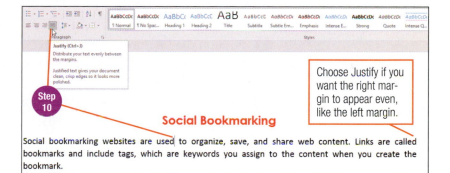

Step 10 — Justify (Ctrl+J)
Distribute your text evenly between the margins.
Justified text gives your document clean, crisp edges so it looks more polished.

Choose Justify if you want the right margin to appear even, like the left margin.

Social Bookmarking
Social bookmarking websites are used to organize, save, and share web content. Links are called bookmarks and include tags, which are keywords you assign to the content when you create the bookmark.

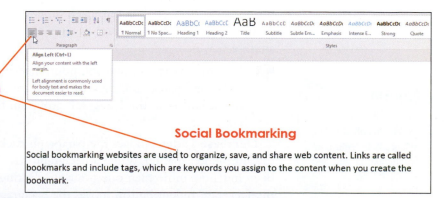

Step 11 — Align Left (Ctrl+L)
Align your content with the left margin.
Left alignment is commonly used for body text and makes the document easier to read.

Social Bookmarking
Social bookmarking websites are used to organize, save, and share web content. Links are called bookmarks and include tags, which are keywords you assign to the content when you create the bookmark.

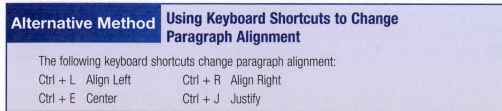

Alternative Method | **Using Keyboard Shortcuts to Change Paragraph Alignment**

The following keyboard shortcuts change paragraph alignment:

Ctrl + L	Align Left	Ctrl + R	Align Right
Ctrl + E	Center	Ctrl + J	Justify

6.6 Indenting Text and Changing Line and Paragraph Spacing

Skills

Indent text

Change line spacing

Change spacing after paragraphs

 Tutorials

Indenting Text

Changing Line Spacing

Changing Spacing Before and After Paragraphs

Paragraphs are indented to set the paragraph apart from the rest of the document. In reports, essays, or research papers, long quotes are indented. A paragraph can be indented for the first line only, or for all lines in the paragraph. Paragraphs can also be indented from the right margin. A paragraph where the first line remains at the left margin but subsequent lines are indented is called a **hanging indent**. Hanging indents are used in bulleted lists, numbered lists, bibliographies, and lists of cited works.

Use the **Line and Paragraph Spacing** button in the Paragraph group to change the spacing between lines of text within a paragraph and to change the spacing before and after paragraphs.

1. With the **SocialMediaProject-YourName** document open, position the insertion point at the left margin of the first paragraph (begins with the text *Social bookmarking websites*).

2. Press the Tab key.

Pressing the Tab key indents the first line of a paragraph 0.5 inch and is referred to as a *first line indent*. The AutoCorrect Options button may also appear. Use the button when it appears to change the first line indent back to a Tab, to stop setting indents when you press Tab, or to change other AutoFormat options.

Step 2

Social Bookmarking

Social bookmarking websites are used to organize, save, and share web content. Links are called bookmarks and include tags, which are keywords you assign to the content when you create the bookmark.

3. Position the insertion point anywhere within the second paragraph (begins with the text *The popular*).

4. Click the Increase Indent button in the Paragraph group on the Home tab.

The **Increase Indent** button indents all lines of a paragraph 0.5 inch. Click the button more than once if you want to move the paragraph further away from the margin. Each time the button is clicked, the paragraph is indented another 0.5 inch.

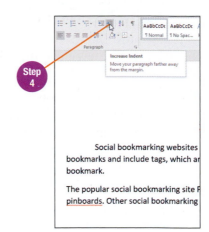

Step 4

5. With the insertion point still positioned in the second paragraph, click the Decrease Indent button in the Paragraph group on the Home tab.

Decrease Indent moves the paragraph back toward the left margin. When a paragraph has been indented more than one position, clicking the Decrease Indent button moves the paragraph back towards the left margin 0.5 inch each time the button is clicked.

6. Position the insertion point anywhere within the third paragraph (begins with the text *Many websites now include*).

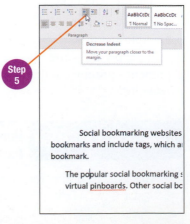

Step 5

App Tip

Use the Clear All Formatting button in the Font group to remove all formatting from selected text.

7. Click the Line and Paragraph Spacing button in the Paragraph group on the Home tab, and then click *Line Spacing Options* at the drop-down list.

8. At the Paragraph dialog box, select the current entry in the *Left* text box and then type 0.5.

9. Select the current entry in the *Right* text box and then type 0.5.

10. Click OK.

The paragraph is indented from both margins by 0.5 inch.

11. With the insertion point still positioned in the paragraph that begins with the text *Many websites*, click the Line and Paragraph Spacing button and then click *Line Spacing Options*.

12. Change the entry in the *Left* and the *Right* text boxes to *0*.

13. Click the *Special* list box arrow and then click *First line*.

14. Click OK.

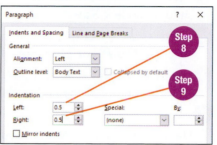

15. Format the second paragraph that begins with the text *The popular* and the last paragraph that begins with the text *Experian* with a first line indent by opening the Paragraph dialog box and then changing the *Special* list box option to *First line*.

16. Click the Select button in the Editing group on the Home tab and then click *Select All* at the drop-down list.

17. Click the Line and Paragraph Spacing button and then click *1.5* at the drop-down gallery.

The line spacing is changed to 1.5 lines for the entire document. Notice the other line spacing options are *1.0*, *1.15*, *2.0*, *2.5*, and *3.0*.

18. With the entire document still selected, click the Line and Paragraph Spacing button and then click *Remove Space After Paragraph* at the drop-down gallery.

19. Click in any section of the document to deselect the text.

20. Save the document using the same name. Leave the document open for the next topic.

Alternative Method | **Changing Indents and Paragraph Spacing Using the Layout and Design tabs**

In the Paragraph group on the Layout tab, use the *Left* and *Right* text boxes to indent paragraphs and the *Before* and *After* text boxes to change the spacing before and after paragraphs.

On the Design tab, use the Paragraph Spacing button in the Document Formatting group to set line and paragraph spacing options for the entire document, including new paragraphs.

6.7 Formatting Using Styles

Skills

Apply styles

Change style set

Tutorials

Applying Styles and
Style Sets

Applying and Modifying
a Theme

A **style** is a set of predefined formatting options that can be applied to selected text or paragraphs with one step. The Styles group on the Home tab shows the styles available in Word. You can also create your own styles. Use the More button at the bottom of the scroll bar at the right of the styles gallery to show more style options and the *Create a Style, Clear Formatting*, and *Apply Styles* options.

Once styles have been applied to text in the document, buttons in the Document Formatting group on the Design tab change the **style set**, which changes the look of a document by applying a group of formatting options to styles. Each style set has different formatting options associated with it.

1. With the **SocialMediaProject-YourName** document open, click the File tab and then click *Save As*.

2. If necessary, select the storage location option where you are saving your work such as *This PC* or *OneDrive - Personal* at the Save As backstage area.

3. Click in the *File name* text box, and then click a second time to place the insertion point at the end of the word *Project* and before *-YourName*.

4. Type WithStyles.

   The file name is now **SocialMediaProjectWithStyles-YourName**.

5. Click the Save button.

6. Position the insertion point anywhere within the title text *Social Bookmarking*.

7. Click the *Title* style in the styles gallery on the Home tab.

8. Position the insertion point anywhere within the first paragraph of text below the title.

9. Click the *Quote* style option in the styles gallery. If the *Quote* style option is not visible, click the More button (▾) at the bottom of the scroll bar located at the right of the styles gallery to display more style options, and then click the *Quote* style option.

Quick Steps

Format Using Styles
1. Select text or position insertion point.
2. Click desired style option.

Change Style Set
1. Click Design tab.
2. Click desired style set.

first paragraph with *Quote* style preview

10 Select *Pinterest.com* in the second paragraph and then click the *Intense Reference* style.

11 Deselect *Pinterest.com*.

Once styles have been applied to text, you can experiment with various style sets in the Document Formatting group on the Design tab. Changing the style set changes font and paragraph formatting options.

12 Click the Design tab.

13 Click the *Basic (Stylish)* style set in the Document Formatting gallery (fourth option from left).

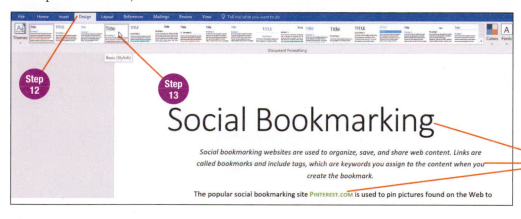

The *Basic (Stylish)* style set causes the look of the document to change.

14 Position the insertion point anywhere within the title text *Social Bookmarking*, click the Home tab, and then click the *Heading 1* style.

The title text formats with the options for the Heading 1 style in the new style set. The formatting applied by a style is determined by the **Theme**. Each theme has a color scheme and a font scheme that affects style options.

App Tip

Change the theme from the Themes button on the Design tab.

15 Save the document using the same name (**SocialMediaProjectWithStyles-YourName**).

16 Click the File tab and then click *Close* to close the document, leaving Word open for the next topic.

View
Model Answer
Compare your completed file with the model answer.

6.8 Creating a New Document from a Template

A **template** is a document that has been created with formatting options applied. Several professional-quality templates for various types of documents are available that you can use rather than creating a new document from scratch. At the Word Start screen, you can browse and preview available templates by category, or you can type in a keyword for the type of document you are looking for in the search text box (contains the text *Search for online templates*) and browse the templates in the search results.

When a Word document has been closed and Word is still active, the New backstage area is used to browse for a template.

1 Click the File tab and then click *New*.

2 At the New backstage area, click Business in the *Suggested searches* section.

3 Scroll down and review the various types of business templates available, then click Home at the top of the New backstage area.

4 Click in the search text box, type time sheet, and then press Enter or click the Start searching button (displays as a magnifying glass at the end of the search text box).

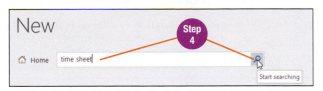

5 Click the first *Time sheet* template in the Templates gallery.

A preview of the template opens with a description that provides information on the template design.

6 Click the Create button.

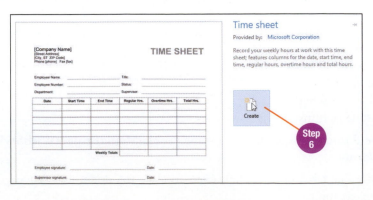

7 Click *[Company Name]* and then type A+ Tutoring Advantage.

8 Click *[Street Address]* and then type 1015 Montreal Way.

9 Click *[City, ST ZIP Code]* and then type St. Paul, MN 55102.

10 Click *[phone]* and then type 888-555-3125.

11 Click *[fax]* and type 888-555-3445.

A+ Tutoring Advantage
1015 Montreal Way
St. Paul, MN 55102
Phone 888-555-3125 Fax 888-555-3445

Steps 7-11

TIME SHEET

App Tip

[Company Name] is referred to as a *text placeholder*. Look for these placeholders in a template to indicate where to personalize the document. Other templates include descriptive text within the template that describes how to use the template.

12 Click next to *Employee Name* and then type your name.

13 Complete the remainder of the time sheet document using the text shown in Figure 6.4 by completing a step similar to Step 12.

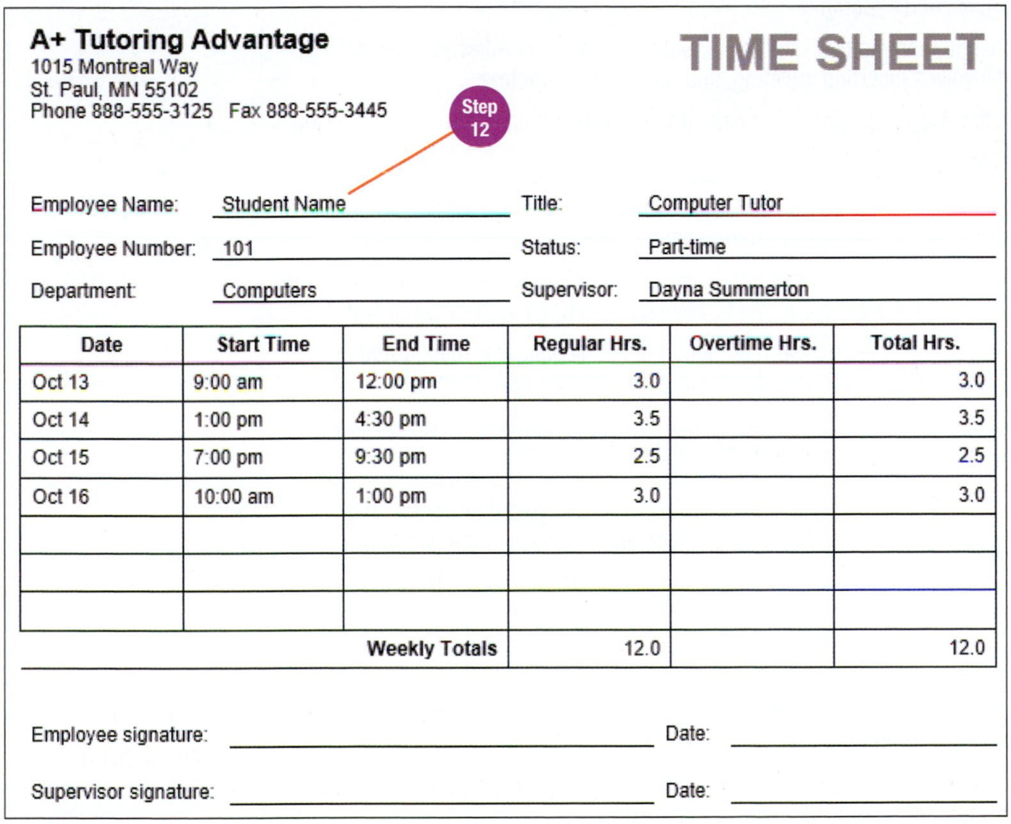

A+ Tutoring Advantage
1015 Montreal Way
St. Paul, MN 55102
Phone 888-555-3125 Fax 888-555-3445

Step 12

TIME SHEET

Employee Name:	Student Name	Title:	Computer Tutor
Employee Number:	101	Status:	Part-time
Department:	Computers	Supervisor:	Dayna Summerton

Date	Start Time	End Time	Regular Hrs.	Overtime Hrs.	Total Hrs.
Oct 13	9:00 am	12:00 pm	3.0		3.0
Oct 14	1:00 pm	4:30 pm	3.5		3.5
Oct 15	7:00 pm	9:30 pm	2.5		2.5
Oct 16	10:00 am	1:00 pm	3.0		3.0
		Weekly Totals	12.0		12.0

Employee signature: _____ Date: _____

Supervisor signature: _____ Date: _____

Figure 6.4
Shown is a completed Oct13to16Timesheet-YourName document.

Quick Steps

Create New Document from a Template
1. Click File tab.
2. Click *New*.
3. Browse or search for template design.
4. Click desired template.
5. Click Create button.
6. Complete document as required.

14 Save the completed time sheet in the Ch6 folder in the CompletedTopicsByChapter folder on your storage medium as **Oct13to16TimeSheet-YourName**. Choose OK when a message displays that the document will be upgraded to the newest file format.

15 Close the document.

View
Model Answer
Compare your completed file with the model answer.

Beyond Basics   **Template Designs**

Templates are available for any type of document. The next time you need to type a letter, memo, report, invitation, announcement, flyer, or labels, look for a template design.

Topics Review

Topic	Key Concepts	Key Terms
6.1 Creating and Editing a New Document	A word processing application is software used to create documents that are composed mostly of text. Start a new blank document from the Word Start screen. Creating a new document includes typing text, editing text, and correcting errors. Wordwrap is the term that describes Word moving text to the next line automatically when you reach the right margin. Double-clicking on the page in blank space and typing is referred to as *click and type*. Text is automatically aligned left, center, or right depending on the location in the line at which click and type occurs. As you type new text, AutoCorrect fixes common misspellings, and AutoFormat automatically converts some text to fractions, ordinals, quotes, hyphens, and hyperlinks. A change made to text that has already been typed is referred to as *editing* and involves inserting, deleting, and replacing characters. The first step in editing is to position the insertion point at the location of the change.	word processing application wordwrap click and type AutoCorrect AutoFormat editing
6.2 Inserting Symbols and Completing a Spelling and Grammar Check	Symbols or special characters, such as a copyright symbol or registered trademark, are entered using the Symbol gallery or Symbol dialog box. The Spelling & Grammar button is used to match words in the document with words in the dictionary; words not found are flagged as potential errors. During a spell check, a word not found in the dictionary is highlighted and suggestions for replacement appear in the Spelling task pane. Ignore, Ignore All, Add, Change, or Change All are buttons in the Spelling task pane used to respond to each potential error.	Symbol gallery Spelling & Grammar
6.3 Finding and Replacing Text	The Find feature highlights all occurrences of a keyword or phrase and provides in the Navigation pane a link to each location in the document. Use Replace if you want Word to automatically change each occurrence of a keyword or phrase with another word or phrase. Find a word with a similar meaning in the Thesaurus.	Find Replace
6.4 Moving Text and Inserting Bullets and Numbering	Turn on the display of hidden formatting symbols using the Show/Hide button in the Paragraph group on the Home tab. Hidden formatting symbols, such as the paragraph symbol, are inserted in a document whenever the Enter key is pressed. Displaying formatting symbols is helpful when moving text to make sure the paragraph symbol is selected before cutting the text. Bullets are items in a list that is set apart from the paragraph and are entered in no particular sequence. A numbered list is a sequential series of tasks or other items that are each preceded by a number.	Bullets button Numbering button

continued…

Topic	Key Concepts	Key Terms
6.5 Formatting Text with Font and Paragraph Alignment Options	Changing the appearance of text is called *formatting*. Character formatting involves applying changes to the appearance of characters, whereas paragraph formatting changes the appearance of an entire paragraph. A font is also called a *typeface* and refers to the design and shape of letters, numbers, and special characters. Change a font, font size, case, font style, font color; highlight text; and add font effects to change character formats. Change a paragraph's alignment from the default Align Left to Center, Align Right, or Justify using the buttons in the Paragraph group on the Home tab. Justified text has extra space within a line so that the left and right margins are even.	formatting character formatting paragraph formatting font Align Left Center Align Right Justify
6.6 Indenting Text and Changing Line and Paragraph Spacing	Press Tab at the beginning of a paragraph to indent only the first line or change *Special* to *First line* at the Paragraph dialog box. A paragraph in which all lines are indented except the first line is called a *hanging indent*. Indent all lines of a paragraph using the Increase Indent button or change the *Left* text box entry at the Paragraph dialog box. A paragraph indents 0.5 inch each time the Increase Indent button is clicked. Use the Decrease Indent button to move a paragraph closer to the left margin; the paragraph moves left 0.5 inch each time the button is clicked. Indent a paragraph from both margins using the *Left* and *Right* text boxes in the Paragraph dialog box. Change line spacing by selecting the desired spacing option from the Line and Paragraph Spacing button. Extra space can be added or removed before or after paragraphs using options from the Line and Paragraph Spacing button or the Paragraph dialog box.	hanging indent Line and Paragraph Spacing Increase Indent Decrease Indent
6.7 Formatting Using Styles	Format text by applying a style, which is a set of predefined formatting options. Change the style set using buttons in the Document Formatting group on the Design tab. Each style set applies a different set of formatting options for the styles on the Home tab, meaning you can change the appearance of a document by changing the style set. A Theme is a set of colors, fonts, and font effects that alter the appearance of a document.	style style set Theme
6.8 Creating a New Document from a Template	A template is a document that is already set up with text and/or formatting options. Browse available templates in the template gallery at the Word Start screen or at the New backstage area. Find a template by browsing the gallery by a category or by typing a keyword in the search text box. Click a template design to preview the template and create a new document based upon the template. Within a template, text placeholders or instructional text is included to help you personalize the document.	template

◆ Recheck
Recheck your understanding of the topics covered in this chapter.

◆ Workbook
Chapter review and assessment resources are available in the *Workbook* ebook.

Chapter 7

Enhancing a Word Document with Special Features

Precheck

Check your understanding of the topics covered in this chapter.

S everal features in Word allow you to add visual appeal, organize information, or format a document for a special purpose, such as a research paper. Word provides different views to work in and navigate a document and includes collaborative tools, such as comments, for working on a document with other people. Several resume and cover letter templates are available in Word to help you build job search documents.

In this chapter, you will enhance documents already typed and finalize an academic research paper by adding formatting, citations, and a Works Cited page. Lastly, you will create a resume and cover letter using templates.

Learning Objectives

7.1 Insert, edit, and label pictures in a document

7.2 Add borders and shading to text and insert a text box in a document

7.3 Insert a table in a document

7.4 Format and modify a table

7.5 Change layout options

7.6 Add text and page numbers in a header for a research paper

7.7 Insert and edit citations in a research paper

7.8 Create a Works Cited page for a research paper and display a document in different views

7.9 Insert and reply to comments in a document

7.10 Create a resume and cover letter from templates

 If you are a SNAP user, go to your SNAP Assignments page to complete the Precheck, Tutorials, and Recheck.

Data Files

Before beginning this chapter, be sure you have copied the student data files for this course to your storage medium. Steps on downloading and extracting the data files are provided in Chapter 1, Topic 1.8, on pages 22–23.

7.1 Inserting, Editing, and Labeling Pictures in a Document

Skills

Insert a picture from a web resource

Insert a picture from a file on your computer

Edit a picture

Insert a caption

Tutorials

Inserting, Sizing, and Positioning an Image

Formatting an Image

Creating and Customizing Captions

Good to Know

According to Microsoft, Bing Image Search applies a copyright filter so that the results displayed are only those tagged with a Creative Commons license. A Creative Commons license means that you have to provide attribution (credit the source). A link to the source of each image is included so that you can properly credit the picture creator.

App Tip

The Online Video button in the Media group on the Insert tab is used to add and play videos within a Word document.

Adding a graphic element, such as a picture, not only adds visual appeal to a document but can also help a reader understand content. You can insert a picture from a web resource or from an image file stored on your computer using the **Online Pictures** button in the Illustrations group on the Insert tab. Once an image has been inserted, you can change the way text wraps around the sides of the image using the **Layout Options** button that appears. Resize the image with the selection handles or drag the image to another position in the document. Use buttons on the Picture Tools Format tab to edit the image.

Inserting a Picture from an Online Source

If you have a picture stored at an online service, such as Flickr, Facebook, or OneDrive, and you have those services connected to Word, you can insert the image from the website within the Word document. You can also search for an image on the web using the search text box to the right of the *Bing Image Search* option at the Insert Pictures dialog box and insert the image into the Word document.

1. Start Word 2016 and open the document **InsulaSummary** from the Ch7 folder in Student_Data_Files.

2. Click the File tab and then click *Save As*. At the Save As backstage area, click *Browse*. At the Save As dialog box, navigate to the CompletedTopicsByChapter folder on your storage medium, create a new folder *Ch7*, and then save a copy of the document within the Ch7 folder as **InsulaSummary-YourName**.

3. Position the insertion point at the beginning of the first paragraph of text.

4. Click the Insert tab and then click the Online Pictures button in the Illustrations group.

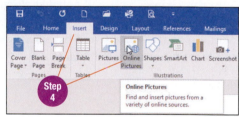

5. At the Insert Pictures dialog box, with the insertion point positioned in the search text box to the right of the *Bing Image Search* option, type brain and then press Enter or click the Search button (displays as a magnifying glass).

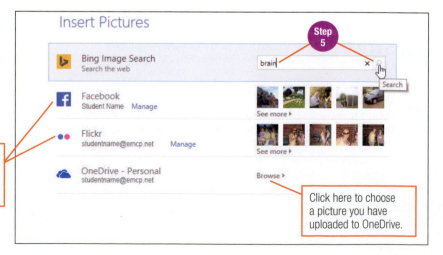

Pictures you have stored at Flickr and Facebook show here if you have connected your Flickr and Facebook accounts to your Microsoft account.

Click here to choose a picture you have uploaded to OneDrive.

6 Scroll down the search results list to the image shown at the right, click to select the image, and then click the Insert button. If you cannot locate the image shown, close the Insert Pictures dialog box and then insert the image using the Pictures button in the Illustrations group (see Steps 13 to 14) using the image file **Brain** in the Ch7 folder in Student_Data_Files.

The image source is shown here. Click the link to visit the web page where the image was located and review the license terms if necessary.

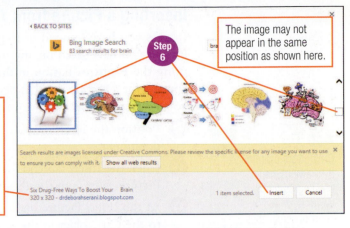

The image may not appear in the same position as shown here.

7 Click the Layout Options button that appears at the top right of the inserted image.

8 Click *Square*, the first option in the *With Text Wrapping* section of the LAYOUT OPTIONS gallery, and then click the Close button.

The LAYOUT OPTIONS gallery provides choices to control how text wraps around the picture object and to control whether the picture should remain fixed at its current position or move with the text. Notice also that the Picture Tools Format tab becomes active when a picture is selected. You will work with buttons in this tab later in this topic.

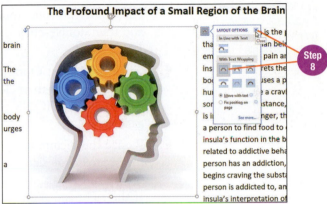

9 Drag the picture to the right until the right edge of the picture is aligned at the right margin.

As you move the picture, green alignment guides help you position the image at the top of the paragraph and at the right margin.

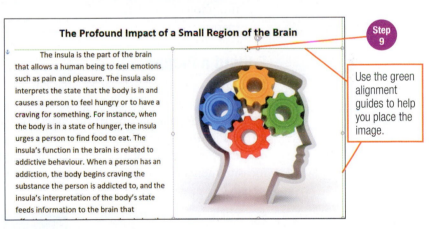

Use the green alignment guides to help you place the image.

10 Drag the bottom left corner selection handle on the image up and toward the right until the picture is resized to the approximate size shown in the Step 10 illustration.

11 Save the revised document using the same name (**InsulaSummary-Your Name**).

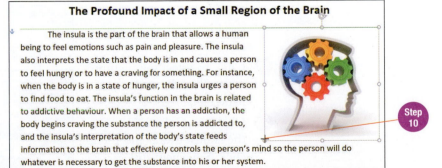

Inserting a Picture from Your Computer

Images you have scanned or imported from your digital camera to your PC or other images you have saved on any storage medium can be inserted into a document using the **Pictures** button in the Illustrations group. The Layout Options button also appears for a picture that has been inserted from your PC for you to specify the way you want document text to wrap around the sides of the picture.

> **App Tip**
>
> Create your own images by drawing Shapes, creating SmartArt or WordArt, or adding a chart or screenshot. Explore these features if you want to add a graphic but do not have a picture available.

12 Position the insertion point at the beginning of the second paragraph in the document.

13 Click the Insert tab and then click the Pictures button in the Illustrations group.

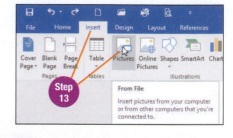

Step 13

From File
Insert pictures from your computer or from other computers that you're connected to.

14 At the Insert Picture dialog box, navigate to the Ch7 folder in the Student_Data_Files folder on your storage medium, and then double-click the image named *USCPhoto*.

15 Click the Layout Options button, click *Square*, and then close the LAYOUT OPTIONS gallery.

16 Resize the picture to the approximate size shown in the image at the right and align the photo at the left margin.

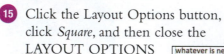

Step 16

whatever is necessary to get the substance into his or her system.

If scientists could alter how the insula works in living people, the insula could be used to treat or cure drug addictive behaviour by focusing on regions that are involved in one's decision making. If the insula could be altered to focus on decision making in areas that involve habits, addiction could be controlled. Scientists could weaken some social functions in the insula that give a person the temptation toward a habit such as drugs or alcohol.

The difficulty is that the insula affects regions of the brain that influence rational

Editing a Picture

Buttons on the Picture Tools Format tab are used to edit an image. Use buttons in the Adjust group to modify the appearance of a picture, such as the sharpness, contrast, or color tone, or to apply an artistic effect. Add a border or picture effect with options in the Picture Styles group. Change the picture's position, text wrapping option, order, alignment, or rotation with buttons in the Arrange group. Crop or specify exact measurements for the picture's height and width with buttons in the Size group.

17 Click to select the picture inserted in the first paragraph.

18 Click *Soft Edge Rectangle* (sixth option) in the Picture Styles group on the Picture Tools Format tab.

Step 18

Soft Edge Rectangle

19 Click to select the picture inserted in the second paragraph.

20 Click the Color button in the Adjust group and then click *Saturation: 400%* (last option) in the *Color Saturation* section of the drop-down gallery.

Step 20

Notice the picture appears brighter than it did before. Saturation refers to the purity of colors in a photo. Some digital cameras use a low saturation level, making the colors seem dull; increasing the saturation level brightens a picture.

Inserting a Caption with a Picture

Adding a caption below a photograph can help a reader understand the context of a picture, or you can use captions to number figures in a report. With the **Insert Caption** feature, Word automatically numbers pictures, inserting the number after the label *Figure*.

21 With the picture in the second paragraph still selected, click the References tab.

22 Click the Insert Caption button in the Captions group.

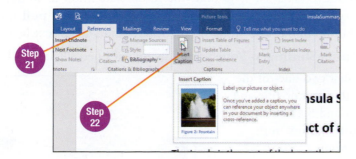

23 At the Caption dialog box, with the insertion point positioned in the *Caption* text box, press the spacebar and then type Insula research is being done at the University of Southern California.

24 Press Enter or click OK.

25 Click in the document outside the caption box to deselect the caption.

26 Save the revised document using the same name (**InsulaSummary-YourName**). Leave the document open for the next topic.

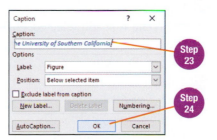

Beyond Basics   **Cropping and Removing the Background of a Picture**

You can remove unwanted portions of a picture with the Crop tool in the Size group on the Picture Tools Format tab. Click the Crop button and then use the crop handles to modify the picture. The portion of the image that will remain appears normal, while the cropped area becomes dark gray. Click outside the image to complete the crop action.

The Remove Background button in the Adjust group on the Picture Tools Format tab is another tool you can use to remove portions of a photo. With this button, you can focus on an object in the foreground of a picture and remove the background. For example, with a photo of an airplane in the sky, you can select the airplane and have Word remove the sky in the background.

Crop

Remove Background

7.2 Adding Borders and Shading, and Inserting a Text Box

Skills

Add a paragraph border

Add shading within a paragraph

Add a border to a page

Insert a text box

Tutorials

Applying Borders

Inserting a Page Border

Applying Shading

Inserting a Text Box

Formatting a Text Box

Add a border and/or add color behind text (called **shading**) to make text stand out from the rest of a document. You can add borders and shading to a single paragraph or to a group of selected paragraphs. With the **Page Borders** button, you can add a border around the edges of the entire page. Add a line that spans the entire page width above the insertion point location by clicking *Horizontal Line* at the **Borders gallery**.

A text box is used to set a short passage of text apart from the rest of a document. Word includes several built-in text box styles that can be used for this purpose. Inserting text inside a box is a way to draw the reader's attention to an important quote or point in a document. A quote inside a text box is called a **pull quote**. Click the **Text Box** button in the Text group on the Insert tab to add a text box.

1. With the **InsulaSummary-YourName** document open, click the Home tab.

2. Select the first two lines of the document that are the title and subtitle text and then click the Borders button arrow in the Paragraph group.

3. Click *Outside Borders*.

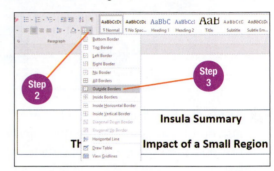

4. With the text still selected, click the Shading button arrow in the Paragraph group.

5. Click *Orange, Accent 6, Lighter 80%* at the Shading color gallery (last color in second row of *Theme Colors* section).

6. Click in any paragraph to deselect the text.

7. Click the Design tab.

8. Click the Page Borders button in the Page Background group.

9. At the Borders and Shading dialog box, with the Page Border tab selected, click the page graphic next to *Shadow* in the *Setting* section.

10. Click the *Width* list box arrow and then click *1 ½ pt* at the drop-down list.

11. Click OK.

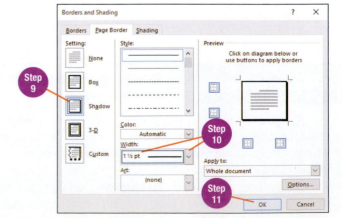

App Tip

Use *Borders and Shading* from the Borders gallery to create a custom border in the Borders and Shading dialog box by changing the border style, color, and width options.

12 Click the Insert tab and then click the Text Box button in the Text group.

13 Scroll down and then click *Whisp Quote* at the drop-down list.

Word adds a text box overlapping other text in the document and with default text already selected inside the text box.

14 Type Some people with damage to the insula were able to quit smoking instantly!.

15 Right-click *[Cite your source here.]* and then choose *Remove Content Control* at the shortcut menu.

16 Point to the border of the box and then drag the text box to the bottom of the page to the approximate location shown in the image below.

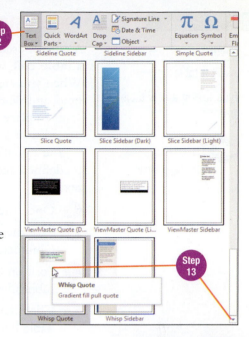

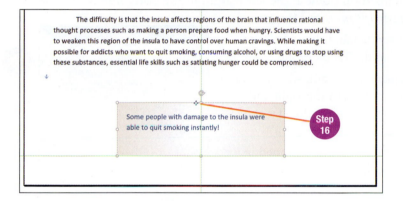

17 Drag the bottom middle selection handle up to reduce the height of the text box as shown in the image below.

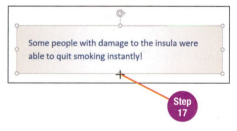

18 Click in any paragraph to deselect the text box and then save the revised document using the same name (**InsulaSummary-YourName**).

19 Close the document.

View
Model Answer
Compare your completed file with the model answer.

Quick Steps

Add a Paragraph Border
1. Select paragraph(s).
2. Click Borders button arrow.
3. Click desired border style.

Add Shading to a Paragraph
1. Select paragraph(s).
2. Click Shading button arrow.
3. Click desired color.

Add a Page Border
1. Click Design tab.
2. Click Page Borders button.
3. Select desired *Setting* and *Style, Color, Width,* or *Art* options.
4. Click OK.

Insert a Text Box
1. Click Insert tab.
2. Click Text Box button.
3. Click desired text box style.
4. Type text.
5. Move and/or resize box as needed.

Beyond Basics   **Formatting Tools to Edit a Text Box**

When a shape, line, drawing, or other graphic object is selected, the Drawing Tools Format tab becomes available. Use buttons on the tab to edit the object. For example, with a selected text box, you can change the text box shape style, fill, or outline and add shape effects. Edit the appearance of the text inside the box by applying a WordArt style, changing the text fill or outline, adding text effects, or changing the alignment or direction of the text.

7.3 Inserting a Table

Skills

Insert a table

Type text in a new table grid

Tutorial
Creating a Table

A table is used to organize and present data in columns and rows. Text is typed within a **table cell**, which is a rectangular box that is the intersection of a column with a row. When you create a table, you specify the number of columns and rows the table will hold, and Word creates a blank grid in which you type the table text. Text that you want to place side by side in columns, or in rows, is ideal for a table. For example, a price list or a catalog with items and descriptions is ideal for a table.

You can also create a table using the Quick Tables feature. A **Quick Table** is a predefined and formatted table with sample data that you can replace with your own text.

1. Open the document **RezMealPlans**.

2. Save the document as **RezMealPlans-YourName** in the Ch7 folder in CompletedTopicsByChapter.

3. Position the insertion point at the left margin in the blank line below the subheading *Meal Plans with Descriptions*.

4. Click the Insert tab and then click the Table button in the Tables group.

5. Click the square in the drop-down grid that is three columns to the right and two rows down (*3x2 Table* displays above grid).

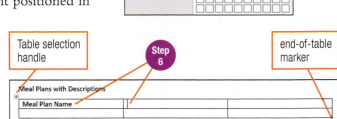

6. With the insertion point positioned in the first table cell, type Meal Plan Name and then press Tab or click in the next table cell.

7. Type Cost and then press Tab or click in the next table cell.

8. Type Description and then press Tab or click in the first table cell in the second row.

9. Type the second row of data as follows. When you finish typing the text in the last column, press Tab to add a new row to the table automatically.

Minimum $2,000   Suitable for students with small appetites who plan to be away from residence most weekends.

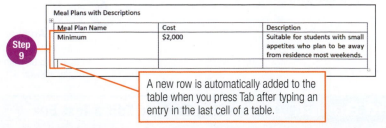

A new row is automatically added to the table when you press Tab after typing an entry in the last cell of a table.

App Tip

The advantage to using a table versus typing information in tabbed columns is that information can wrap around within a table cell.

Oops!

Added a new row by mistake? Click the Undo button to remove the extra row.

10. Type the remainder of the table as shown in Figure 7.1 on page 165 by completing steps similar to those in Steps 6 to 9, except do not press Tab after typing the last table cell entry.

Meal Plans with Descriptions

Meal Plan Name	Cost	Description
Minimum	$2,000	Suitable for students with small appetites who plan to be away from residence most weekends.
Light	$2,200	Best plan for students with a lighter appetite who spend occasional weekends on campus.
Full	$2,400	Full is the most popular plan. This plan is for students with an average appetite who will stay on campus most weekends.
Plus	$2,600	Students with a hearty appetite who will stay on campus most weekends choose the Plus plan.

> The column alignment is Justified because of the style set in the document. Generally, cells are aligned left in new tables.

Figure 7.1

Text for first table with the design and layout options for the current document style is shown.

11 Position the insertion point at the left margin in the blank line below the subheading *Meal Plan Fund Allocations*.

12 Click the Insert tab, click the Table button, and then click *Insert Table*.

You can also insert a new table using a dialog box in which you specify the number of columns and rows.

13 At the Insert Table dialog box, with the value in the *Number of columns* text box already set to *5*, press Tab or select the value in the *Number of rows* text box, type *5*, and then press Enter or click OK.

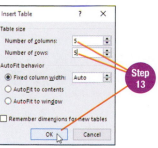

14 Type the data in the new table as shown in Figure 7.2. Click in the paragraph below the table after typing the text in the last table cell.

15 Save the document using the same name (**RezMealPlans–YourName**). Leave the document open for the next topic.

Meal Plan Fund Allocations

Meal Plan Name	Total Cost	Operating Fund	Basic Fund	Flex Fund
Minimum	$2,000	$200	$1,575	$225
Light	$2,200	$200	$1,725	$275
Full	$2,400	$200	$1,775	$425
Plus	$2,600	$200	$1,850	$550

Note that the Basic fund is tax exempt and is designed for use at all on-campus restaurants. Flex fund purchases are taxable.

> Click outside the table grid after typing the last table cell entry to avoid adding a new row to the table.

Figure 7.2

Text for second table is shown above.

7.4 Formatting and Modifying a Table

Skills

Apply and customize a table style

Insert and delete rows and columns

Change column width

Change cell alignment

Merge cells

Tutorials

Changing the Table Layout

Changing the Table Design

Customizing Cells in a Table

Once a table has been inserted into the document, use buttons in the Table Tools Design and Layout tabs to format the table and to add or delete rows and columns. Choose from a predesigned collection of formatting options to add borders, shading, and color to a table from the **Table Styles** gallery.

1 With the **RezMealPlans–YourName** document open, position the insertion point in any table cell within the first table.

2 Click the Table Tools Design tab.

3 Click the More button located at the bottom of the scroll bar at the right of the Table Styles gallery.

4 Click *Grid Table 4 – Accent 2* (tenth option in second row in *Grid Tables* section; the location may vary on your screen depending on your screen size and resolution setting).

Notice the formatting applied to the column headings and text in the first column. A row with shading and other formatting applied uniformly to every other row to make the data easier to read is referred to as a **banded row**. The border around each cell is now colored orange. The check boxes in the Table Style Options group, the Shading button in the Table Styles group, and the buttons in the Borders group are used to further modify the table formatting.

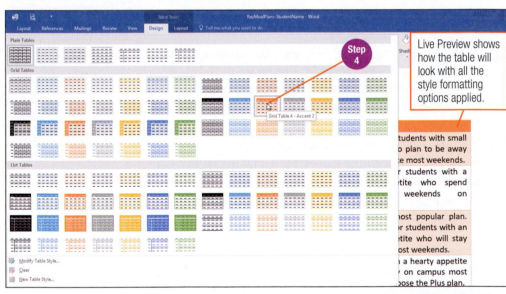

Quick Steps

Format a Table
1. Position insertion point within table cell.
2. If necessary, click Table Tools Design tab.
3. Click desired Table Style option.
4. Customize Table Style Options, Shading, or Borders as desired.

Insert Columns or Rows
1. Position insertion point in desired table cell.
2. If necessary, click Table Tools Layout tab.
3. Click Insert Above, Insert Below, Insert Left, or Insert Right button.

Delete Columns or Rows
1. Position insertion point in desired table cell.
2. If necessary, click Table Tools Layout tab.
3. Click Delete button.
4. Click desired option at drop-down list.

5 Click the *First Column* check box in the Table Style Options group to clear the check mark.

Notice the bold formatting is removed from the text in the first column.

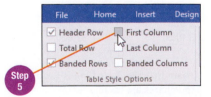

6 Select the column headings in the first row of the table, click the Shading button arrow on the Mini toolbar or in the Table Styles group on the Table Tools Design tab, and then click *Orange, Accent 2, Darker 50%* (sixth option in last row of *Theme Colors* section).

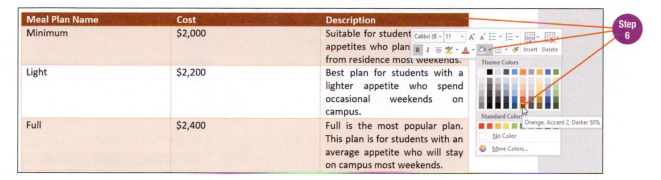

Inserting and Deleting Columns and Rows

Buttons in the Rows & Columns group on the Table Tools Layout tab are used to insert or delete columns or rows. Position the insertion point within a table row and click the Insert Above or Insert Below button to add a new row to the table. The Insert Left and Insert Right buttons are used to add a new column to the table.

Position the insertion point within a table cell, select multiple rows or columns or select the entire table, and then click the Delete button to delete cells, a column, a row, or the table.

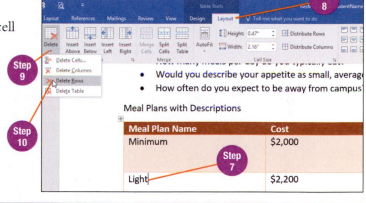

7 Position the insertion point within any table cell in the third row of the first table (begins with *Light*).

8 Click the Table Tools Layout tab.

9 Click the Delete button in the Rows & Columns group.

10 Click *Delete Rows*.

11 Position the insertion point within any table cell in the third row of the second table (begins with *Light*), click the Delete button, and then click *Delete Rows*.

12 Position the insertion point within any table cell in the last column of the first table.

13 Click the Insert Left button in the Rows & Columns group.

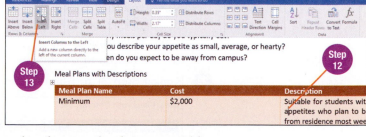

A new column is created between the *Cost* and *Description* columns. Notice that Word adjusts each column to be the same width. New rows are inserted by following a similar process.

14 Position the insertion point within the table cell in the first row of the new column (between *Cost* and *Description*) and then type Daily Spending.

15 Type the values below the column heading in rows 2, 3, and 4 as follows:

$18.35

$22.00

$23.85

Modifying Column Width and Alignment and Merging Cells

Adjust the width of a column by dragging the border line between columns left or right. You can also enter precise width measurements in the *Width* text box in the Cell Size group on the Table Tools Layout tab. Use the buttons in the Alignment group to align text within cells horizontally and vertically. Combine two or more cells into one cell using the Merge Cells button or divide a cell into two or more cells using the Split Cells button in the Merge group.

16 Position the insertion point within any table cell in the second column of the first table (column heading is *Cost*).

17 Click the *Width* down-pointing arrow in the Cell Size group until the value is *1"*.

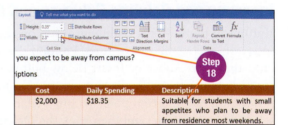

18 Position the insertion point within any table cell in the last column of the first table (column heading is *Description*) and then click the *Width* up-pointing arrow until the value is *2.3"*.

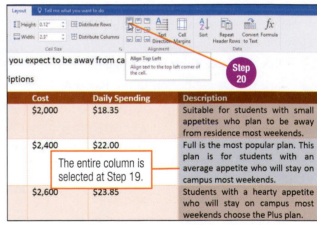

19 With the insertion point still positioned within the last column of the first table, click the Select button in the Table group and then click *Select Column* at the drop-down list.

20 Click the Align Top Left button in the Alignment group (first button).

21 Select the first column in the first table (column heading is *Meal Plan Name*) and then click the Align Center button in the Alignment group (second button in second row).

The entire column is selected at Step 19.

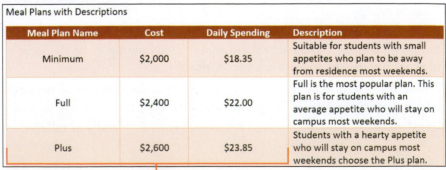

22 Repeat Step 21 to Align Center the second and third columns in the first table.

Meal Plans with Descriptions			
Meal Plan Name	**Cost**	**Daily Spending**	**Description**
Minimum	$2,000	$18.35	Suitable for students with small appetites who plan to be away from residence most weekends.
Full	$2,400	$22.00	Full is the most popular plan. This plan is for students with an average appetite who will stay on campus most weekends.
Plus	$2,600	$23.85	Students with a hearty appetite who will stay on campus most weekends choose the Plus plan.

Steps 21-22

23. Position the insertion point within any table cell in the first row of the second table, click the Table Tools Layout tab if necessary, and then click the Insert Above button in the Rows & Columns group.

24. With the new row already selected, click the Merge Cells button in the Merge group.

25. With the new row still selected, type Breakdown of Meal Plan Cost by Fund and then click the Align Top Center button in the Alignment group (second button in first row).

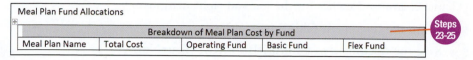

Meal Plan Name	Total Cost	Operating Fund	Basic Fund	Flex Fund

Steps 23-25

26. Select the column headings and all the values in columns 2, 3, 4, and 5 in the second table and then click the Align Center button.

27. With the insertion point positioned within any table cell in the second table, click the Select button and then click *Select Table*.

28. Click the Table Tools Design tab.

29. Click the Borders button arrow in the Borders group and then click *No Border*.

App Tip

You can also select the table with a mouse by clicking the Table selection handle at the top left corner of the table.

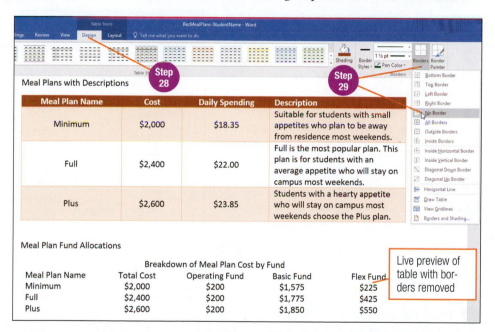

Live preview of table with borders removed

30. Click in the paragraph below the table to deselect the table.

31. Apply the Heading 2 style to the text *Meal Plans with Descriptions* and *Meal Plan Fund Allocations* above the first and second tables.

32. Save the document using the same name (**RezMealPlans-YourName**) and then close the document.

App Tip

Create a table for any type of columnar text instead of setting tabs—tables are simpler to create and have more formatting options.

View
Model Answer
Compare your completed file with the model answer.

Alternative Method | **Modifying a Table Using the Shortcut Menu or Mini Toolbar**

Right-click within a table cell or with table cells selected to display a context-sensitive shortcut menu and Mini toolbar. Use options from the shortcut menu or Mini toolbar to insert or delete cells, columns, or rows; merge or split cells; change a border style; or modify table properties.

7.5 Changing Layout Options

Skills

Change page orientation

Change margins

Insert a page break

 Tutorials

Changing Page Orientation

Changing Paper Size

Changing Margins

Inserting and Removing a Page Break

Inserting and Deleting a Section Break

By default, new documents in Word are set up for a letter-sized page (8.5 x 11 inches) in portrait orientation with 1-inch margins at the left, right, top, and bottom. **Portrait** orientation means the page is vertically oriented (taller than it is wide). In portrait mode, a page has a standard 6.5-inch line length (8.5 inches minus 2 inches for the left and right margins). This is the orientation commonly used for documents and books. You can change to **landscape** orientation (the page is rotated to make it wider than it is tall). In landscape orientation, the text has a 9-inch line length (11 inches minus 2 inches for the left and right margins).

1. Open the document **ChildLitBookRpt**.

2. Save the document as **ChildLitBookRpt-YourName** in the Ch7 folder in CompletedTopicsByChapter.

3. Click the Layout tab.

4. Click the Orientation button in the Page Setup group and then click *Landscape*.

Step 3

Step 4

Notice the width of the page is extended, and the page is now wider than it is tall.

5. Scroll down to view the document in landscape orientation.

6. With the insertion point positioned at the top of the document, click the Margins button in the Page Setup group.

7. Click *Custom Margins*.

8. With the insertion point positioned in the *Top* text box in the *Margins* section of the Page Setup dialog box, press Tab twice or select the current value in the *Left* text box and then type 1.2.

9. Press Tab or select the current value in the *Right* text box, type 1.2, and then press Enter or click OK.

10. Scroll down to view the document with the new left and right margin settings.

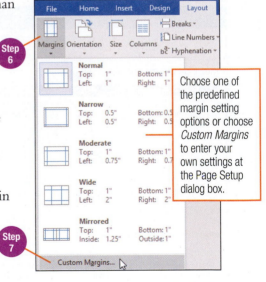

Step 6

Step 7

Choose one of the predefined margin setting options or choose *Custom Margins* to enter your own settings at the Page Setup dialog box.

Step 8

Step 9

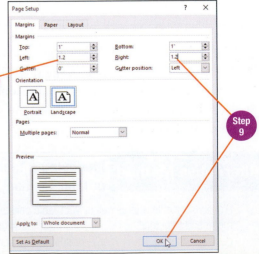

App Tip

Use the Size button in the Page Setup group to change the paper size to legal (8.5 x 14), envelope, or several other predefined photo or index card sizes.

App Tip

The Columns button in the Page Setup group formats a document into two or more newspaper-style columns of text.

Sometimes you want to end a page before the point at which Word ends a page automatically and starts a new page (referred to as a **soft page break**). Soft page breaks occur when the maximum number of lines that can fit within the current page size and margins has been reached. A page break that you insert at a different location is called a **hard page break**.

11 Position the insertion point at the left margin next to the subtitle *The Allegories* near the bottom of page 1.

12 Click the Insert tab.

13 Click the Page Break button in the Pages group.

Notice that all the text from the insertion point onward is moved to page 2.

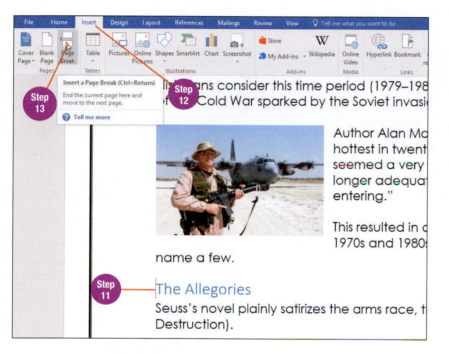

14 Scroll up and down to view the book report with the new page break.

15 Save the document using the same name (**ChildLitBookRpt-YourName**) and then close the document.

Alternative Method	**Creating a Page Break Using the Layout Tab or with a Keyboard Shortcut**

You can also insert a page break using the Breaks button in the Page Setup group on the Layout tab or by using the keyboard command Ctrl + Enter.

App Tip

Insert hard page breaks as your last step in preparing a document, because hard page breaks do not adjust if you add or delete text.

Oops!

Page break at wrong location? Press Backspace until the page break is deleted or use Undo to remove the page break. Position the insertion point at the correct location and try Step 13 again.

Quick Steps

Change to Landscape Orientation
1. Click Layout tab.
2. Click Orientation button.
3. Click *Landscape*.

Change Margins
1. Click Layout tab.
2. Click Margins button.
3. Click predefined margin option.

OR

1. Click Layout tab.
2. Click Margins button.
3. Click *Custom Margins*.
4. Enter custom measurements.
5. Click OK.

Insert a Hard Page Break
1. Position insertion point.
2. Click Insert tab.
3. Click Page Break button.

View

Model Answer
Compare your completed file with the model answer.

Beyond Basics | **Changing Page Layout for a Section of a Document**

By default, changes such as margins or orientation affect the entire document. A section break is inserted to change page layout options for a portion of a document. The Breaks button in the Page Setup group on the Layout tab is used to insert a section break. Choose *Next Page* to insert a section break that also starts a new page or choose *Continuous* to have the section break start at the insertion point position without starting a new page. For example, use section breaks if you want one page in a document to be landscape while the other pages are portrait. To do this, insert a section break where you want a landscape page, change the orientation to landscape, then insert another section break after the landscape page and return the orientation to portrait.

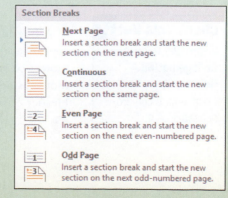

Section Breaks

Next Page
Insert a section break and start the new section on the next page.

Continuous
Insert a section break and start the new section on the same page.

Even Page
Insert a section break and start the new section on the next even-numbered page.

Odd Page
Insert a section break and start the new section on the next odd-numbered page.

7.6 Adding Text and Page Numbers in a Header for a Research Paper

Skills

Add text in a header

Insert page numbers in a header in MLA style

Tutorials

Inserting and Removing Page Numbers

Inserting and Removing a Predesigned Header and Footer

Editing a Header and Footer

Formatting a Report in MLA Style

Creating a Different First Page Header and Footer

Creating Odd and Even Page Headers and Footers

Formatting a Report in APA Style

During the course of your education, chances are you will have to submit a research paper or essay that is formatted for a specific **style guide** (a set of rules for paper formatting and referencing). Style guides are used in academic and professional writing; MLA (Modern Language Association) and APA (American Psychological Association) are two guides used often. See Table 7.1 for general MLA and APA guidelines. Always check an instructor's assignment instructions in case he or she has requirements that may be in addition to the formatting guidelines in Table 7.1.

Table 7.1

Formatting and Page Layout Guidelines for MLA and APA

Layout Item	MLA	APA
Paper size and margins	8.5 x 11 with 1-inch margins	8.5 x 11 with 1-inch margins
Font size	12-point; typeface is not specified but should be easily readable	12-point, with Times New Roman the preferred typeface
Line and paragraph spacing	2.0 with no spacing between paragraphs	2.0 with no spacing between paragraphs
Paragraph indent	Indent first line 0.5 inch	Indent first line 0.5 inch
Page numbering	Top right of each page one space after your last name	Top right of each page with the title of the paper all uppercase at the left margin on the same line
Title page	No (unless specifically requested by your instructor)	Yes Running Head: title of the paper at the left margin all uppercase with page number at the right margin 1 inch from the top. In the upper half of the page centered horizontally include: Title of the paper Your name School name
First page	Top left corner (double-spaced): Your name Instructor's name Course title Date A double-space below the above headings center the title (title case) and then begin the paper.	Center the word *Abstract* at the top of the page. Type a brief summary of the paper in a single paragraph in block format (no indents). Limit yourself to approximately 150 words. Start paper on a new page after the Abstract with the title of the paper centered (title case) at the top of the page.
Bibliography	Create separate Works Cited page at the end of the document organized alphabetically by author.	Create separate References page at the end of the document organized alphabetically by author.

Check This Out ✔

http://CA2.Paradigm College.net/MLA Guide

Go here for a comprehensive MLA guide.

A **header** is text that appears at the top of each page, and a **footer** is text that appears at the bottom of each page. Word provides several predefined headers and footers or you can create your own. Page numbering is added within a header or footer using the **Page Number** button in the Header & Footer group on the Header & Footer Tools Design tab.

 Open the document **China&TibetEssay**.

2 Save the document as **China&TibetEssay–YourName** in the Ch7 folder in CompletedTopicsByChapter.

3 Scroll down and review the formatting in the essay. Notice the paper size, font, margins, line and paragraph spacing, and first line indents are already formatted.

4 Position the insertion point at the beginning of the document and replace the text *Michael Seguin* with your first and last name.

The first four lines of this report are set up in MLA format for a first page; however, you need to add the page numbering for an MLA report.

5 Click the Insert tab and then click the Header button in the Header & Footer group.

Quick Steps

Insert a Header or Footer
1. Click Insert tab.
2. Click Header or Footer button.
3. Click built-in option or choose *Edit Header* or *Edit Footer*.
4. Type text and/or add options as needed.
5. Click Close Header and Footer button.

Insert Page Numbering
1. Within Header or Footer pane, click Page Number button.
2. Point to *Current Position*.
3. Click desired page number option.
4. Click Close Header and Footer button.

Step 5

6 Click *Edit Header* at the drop-down list.

7 Press Tab twice to move the insertion point to the right margin, type your last name, and then press the spacebar.

8 Click the Page Number button in the Header & Footer group.

9 Point to *Current Position* and then click *Plain Number*.

10 Select your last name and the page number in the Header pane.

11 Click the *Font* option box arrow on the Mini toolbar, scroll down the font list, and then click *Times New Roman*.

12 Click the *Font Size* option box arrow on the Mini toolbar and then click *12*.

13 Click within the Header pane to deselect the text.

14 Click the Close Header and Footer button in the Close group on the Header & Footer Tools Design tab.

15 Scroll down through the document to view your last name and the page number at the top of each page.

16 Save the document using the same name (**China&TibetEssay–YourName**). Leave the document open for the next topic.

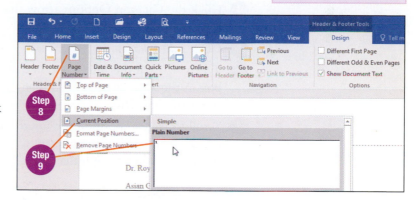

Step 8

Step 9

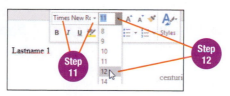

Step 11   Step 12

Oops!

Mini toolbar not visible? Click the Home tab and change the font and font size using the buttons in the Font group. Click the Header & Footer Tools Design tab at Step 13.

Step 14

Check This Out ✓

http://CA2.Paradigm
College.net/APAGuide

Go here for a comprehensive APA guide.

Beyond Basics   **Removing Page Number from First Page**

In many reports or books, a header and/or page number is not shown on the first page. Click the *Different First Page* check box on the Header & Footer Tools Design tab to create a First Page header that you can leave blank.

7.7 Inserting and Editing Citations

Skills

Edit a citation

Insert a citation

Tutorials

Inserting Sources and Citations

Editing a Citation and a Source

Direct quotations copied from sources or material you have paraphrased from another source needs to be referenced in a **citation** (source of the information used). Word provides tools to manage sources, insert citations, and edit citations.

1. With the **China&TibetEssay-YourName** document open, click the References tab.

2. Look at the *Style* option in the Citations & Bibliography group. If the *Style* is not *MLA*, click the *Style* option box arrow and then click *MLA*. (You may need to slide or scroll down the list.)

    You will begin by editing an existing citation.

3. Scroll down to page 3 and then click the insertion point within the *(Tsering)* citation at the end of the third sentence in the first paragraph (sentence begins with *When delegates disagreed . . .*) to display the citation placeholder.

4. Click the Citation Options arrow that appears.

5. Click *Edit Citation*.

6. Type par. 12 at the Edit Citation dialog box in the *Pages* text box and then press Enter or click OK.

7. Scroll to page 1, position the insertion point left of the period at the end of the quotation in the first sentence in the second paragraph that begins *China currently claims . . .* and then press the spacebar.

8. Click the Insert Citation button in the Citations & Bibliography group.

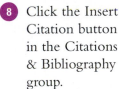

9. Click *Add New Source* at the drop-down list.

10. At the Create Source dialog box, if *Type of Source* is not *Web site*, click the *Type of Source* list box arrow and then click *Web site*. (You may need to slide or scroll down the list.)

11. Click in the *Author* text box and then type Bajoria, Jayshree.

12. Click in the *Name of Web Page* text box and then type The Question of Tibet.

13. Press Tab or click in the *Year* text box and then type 2008.

14. Continue to press Tab or click in the designated text boxes and then type the information as shown below:

Month	December	*Month Accessed*	November
Day	5	*Day Accessed*	15
Year Accessed	2018	*Medium*	Web

Good to Know

In the seventh edition of the MLA handbook, URLs are no longer required. MLA advises writers to include URLs only if a reader is unlikely to find the source without the web address.

15 Click OK.

16 Click in the *(Bajoria)* citation, click the Citation Options arrow, and then click *Edit Citation.*

17 Type par. 2 in the *Pages* text box at the Edit Citation dialog box and then click OK.

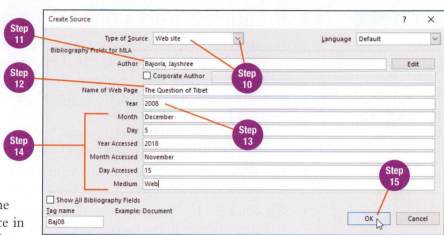

Step 11 — Type of Source

Step 12 — Author

Step 10

Step 13

Step 14

Step 15

18 Position the insertion point left of the period at the end of the quoted text in the fourth sentence in the second paragraph (sentence that begins *Praag points out. . .*), press the spacebar, click the Insert Citation button, and then click *Add New Source.*

19 Enter the information in the designated text boxes in the Create Source dialog box with the *Type of Source* set to *Web site* as follows.

Author	Praag, Michael C.
Name of Web Page	The Historical Status of Tibet: A Summary
Year	1996
Month	June
Day	26
Year Accessed	2018
Month Accessed	November
Day Accessed	14
Medium	Web

20 Edit the Praag citation to add *par. 18* in the citation by completing steps similar to Steps 16 to 17.

21 Position the insertion point left of the period at the end of the fourth sentence in the second paragraph on page 2 that begins *While independence was . . .* and then press the spacebar.

> In addition to China's suspicious historical claim to Tibet since the thirteenth century, many point to Tibet's period of independence following the fall of the Qing Dynasty as evidence of Tibet's autonomy from China. When the Qing Dynasty fell in 1912, the thirteenth Dalai Lama declared Tibet an independent nation a year later. Many refer to this period in Tibet's history as a period of de facto independence. While independence was declared, western countries including Britain and the United States, did not recognize Tibet as fully independent, resulting in the "de facto" title (Bajoria par. 3). According to Praag, from 1911 to 1950, Tibet successfully avoided

Steps 21–23

22 Click the Insert Citation button and then click *Bajoria, Jayshree.*

23 Edit the citation to add *par. 3* in the citation.

24 Save the document using the same name (**China&TibetEssay–YourName**). Leave the document open for the next topic.

Quick Steps

Insert a Citation with a New Source
1. Position insertion point.
2. Click References tab.
3. Click Insert Citation button.
4. Click *Add New Source.*
5. If necessary, change *Type of Source.*
6. Enter required information.
7. Click OK.

Insert a Citation with an Existing Source
1. Position insertion point.
2. Click References tab.
3. Click Insert Citation button.
4. Click required source.

Edit a Citation
1. Position insertion point in citation.
2. Click Citation Options button.
3. Click *Edit Citation.*
4. Type page or other reference.
5. Click OK.

Good to Know

MLA recommends the abbreviations *n. pag.* for a source without page numbers, *n.d.* for a source without a date, and *n.p.* for a source without a publisher.

Beyond Basics   **Editing a Source**

To change the source information for a citation, position the insertion point within the citation, click the Citation Options arrow, and then click *Edit Source.* This action opens the Edit Source dialog box in which you can make changes to the bibliography fields for the reference.

7.8 Creating a Works Cited Page and Using Word Views

Skills

Insert a page break

Create a Works Cited page

Browse a document in different views

Tutorials

Inserting a Works Cited Page

Managing Sources

Inserting Footnotes and Endnotes

Changing Document Views

The **Bibliography** button in the Citations & Bibliography group on the References tab is used to generate a **Works Cited** page for MLA papers or a References page for APA papers. The MLA style guide requires a Works Cited page to be on a separate page at the end of the document organized alphabetically by author's name or by title when an author's name is absent.

Word provides various views in which to review a document, including Read Mode view that provides maximum screen space for reading longer documents.

1. With the **China&TibetEssay-YourName** document open, move the insertion point to the left margin on the blank line below the last paragraph.

2. Click the Insert tab and then click the Page Break button in the Pages group to start a new page.

3. Click the References tab.

4. Click the Bibliography button in the Citations & Bibliography group.

5. Click *Works Cited* in the drop-down list.

Word automatically generates the Works Cited page. In the next steps, you will format the text to match the font, size, and spacing of the rest of the document.

6. Select all the text in the Works Cited page.

Word surrounds the entire text on the page with a border and displays a Bibliographies button and an Update Citations and Bibliography button along the top of the placeholder.

7. Click the Home tab and then make the following changes to the selected text:

   a. Change the font to 12-point Times New Roman.

   b. Change the line spacing to 2.0.

   c. Open the Paragraph dialog box and change *Before* and *After* in the *Spacing* section to *0 pt*.

8. Select the title text *Works Cited*, change the font color to Automatic (black), and center the title.

App Tip

Do not change formatting until you are sure your Works Cited page is complete, because the page will revert to predefined formats if you make changes to sources and then update the Works Cited page.

Good to Know

The seventh edition of the MLA guide now requires all entries in a reference list to include the medium in which the reference has been published, such as film, print, or web.

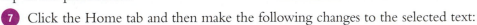

Use the Bibliographies button to change to a different bibliography style or to convert the Works Cited page to static text that can be edited.

Regenerate the Works Cited page using this button if you make a change to any of the source information.

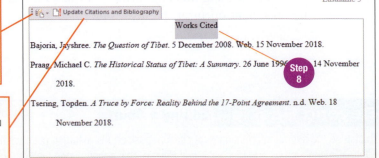

The default view for new documents is **Print Layout view**, which displays the document as it will appear when printed. **Read Mode view** displays a document full screen in columns, allowing you to read longer documents more easily without screen elements such as the Quick Access Toolbar and ribbon. **Draft view** hides print elements, such as headers and footers. **Web Layout view** displays a document as it would appear as a web page, and **Outline view** displays content as bulleted points.

9 Position the insertion point at the beginning of the document.

10 Click the View tab and then click the Read Mode button in the Views group.

Use Read Mode to view a document without editing.

App Tip

The *Welcome back!* balloon appears at the right side of the screen when you reopen a document. Click the balloon to scroll to where you left the document when you closed it.

11 Click the View tab, point to *Layout*, and then click *Paper Layout* to display the document as single pages. Skip this step if your screen is already displaying the document as single pages.

12 Scroll down to the end of the document to view the document in Read Mode, then click the View tab, point to *Layout*, and then click *Column Layout*.

App Tip

In Print Layout view, turn on the Navigation pane (View tab, Navigation Pane check box in the Show group) and click *Pages* at the top of the pane to move through a document by clicking miniature page thumbnails.

13 Click the Previous Screen button (left-pointing arrow inside circle at the middle right of the screen) to move back to the previous screen until you have returned to the beginning of the document.

14 Click the View tab and then click *Edit Document* to return to Print Layout view.

15 Click the Draft button in the Views group on the View tab to view just the text in the document and then browse the document.

16 Click the Print Layout view button near the right end of the Status bar.

Quick Steps

Generate a Works Cited Page
1. Position insertion point at end of document.
2. Insert page break.
3. Click References tab.
4. Click Bibliography button.
5. Click *Works Cited*.
6. Format as required.

Change Document View
1. Click View tab.
2. Click desired view button.

17 Save the document using the same name (**China&TibetEssay–YourName**). Leave the document open for the next topic.

Beyond Basics   Footnotes and Endnotes

In some academic papers, you need to insert footnotes or endnotes. Footnotes are explanatory comments or source information placed at the bottom of a page. Endnotes are explanatory comments or source information that appear at the end of a section or document. Position the insertion point and use the Insert Footnote or Insert Endnote button on the References tab to add these elements.

7.9 Inserting and Replying to Comments

Skills

Insert comments

Change the markup view

Reply to a comment

Mark a comment done

▶ **Tutorials**

Inserting Comments

Managing Comments

Tracking Changes in a Document

Comments is a collaborative tool in Word that is useful when working on a document with another person or team. A **comment** is a short note associated with text that provides explanatory information to a reader. A comment can also be used to pose a question to document reviewers. You can reply to a comment made by someone else and mark a comment as done.

When working in teams or on group projects, consider using comments in documents to explain portions of your text, ask questions of your teammates, or add general feedback.

1. With the **China&TibetEssay-YourName** document open, position the insertion point at the beginning of the document.

2. Select the words *Sakya Lama* in the second sentence in the first paragraph.

3. Click the Review tab.

4. Click the New Comment button in the Comments group.

Word opens a new comment box (referred to as a *comment balloon*) in the Markup Area. The **Markup Area** is at the right side of the screen, where comments and other document changes, such as insertions and deletions, are shown when the Track Changes feature is turned on.

5. Type Consider adding more explanation about Sakya Lama and the Yuan dynasty and then click in the document outside the comment box.

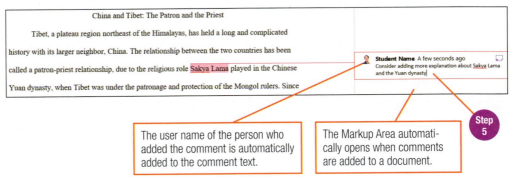

The user name of the person who added the comment is automatically added to the comment text.

The Markup Area automatically opens when comments are added to a document.

6. Select *modern historians* in the third sentence in the second paragraph.

7. Click the New Comment button.

8. Type Provide a few names of modern historians? and then click in the document.

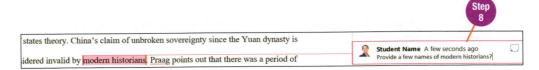

9 Position the insertion point at the beginning of the document.

10 If necessary, click the Review tab.

11 Click the Display for Review button arrow in the Tracking group (currently displays *All Markup* or *Simple Markup*) and then click *No Markup* at the drop-down list.

Notice the two comments on page 1 are removed from the document display.

12 Click the Display for Review button arrow and then click *Simple Markup* at the drop-down list.

13 If the Show Comments button in the Comments group is not active, click Show Comments to turn on the display of comment text, then point to the first comment balloon on page 1.

Notice Word shows a callout line pointing to the text with which the comment is associated. The comment box also displays a Reply button.

Your time may vary.

14 Click the Reply button (displays as a small white page with a left-pointing arrow) in the first comment box.

15 Type Asked Dr. Smith and he said this is not necessary and then click in the document.

Notice the reply comment text is indented below the original comment in a conversation-style dialogue.

16 Point to the second comment balloon on page 1 and then click the Reply button.

17 Type I'll check with Jose to see if he wants to include a few names.

18 Right-click the first comment balloon on page 1 and then click *Mark Comment Done* at the shortcut menu.

19 Click in the document outside the comment box.

Notice the comment text is dimmed for the comment marked as done. You can also delete a comment instead of marking the comment done.

20 Save the document using the same name (**China&TibetEssay-YourName**) and then close the document.

Quick Steps

Insert a Comment
1. Position insertion point or select text.
2. Click Review tab.
3. Click New Comment button.
4. Type comment text.
5. Click in document.

Reply to a Comment
1. Point to comment balloon.
2. Click Reply button.
3. Type reply text.
4. Click in document.

Mark a Comment Done
1. Right-click comment.
2. Click *Mark Comment Done*.

Change Display for Review
1. Click Review tab.
2. Click Display for Review button arrow.
3. Click desired markup view.

View
Model Answer
Compare your completed file with the model answer.

Beyond Basics   **Tracking Changes Made to a Document**

In situations when a document will be circulated to multiple readers for revisions, turning on track changes is a good idea. Track changes (Review tab) logs each person's insertions, deletions, and formatting changes. Changes can be reviewed, accepted, and rejected in the Revisions pane.

7.10 Creating a Resume and Cover Letter from Templates

Word provides several professionally designed and formatted resume and cover letter templates that take the work out of designing and formatting these two crucial documents, letting you focus your efforts on writing documents that will win you a job interview!

Check This Out ✓

http://CA2.Paradigm College.net/Career Advice

Go here for articles and examples on how to write effective resumes and cover letters.

Good to Know 🎓

Most recruiters advise job seekers to begin a job search with a reverse chronological resume style which lists your work experience from most to least recent.

App Tip ▶

Print multiple copies of a resume by changing the *Copies* value at the Print backstage area.

Oops! !

Having trouble removing spaces in the resume? The resume is formatted as a table. Use the skills you learned in Topic 7.4 to delete table rows to remove space between sections where you deleted template content.

1. Click the File tab and then click *New*.

2. At the New backstage area, click in the search text box, type entry level resume, and then press Enter or click the Start searching button (displays as a magnifying glass).

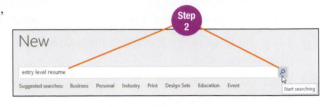

3. Click the thumbnail for *Resume for recent college graduate* in the Templates gallery.

4. Click the Create button at the preview window for the selected resume.

5. Select the text *STUDENT NAME* in the *Author* placeholder and then type Dana Jelic. Note that the text will appear all uppercase.

6. Right-click the *Street Address* placeholder and then click *Remove Content Control* at the shortcut menu.

7. Delete the *City, ST ZIP Code* placeholder by completing a step similar to Step 6.

8. Click to select the *Phone Number* placeholder and then type 800-555-4577.

9. Click to select the *E-mail Address* placeholder and then type jelic@emcp.net.

10. Type the remainder of the resume as shown in Figure 7.3 by selecting and editing text in placeholders or by deleting placeholders or text.

Figure 7.3

Shown is the resume text for Topic 7.10.

11. Save the document as **JelicResume-YourName** in the Ch7 folder in CompletedTopicsByChapter. Click OK if a message appears about upgrading to the newest file format.

12. Close the resume document and then display the New backstage area.

13. Type cover letter in response to technical position in the search text box and then press Enter or click the Start searching button.

14. Select and create a document using the template *Sample cover letter in response to a technical position advertisement.*

15. Create the letter as shown in Figure 7.4 by selecting and editing text in placeholders, adding more space between text, or by deleting placeholders or text. (Note that the letter closing in the template is not shown in Figure 7.4. Use the default text in this section.)

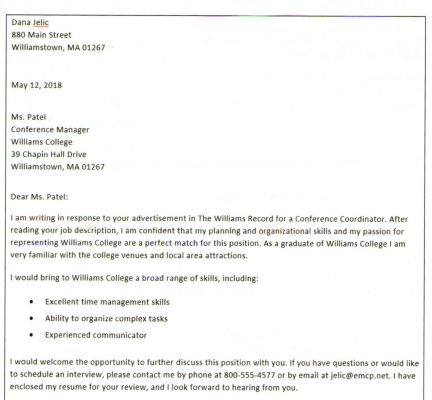

Dana Jelic
880 Main Street
Williamstown, MA 01267

May 12, 2018

Ms. Patel
Conference Manager
Williams College
39 Chapin Hall Drive
Williamstown, MA 01267

Dear Ms. Patel:

I am writing in response to your advertisement in The Williams Record for a Conference Coordinator. After reading your job description, I am confident that my planning and organizational skills and my passion for representing Williams College are a perfect match for this position. As a graduate of Williams College I am very familiar with the college venues and local area attractions.

I would bring to Williams College a broad range of skills, including:

- Excellent time management skills
- Ability to organize complex tasks
- Experienced communicator

I would welcome the opportunity to further discuss this position with you. If you have questions or would like to schedule an interview, please contact me by phone at 800-555-4577 or by email at jelic@emcp.net. I have enclosed my resume for your review, and I look forward to hearing from you.

Figure 7.4

Use this cover letter text for Topic 7.10.

Quick Steps

Create a Resume
1. Click File tab.
2. Click *New*.
3. Type entry in the search text box and press Enter.
4. Click desired template.
5. Click Create button.
6. Edit as required.

Create a Cover Letter
1. Click File tab.
2. Click *New*.
3. Type entry in the search text box and press Enter.
4. Click desired template.
5. Click Create button.
6. Edit as required.

16. Click the Layout tab and then click the Page Setup dialog box launcher.

17. If necessary, click the Layout tab in the Page Setup dialog box.

18. Click the *Vertical alignment* list box arrow, click *Center* at the drop-down list, and then click OK.

19. Save the document as **JelicCoverLetter-YourName** in the Ch7 folder in CompletedTopicsByChapter and then close the document.

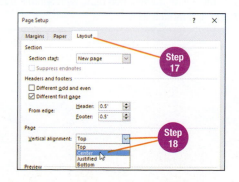

View

Model Answer
Compare your completed file with the model answer.

Topics Review

Topic	Key Concepts	Key Terms
7.1 Inserting, Editing, and Labeling Pictures in a Document	Graphic elements assist with comprehension and/or add visual appeal to documents.	Online Pictures Layout Options Pictures Insert Caption
	Insert a picture from a web resource or an online service, such as Flickr, Facebook, or OneDrive, using the Online Pictures button in the Illustrations group on the Insert tab.	
	Buttons in the Layout Options gallery are used to control how text wraps around the picture and the position of the picture on the page.	
	The Pictures button on the Insert tab is used to insert pictures from image files stored on your computer.	
	Edit the appearance of an image and/or add special effects using buttons on the Picture Tools Format tab.	
	A caption is explanatory text above or below a picture that is added using the Insert Caption button on the References tab.	
	Word automatically numbers pictures as Figures.	
7.2 Adding Borders and Shading, and Inserting a Text Box	Add a border or shading to paragraphs to make text stand out on a page.	shading Page Borders Borders gallery pull quote Text Box
	Shading is color applied to the page behind the text.	
	Apply a border to selected text using the Borders gallery from the Borders button arrow in the Paragraph group on the Home tab.	
	Add shading using the Shading button arrow in the Paragraph group on the Home tab.	
	A page border surrounds the entire page and is added from the Page Borders button in the Page Background group on the Design tab.	
	A pull quote is a quote placed inside a text box.	
	Insert text inside a box using the Text Box button in the Text group on the Insert tab.	
7.3 Inserting a Table	A table is a grid of columns and rows in which you type text and is used when you want to arrange text side by side or in rows.	table cell Quick Table
	A table cell is a rectangular-shaped box in the table grid that is the intersection of a column and a row into which you type text.	
	Create a table by clicking a square in a drop-down grid or by entering the number of columns and rows in the Insert Table dialog box.	
	Pressing Tab in the last table cell automatically adds a new row to the table.	
	A Quick Table is a predesigned table with sample data, such as calendars and tabular lists.	

continued…

Topic	Key Concepts	Key Terms
7.4 Formatting and Modifying a Table	Apply a predesigned collection of borders, shading, and color to a table using an option from the Table Styles gallery.	Table Styles banded row
	Shading or other formatting applied to every other row to make the table data easier to read is called a *banded row*.	
	Check boxes in the Table Style Options group are used to customize the formatting applied from a Table Styles option.	
	Apply shading or borders using buttons in the Table Styles and Borders group on the Table Tools Design tab.	
	New rows and columns are inserted above, below, left, or right of the active table cell using buttons in the Rows & Columns group on the Table Tools Layout tab.	
	Remove selected table cells, rows, or columns or the entire table using options from the Delete button in the Rows & Columns group.	
	Adjust the width of a column by changing the *Width* text box value in the Cell Size group on the Table Tools Layout tab or by dragging the column border.	
	Buttons to change alignment options for selected table cells are found in the Alignment group on the Table Tools Layout tab.	
	Cells in a table can be merged or split using buttons in the Merge group on the Table Tools Layout tab.	
7.5 Changing Layout Options	Portrait orientation means the text on the page is oriented to the taller side (8.5-inch width), while landscape orientation rotates the text to the wider side of the page (11-inch measurement becomes the page width).	portrait landscape soft page break hard page break
	Change the margins by choosing one of the predefined margin options or by entering measurements for the top, bottom, left, and right margins at the Page Setup dialog box.	
	A soft page break is a page break that Word inserts automatically when the maximum number of lines that can fit on a page has been reached.	
	A hard page break is a page break inserted by you in a different location than where the soft page break occurred.	
	A section break is inserted from the Breaks button in the Page Setup group on the Layout tab and is used to format a portion of a document with different page layout options.	
7.6 Adding Text and Page Numbers in a Header for a Research Paper	A style guide is a set of rules for formatting and referencing academic papers.	style guide header footer Page Number
	A header is text that appears at the top of each page, while a footer is text that appears at the bottom of each page.	
	Click the Insert tab and choose the Header or Footer button to create a header or footer in the Header or Footer pane.	
	Page numbers are added to a document at the top or bottom of a page within a Header or Footer pane using the Page Number button on the Header & Footer Tools Design tab.	

continued…

Topic	Key Concepts	Key Terms
7.7 Inserting and Editing Citations	A citation provides a reader with the reference for information quoted or paraphrased within an academic paper. Position the insertion point where a citation is needed and use the Insert Citation button on the References tab to create a reference. Click *Add New Source* at the Insert Citation drop-down list to enter information for a new reference for a citation. Edit a citation to add a page number or paragraph number to the reference.	citation
7.8 Creating a Works Cited Page and Using Word Views	A Works Cited page is a separate page at the end of the document with the references used for the paper. Use the Bibliography button in the Citations & Bibliography group on the References tab to generate a Works Cited page formatted for the MLA style guide. Print Layout view displays the document as it will appear when printed. Read Mode view displays a document full screen in columns or pages without editing tools. Draft view hides print elements, such as headers, footers, and page numbering. Web Layout view displays the document as a web page. Outline view displays content as bullet points. Footnotes are sources or explanatory comments placed at the bottom of a page, while endnotes are sources or explanatory comments placed at the end of a section or document in an academic paper.	Bibliography Works Cited Print Layout view Read Mode view Draft view Web Layout view Outline view
7.9 Inserting and Replying to Comments	A comment is a short note associated with text that provides explanatory information or poses a question to a reader. Select text that you want to associate with a comment and type the comment text inside a comment balloon by clicking the New Comment button on the Review tab. Comment balloons display in the Markup Area, which is a pane that opens at the right side of the document when comments are added. A document with comments can be shown with *No Markup*, *Simple Markup*, or *All Markup*, which refers to the way comment boxes are displayed. Point at a comment balloon and use the Reply button to enter reply text that responds to a comment. Mark a comment as done to retain the comment text but display the comment dimmed in the Markup Area.	comment Markup Area
7.10 Creating a Resume and Cover Letter from Templates	At the New backstage area, search for a resume template and a cover letter template in various styles, themes, and purposes. Personalize a resume and cover letter by adding content and deleting content controls not needed using the placeholders in a resume or cover letter template.	

Recheck
Recheck your understanding of the topics covered in this chapter.

Workbook
Chapter review and assessment resources are available in the *Workbook* ebook.

Chapter 8

Creating, Editing, and Formatting Excel Worksheets

Precheck
Check your understanding of the topics covered in this chapter.

Microsoft Excel (referred to as *Excel*) is a **spreadsheet application** used to create, process, and present information that is organized in a grid of columns and rows. In Excel, data can be calculated, analyzed, and graphed in a chart. The ability to do "what-if" analysis is a popular feature of the application; in this type of analysis, one or more values are changed to view the effect on other values. Excel is used for budget, income, expense, investment, loan, schedule, grading, attendance, inventory, and research data. Any information that can be set up in a grid-like structure is suited to Excel.

A file that you save in Excel is called a **workbook**. A workbook contains a collection of worksheets; a **worksheet** is the structure into which you enter, edit, and manipulate data. Think of a workbook as a binder and a worksheet as a page within the binder. Initially, a workbook has only one worksheet (page), but you can add more as needed.

Many of the features that you learned about in Word operate the same or similarly in Excel, which will make learning Excel faster and easier. You will begin by creating new worksheets in a blank workbook and then open other worksheets to practice navigating, editing, sorting, and formatting tasks.

Note: If you are using a tablet, consider connecting it to a USB or wireless keyboard because parts of this chapter involve a fair amount of typing.

 SNAP If you are a SNAP user, go to your SNAP Assignments page to complete the Precheck, Tutorials, and Recheck.

 Data Files
Before beginning this chapter, be sure you have copied the student data files for this course to your storage medium. Steps on downloading and extracting the data files are provided in Chapter 1, Topic 1.8, on pages 22–23.

Learning Objectives

8.1 Create and edit a new worksheet

8.2 Format cells with font, number, border, and merge and center options

8.3 Adjust column width and row height, and change cell alignment

8.4 Use the Fill feature to enter and copy data, and use AutoSum to add a column or row of values

8.5 Insert and delete rows and columns

8.6 Sort data and apply cell styles

8.7 Change page orientation, scale a worksheet, and display formulas in cells

8.8 Insert and rename a worksheet, copy cells, and indent data within a cell

8.9 Use Go To to move to a cell, freeze panes, apply shading options, and wrap text and rotate cell entries

8.1 Creating and Editing a New Worksheet

Skills

Open a new blank workbook

Enter text and values

Create formulas

Edit cells

Tutorials

Opening a Blank Workbook

Entering Data

Changing the Zoom

Entering Formulas Using the Keyboard

Entering Formulas Using the Mouse

Editing Data

Clearing Cell Contents and Formatting

Checking Spelling

Determining the Order of Operations

When you start a new blank workbook, you begin at a worksheet, which is divided into columns and rows. The intersection of a column and a row is called a **cell**, and in each cell you can type text, a value, or a formula. The cell with the green border is called the **active cell**. The active cell is the location in which the next entry you type is stored. The active cell is also the cell affected by the next command (action) you perform. Each cell is identified with the letter of the column and the number of the row that intersect to form the cell. For example, A1 refers to the cell in column A, row 1. A new workbook starts with one worksheet labeled *Sheet1* that has columns labeled A to Z, AA to AZ, BA to BZ, and so on to the last column, which is labeled XFD. Rows are numbered 1, 2, 3, up to 1,048,576.

Creating a New Worksheet

Begin a new worksheet by entering titles, column headings, and row headings to give the worksheet an organizational layout and provide context for the reader. Next, enter values in the columns and rows. Complete the worksheet by inserting formulas that perform calculations or otherwise summarize data.

1. Start Excel 2016.

2. At the Excel Start screen, click *Blank workbook* in the Templates gallery. Compare your screen with the one shown in Figure 8.1, and read the descriptions of screen elements in Table 8.1 on page 187.

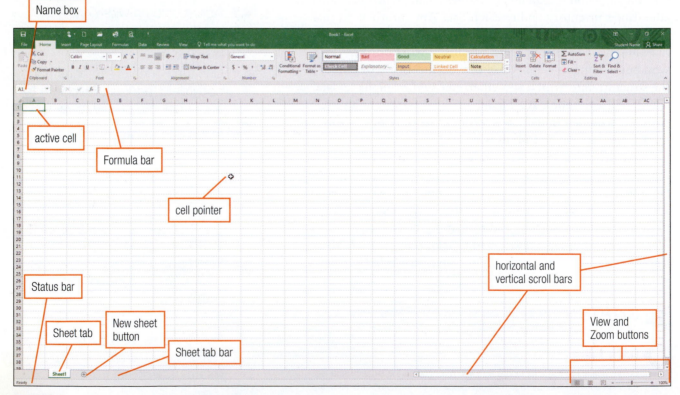

Figure 8.1

A new blank worksheet in Excel is shown here. The worksheet area below the ribbon is divided into columns and rows, creating cells into which data and formulas are typed and stored.

Table 8.1

Excel Features

Feature	Description
Active cell	Location in which the next typed data will be stored and that will be affected by the next command. Make a cell active by clicking it or by moving to it using the Arrow keys.
Cell pointer	Icon that displays when you are able to select cells with the mouse by clicking or dragging. On a touch device with no mouse attached, tap a cell to display selection handles (round circles) at the top left and bottom right corners.
Formula bar	Bar that displays contents stored in the active cell and is also used to create formulas.
Horizontal and vertical scroll bars	Tools used to view parts of a worksheet not shown in the current viewing area.
Name box	Box that displays the address or name of the active cell.
New Sheet button	Button on the Sheet tab bar used to insert a new worksheet.
Sheet tab	Tab that displays the name of the active worksheet. By default, new sheets are named Sheet# where # is the number of the sheet in the workbook.
Sheet tab bar	Bar that displays sheet tabs and is used to navigate between worksheets.
Status bar	Bar that displays messages indicating the current mode of operation; Ready indicates the worksheet is ready to accept new data.
View and Zoom buttons	Buttons used to change the appearance of the worksheet. Excel opens in Normal view. Other view buttons include Page Layout and Page Break Preview. Zoom buttons are used to enlarge or shrink the display.

Entering Text and Values

When you start a new worksheet, the active cell is A1 at the top left corner of the worksheet. Entries are created by activating a cell and typing text, a value, or a formula.

3 With A1 the active cell, type Car Purchase Cost and then press Enter or click A2 to make A2 the active cell.

4 Type Preowned Ford Focus Sedan and then press Enter twice or click A4.

5 Type Total Purchase Price and then click A6.

6 Type Loan Details: and then click B7.

7 Type the remaining row headings by moving the active cell and typing the text shown in the image at right.

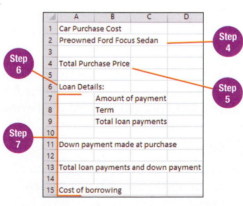

8 Click F4, type 16700.00, and then click E7.

Zeros to the right of a decimal are not stored or shown. You will learn how to format a cell to show the decimal place values in Topic 8.2.

App Tip

You can use Arrow keys to move up, down, left, or right from cell to cell.

App Tip

Excel's AutoComplete matches an entry in the same column with the first few characters that you type. Accept an AutoComplete entry by pressing Tab, Enter, or an Arrow key, or continue typing to ignore the suggestion.

App Tip

By default, text entries align at the left edge of a cell and numeric entries are right aligned.

9   Type the remaining values by activating the cell and typing the numbers shown below.

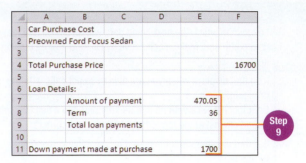

Creating Formulas to Perform Calculations

A **formula** is used to perform mathematical operations on values. A formula entry begins with the equals sign (=) to indicate to Excel the entry that follows is a calculation. Following the equals sign, type the first cell address that contains a value you want to use, type a mathematical operator, and then type the second cell address. Continue typing mathematical operators and cell addresses until finished. The mathematical operators are + (addition), – (subtraction), * (multiplication), / (division), and ^ (exponentiation).

10   Click E9 to make it the active cell and then type =e7*e8.

11   Click the Enter button in the Formula bar.

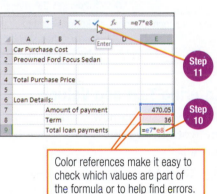

Color references make it easy to check which values are part of the formula or to help find errors.

Excel calculates the result and displays the value in E9. Notice that the cell in the worksheet area displays the formula result, while the Formula bar displays the formula used to calculate the result. Notice also that Excel capitalizes column letters in cell addresses within formulas.

Another way to enter a formula is to use the pointing method, in which you click the desired cells instead of typing their cell addresses.

12   Make F13 the active cell and then type =.

13   Click E9.

A moving dashed border surrounds E9, the cell is color coded, and the address *E9* is inserted in both cell F13 and the Formula bar.

14   Type +.

15   Click E11 and then click the Enter button in the Formula bar or press Enter.

16   Make F15 the active cell, type the formula =f13-f4 or enter the formula using the pointing method, and then click the Enter button in the Formula bar or press Enter.

The result, *1921.8*, displays in the cell.

> **Oops!**   !
>
> Clicked the wrong cell? Just click the correct cell—the cell reference is not fixed in the formula until you type an operator. You can also press the Esc key to start over.

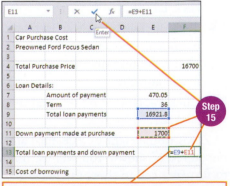

The formula created in Steps 12 to 15 by typing the equals sign (Step 12), clicking cells (Steps 13 and 15), and typing the operator (Step 14).

Editing Cells

An entire cell entry can be changed by making the cell active and typing a new entry to replace the existing contents. Double-click a cell to open it for editing in the worksheet area; this will allow you to change the entry rather than replace it. You can also edit a cell's contents by making the cell active and then inserting or deleting characters or spaces in the Formula bar.

To delete the contents of an active cell, press the Delete key or click the **Clear button** in the Editing group on the Home tab and then click *Clear All* at the drop-down list.

17 Make E7 the active cell, type *480.95*, and then press Enter.

Notice that the new payment amount causes the values in E9, F13, and F15 to update.

18 Double-click F4, position the insertion point between *6* and *7*, press Backspace to remove *6*, type *5*, and then click any other cell.

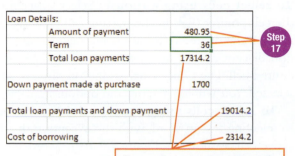

These values are updated automatically when the new loan payment is entered in E7.

19 Make E11 the active cell, click in the Formula bar, position the insertion point between *1* and *7*, press Backspace to remove *1*, and then click any other cell.

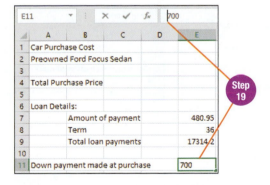

20 Save the new workbook as **CarCost-YourName** in a new folder named *Ch8* in the CompletedTopicsByChapter folder on your storage medium. Leave the workbook open for the next topic.

Quick Steps

Enter a Formula
1. Activate formula cell.
2. Type =.
3. Type first cell address.
4. Type operator symbol.
5. Type next cell address.
6. Continue Steps 4–5 until finished.
7. Press Enter.

Edit a Cell
1. Double-click cell.
2. Position insertion point.
3. Insert or delete characters or spaces as needed.
4. Press Enter.

App Tip

F2 is the keyboard command to edit a cell.

App Tip

AutoCorrect operates in Excel; however, red wavy lines do *not* appear below misspelled words. Consider using the Spelling feature in the Proofing group on the Review tab to spell check all worksheets.

Beyond Basics   Order of Operations in Formulas

If you combine operations in a formula, Excel will automatically calculate exponentiation, multiplication, and division before addition and subtraction. You can tell Excel to perform a particular operation first by using parentheses around that part of the formula. For example, in the formula *=(A1+A2)*A3*, Excel adds the values in A1 and A2 first and then multiplies the result by the value in A3.

8.2 Formatting Cells

Skills

Select cells

Change the font

Apply bold formatting

Format values

Add borders

Merge cells

Tutorials

Selecting Cells

Merging and Centering Cells

Applying Font Formatting

Adding Borders to Cells

Applying Number Formatting

Applying Formatting Using the Format Cells Dialog Box

Much like the Home tab in Word, the Home tab in Excel contains the formatting options for changing the appearance of text, values, or formula results. The Font group contains buttons to change the font, font size, and font color, and to apply bold, italic, underline, borders, and shading. The Alignment group contains buttons to align text or values within the cell edges.

Selecting Cells and Applying Font Formatting

To select cells using a mouse: Select adjacent cells with a mouse by positioning the cell pointer (✛) in the starting cell and dragging in the required direction until all the desired cells have been included in a shaded selection rectangle. Select nonadjacent cells by holding down the Ctrl key while you click the mouse in each desired cell. The Quick Analysis button displays when a group of cells has been selected. You will learn about the options available from this button in Chapter 9.

   To select cells using touch: Selecting cells on a touch device in Excel is similar to selecting text in Word, with the exception that two selection handles display in the active cell, as shown in Figure 8.2. As in Word, tap inside the selection area to display the Mini toolbar.

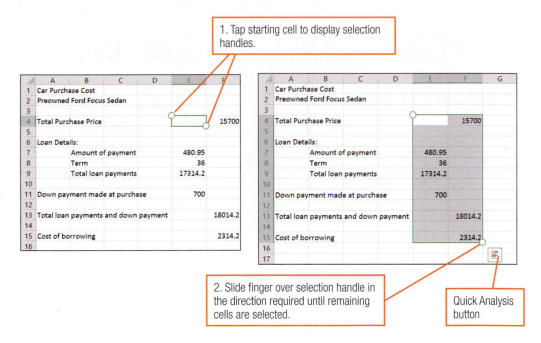

1. Tap starting cell to display selection handles.

2. Slide finger over selection handle in the direction required until remaining cells are selected.

Quick Analysis button

Figure 8.2

Selecting cells on a touch device is not much different from selecting cells using a mouse.

1 With the **CarCost–YourName** workbook open, starting at cell A1, select all the cells down and right to F15.

   A rectangular-shaped group of cells is referred to as a **range**. A range is referenced with the address of the cell at the top left corner, a colon (:), and the address of the cell at the bottom right corner. For example, the reference for the range selected in Step 1 is *A1:F15*.

2 Click the Font list box arrow in the Font group on the Home tab, scroll down the gallery, and then click *Century Gothic*.

3 Click any cell to deselect the range.

4 Select A1:A2 and then click the Bold button in the Font group.

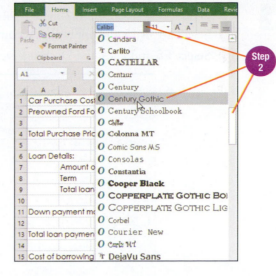

Notice the entire text in A1 and A2 is bold, including the characters that spill over the edge of column A into columns B, C, and D. This is because the entire text entry is stored in the cell that was active when the text was typed. Overflow text that displays in adjacent columns is not problematic when the adjacent columns are empty. You will learn how to widen a column to fit overflow text in the next topic.

5 Select cell F15 and then click the Bold button.

Formatting Numbers

By default, cells in a new worksheet are all in the General format, which has no specific appearance options. Buttons in the Number group on the Home tab are used to format the appearance of numeric entries in a worksheet. Add a dollar symbol, insert commas to indicate thousands, and/or adjust the number of decimal places to improve the appearance of values. Use the Percent Style format to convert decimal values to percentages and include the percent symbol.

Use the *Number Format* option box arrow (next to *General*) to choose other formats for dates, times, fractions, or scientific values, or to open the Format Cells dialog box from the *More Number Formats* option.

6 Select E4:F15.

7 Click the Comma Style button in the Number group.

Comma Style formats values with a comma in thousands and two decimal places.

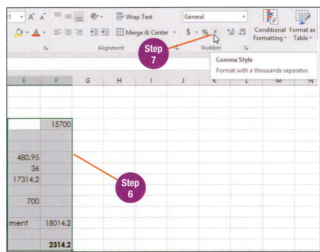

8 Select F4 and then click the Accounting Number Format button in the Number group. (Do not click the down-pointing arrow on the button.)

Accounting Number Format adds a currency symbol ($ for the United States and Canada), commas to indicate thousands, and two decimal places. Use the Accounting Number Format button arrow to choose a format that uses a currency symbol other than the dollar symbol (such as €, which stands for Euro).

> **App Tip**
>
> Excel automatically widens columns as needed when you apply a format that adds more characters to a column, such as Comma Style.

App Tip

Accounting Number Format aligns the currency symbol at the left edge of the cell. The *Currency* option (in the Number Format drop-down list) places the currency symbol immediately to the left of the value. Use Accounting Number Format if you want all dollar symbols to align at the same position.

9 Select E7 and apply the Accounting Number Format.

10 Select F13:F15 and apply the Accounting Number Format.

11 Select E8 and click the Decrease Decimal button twice.

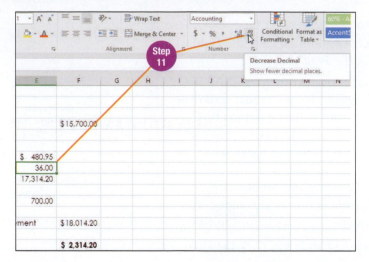

One decimal place is removed from the active cell or range each time you click the **Decrease Decimal button**. Click the **Increase Decimal button** to add one decimal place.

Adding Borders

Borders in various styles and colors can be added to the top, left, bottom, or right edge of a cell. Borders are used to underscore column headings or totals, or to otherwise emphasize cells.

Oops! !

Apply the wrong format to a cell or range? Use the Undo command, or simply apply the correct format to the cell or range.

12 Select F13.

13 Click the Bottom Border button arrow in the Font group.

14 Click *Top and Bottom Border* at the drop-down list.

15 Select F15.

16 Click the Top and Bottom Border button arrow and then click *Bottom Double Border* at the drop-down list.

17 Click any other cell to view the border style applied to F15.

App Tip

The Borders button updates to the most recently selected border style so that you can apply the same border to another cell or range by simply clicking the button (not the arrow).

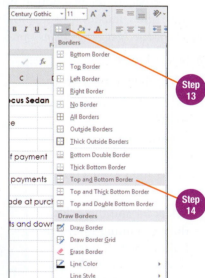

Merging and Centering Cells

A worksheet title is often centered across the columns used in the worksheet. The **Merge & Center button** in the Alignment group is used to combine a group of cells into one large cell and center its contents. Use the Merge & Center button arrow to merge without centering, or to unmerge a merged cell.

18 Select A1:F1.

19 Click the Merge & Center button in the Alignment group.

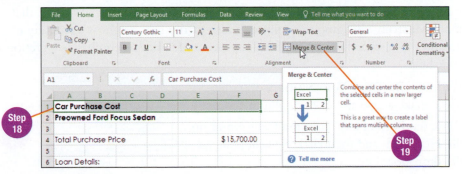

20 Select A2:F2 and then click the Merge & Center button.

21 Save the workbook using the same name (**CarCost-YourName**). Leave the workbook open for the next topic.

Alternative Method	**Formatting with the Mini Toolbar or Keyboard Shortcuts**

Consider applying formatting with the Mini toolbar, which you display by right-clicking inside selection area. Alternatively, you can use these keyboard shortcuts:

Ctrl + B	Apply bold formatting
Ctrl + I	Apply italic formatting
Ctrl + 1 (one)	Open the Format Cells dialog box
Ctrl + Shift + $	Apply US currency formatting
Ctrl + Shift + %	Apply percent style formatting

Quick Steps

Change a Font
1. Select cell(s).
2. Click Font list box arrow.
3. Click desired font.

Change Number Format
1. Select cell(s).
2. Click *Number Format* option box arrow, Accounting Number Format button, Percent Style button, or Comma Style button.
3. Click format option if needed.

Adjust Decimal Places
1. Select cell(s).
2. Click Increase Decimal or Decrease Decimal button.

Add Borders
1. Select cell(s).
2. Click Borders button arrow.
3. Click border style.

Merge Cells
1. Select cells.
2. Click Merge & Center button.

Beyond Basics   **Format Cells Dialog Box**

Click the dialog box launcher located at the bottom right of the Font, Alignment, or Number group to open the Format Cells dialog box. Use the dialog box to apply multiple formats in one operation, to further customize format options, or to apply font effect options *Strikethrough*, *Superscript*, and *Subscript*.

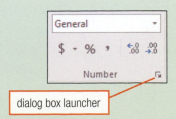

dialog box launcher

8.3 Adjusting Column Width and Row Height, and Changing Cell Alignment

Skills

Adjust column width

Adjust row height

Change cell alignment

Tutorials

Adjusting Column Width and Row Height

Applying Alignment Formatting

In a new worksheet, each column width is 8.43 and each row height is 15. The column width value is the number of characters at the default font that can be displayed in the column. The row height value is a points measurement, with 1 point being approximately 1/72 of an inch. Make cells larger by widening a column's width, or increasing a row's height. In many instances, Excel automatically makes columns wider and rows taller to accommodate the cell entry, formula result, or format that you apply. Manually changing the column width or the row height is a technique used to add more space between cells to improve readability or emphasize a section of the worksheet.

1 With the **CarCost-YourName** workbook open, make active any cell in column E.

2 Click the Format button in the Cells group on the Home tab and then click *Column Width* at the drop-down list.

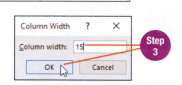

3 Type 15 in the *Column width* text box at the Column Width dialog box and then press Enter or click OK.

4 Make active any cell in column F, click the Format button, and then click *Column Width* at the drop-down list.

5 Type 10 at the Column Width dialog box and then press Enter or click OK.

Notice the cells with values in column F have been replaced with a series of pound symbols (#####). This occurs when the column's width is too narrow to show all the characters.

6 Make F4 the active cell.

7 Click the Format button and then click *AutoFit Column Width* at the drop-down list.

AutoFit changes the width of the column to fit the contents of the active cell. F4 was made active at Step 6 because this cell has the largest number in the column. Notice the pound symbols have disappeared, and the values are redisplayed now that the column is wide enough.

> **App Tip**
>
> The maximum width for a column is 255 characters and spaces, and the maximum height for a row is 409 points.

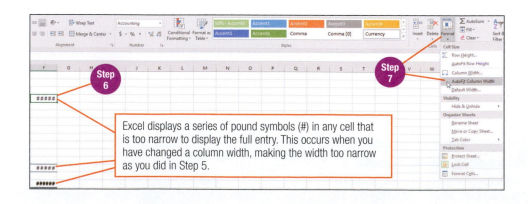

Excel displays a series of pound symbols (#) in any cell that is too narrow to display the full entry. This occurs when you have changed a column width, making the width too narrow as you did in Step 5.

8 Select A1:A2.

Select cells in multiple rows or columns to change the height or width of more than one row or column at the same time.

9 Click the Format button and then click *Row Height* at the drop-down list.

10 Type 26 in the *Row height* text box at the Row Height dialog box, and then press Enter or click OK.

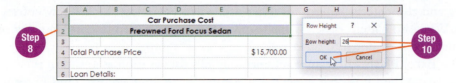

The Alignment group on the Home tab contains buttons to align the entry of a cell horizontally and/or vertically. You can align at the left, center, or right horizontally, or at the top, middle, or bottom vertically.

11 With A1:A2 still selected, click the Middle Align button in the Alignment group.

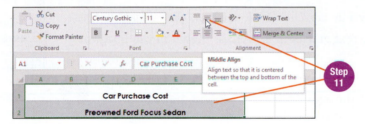

Middle Align centers text vertically between the top and bottom edges within a cell.

12 Make active any cell in row 15.

13 Click the Format button, click *Row Height*, type 26, and then press Enter or click OK.

14 Select A15:F15 and then click the Middle Align button.

15 Click any cell to deselect the range.

16 Save the workbook using the same name (**CarCost-YourName**).

17 Click the File tab and then click *Close* to close the workbook.

Quick Steps

Change Column Width
1. Activate any cell within column.
2. Click Format button.
3. Click *Column Width*.
4. Type width.
5. Click OK.

Change Row Height
1. Activate any cell within row.
2. Click Format button.
3. Click *Row Height*.
4. Type height.
5. Click OK.

AutoFit Column Width or Row Height
1. Activate cell with contents that are longest in the column or tallest in the row.
2. Click Format button.
3. Click *AutoFit Row Height* or *AutoFit Column Width*.

View
Model Answer
Compare your completed file with the model answer.

Alternative Method **Changing Column Width or Row Height Using a Mouse**

Change column widths using a mouse by dragging the column boundary at the right of the column letter to the right to increase the width or to the left to decrease the width. Change row height using a mouse by dragging the row boundary below the row number up to decrease the height or down to increase the height. Double-click the right column boundary or the bottom row boundary to AutoFit the column width or row height.

8.4 Entering or Copying Data with the Fill Command and Using AutoSum

Skills

Use Auto Fill to enter a series

Use Fill Right to copy a value

Use the fill handle to enter data

Add a column with AutoSum

Tutorials

Populating Data Using Flash Fill

Entering Formulas Using the AutoSum Button

Entering Data Using the Fill Handle

Copying Formulas

Using Undo and Redo

The **Auto Fill** feature in Excel is used to enter data automatically based on a pattern or series that exists in an adjacent cell or range. For example, if *Monday* is entered in cell A1, Auto Fill can enter *Tuesday*, *Wednesday*, and so on automatically in the cells immediately to the right of or below A1. Excel fills many common text or number series, and also detects patterns for other data when you select the first few entries in a list. When no pattern or series applies, the **Fill** feature is used to copy an entry or formula across or down to other cells.

The Excel **Flash Fill** feature automatically fills data as soon as a pattern is recognized. When Flash Fill presents a suggested list in dimmed text, press Enter to accept the suggestions, or ignore the suggestions and continue typing. A Flash Fill Options button appears when a list is presented with options to undo Flash Fill, accept the suggestions, or select changed cells.

Using Auto Fill and Fill Right

1. Click the File tab, click *New*, and then click *Blank workbook*.

2. Type the text entries in A2:A13 as shown in the image at right.

3. Change the width of column A to 18.

4. Make B1 the active cell, type Sep, and then click the Enter button on the Formula bar.

5. Select B1:I1.

6. Click the Fill button in the Editing group, and then click *Series* at the drop-down list.

	A	B
1		
2	Expenses	
3	Housing	
4	Food	
5	Transportation	
6	Smartphone	
7	Internet	
8	Entertainment	
9	Total Expenses	
10		
11	Income	
12		
13	Cash left over	

Step 2

7. Click *AutoFill* in the *Type* section of the Series dialog box and then click OK.

   Auto Fill enters the column headings *Oct* through *Apr* in the selected range.

8. Make B3 the active cell, type 875, and then click the Enter button.

9. Select B3:I3, click the Fill button, and then click *Right* at the drop-down list.

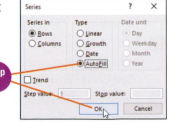

Fill Right copies the entry in the first cell to the other cells within the selected range.

10 Enter the remaining values as shown in the image at right. In rows 5, 6, and 7, use Fill Right to enter the data by completing tasks similar to those in Steps 8 and 9.

	A	B	C	D	E	F	G	H	I
1		Sep	Oct	Nov	Dec	Jan	Feb	Mar	Apr
2	Expenses								
3	Housing	875	875	875	875	875	875	875	875
4	Food	260	340	310	295	320	280	300	345
5	Transportation	88	88	88	88	88	88	88	88
6	Smartphone	48	48	48	48	48	48	48	48
7	Internet	42	42	42	42	42	42	42	42
8	Entertainment	150	110	95	175	100	85	95	120
9	Total Expenses								

Step 10

Using the Fill Handle to Copy Cells

The **fill handle** is a small, green square at the bottom right corner of the active cell or range. The fill handle can be used to copy data from a cell or range to adjacent cells.

To use the fill handle with a mouse: When you point at the square with a mouse, the cell pointer changes appearance from the large white cross to the fill handle (**+**). Drag right or down when you see the fill handle icon to copy data or a formula, or to extend a series from the active cell or range to adjacent cells.

To use the fill handle on a touch device: Tapping a cell on a touch device displays the active cell with two selection handles instead of the fill handle. See Figure 8.3 for instructions on how to use the fill handle on a touch device.

App Tip

The Auto Fill Options button appears when you release the mouse after dragging the fill handle to copy a cell. Use the button to instruct Excel to fill the cells with the formatting options only from the copied cell (no data is copied), or to copy the data with no formatting options (only data is copied).

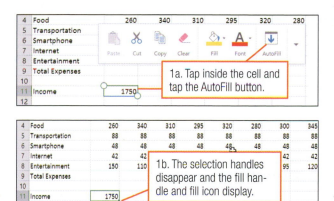

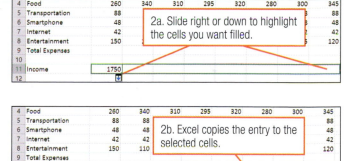

Figure 8.3

How to use the fill handle on a touch device.

11 Make B11 the active cell, type 1750, and then click the Enter button.

12 Drag the fill handle right to I11.

The value *1750* is copied to the cells in the selected range. See Beyond Basics at the end of this topic for more information about using the versatile fill handle.

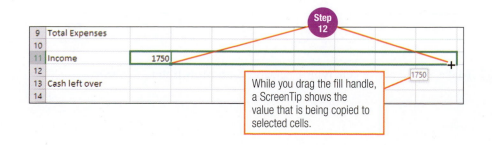

Step 12

While you drag the fill handle, a ScreenTip shows the value that is being copied to selected cells.

Oops! !

Don't see the fill handle? The feature may have been turned off. Open the Advanced panel at the Excel Options dialog box (click File, *Options*, *Advanced*), and then click *Enable fill handle and cell drag-and-drop* to insert a check mark.

Using the SUM function and AutoSum

To add the expenses in B9, you could type the formula =B3+B4+B5+B6+B7+B8; however, Excel includes a built-in preprogrammed function called SUM that can be used to add a column or row of numbers. The SUM function is faster and easier to use. To add the expenses in B9 using SUM, you would type the formula =SUM(B3:B8). Notice after =SUM you need to provide only the range of cells to add enclosed in parentheses, rather than all the individual cell references and addition symbols. Because the SUM function is used frequently, an **AutoSum button** is included in the Editing group on the Home tab and the AutoSum feature automatically detects the range to be added.

13 Make B9 the active cell.

14 Click the AutoSum button in the Editing group on the Home tab. (Do not click the down-pointing arrow at the right of the AutoSum button.)

Excel inserts the formula =SUM(B3:B8) in B9 with the suggested range B3:B8 selected.

15 Click the Enter button or press Ctrl + Enter to complete the formula. (Ctrl + Enter completes the entry and keeps B9 as the active cell.)

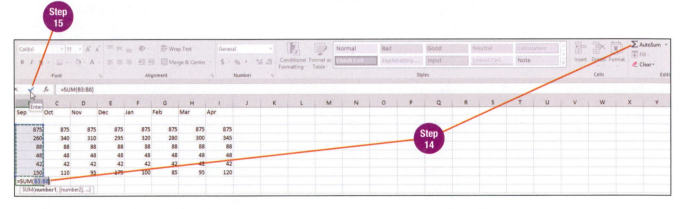

16 With B9 the active cell, drag the fill handle right to I9.

In this instance, using the fill handle copies the formula in B9 to the selected cells.

17 Make B13 the active cell, type =b11-b9, and then click the Enter button or press Ctrl + Enter.

18 With B13 the active cell, drag the fill handle right to I13.

8	Entertainment	150	110	95	175	100	85	95	120
9	Total Expenses	1463	1503	1458	1523	1473	1418	1448	1518
10									
11	Income	1750	1750	1750	1750	1750	1750	1750	1750
12									
13	Cash left over	287	247	292	227	277	332	302	232
14									

Step 18

19 Make J1 the active cell, type Total, and then press Enter.

20 Make J3 the active cell and then click the AutoSum button.

In this instance, Excel suggests the range B3:I3 in the SUM function. Excel looks for values immediately above or to the left of the active cell. Because no value exists above J3, Excel correctly suggests adding the values to the left in the same row.

21 Click the Enter button to accept the formula.

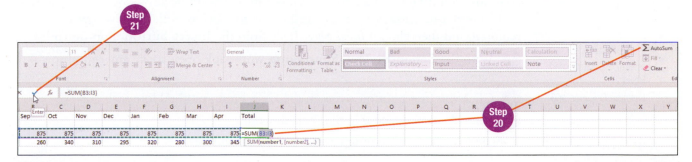

22 With J3 the active cell, drag the fill handle down to J13.

23 Make J10 the active cell, and then either press the Delete key or click the Clear button in the Editing group and then click *Clear All* at the drop-down list.

24 Make J12 the active cell and repeat the instruction in Step 23.

◢	A	B	C	D	E	F	G	H	I	J
1		Sep	Oct	Nov	Dec	Jan	Feb	Mar	Apr	Total
2	Expenses									
3	Housing	875	875	875	875	875	875	875	875	7000
4	Food	260	340	310	295	320	280	300	345	2450
5	Transportation	88	88	88	88	88	88	88	88	704
6	Smartphone	48	48	48	48	48	48	48	48	384
7	Internet	42	42	42	42	42	42	42	42	336
8	Entertainment	150	110	95	175	100	85	95	120	930
9	Total Expenses	1463	1503	1458	1523	1473	1418	1448	1518	11804
10										
11	Income	1750	1750	1750	1750	1750	1750	1750	1750	14000
12										
13	Cash left over	287	247	292	227	277	332	302	232	2196

25 Save the new workbook as **SchoolBudget-YourName** in the Ch8 folder in CompletedTopicsByChapter on your storage medium. Leave the workbook open for the next topic.

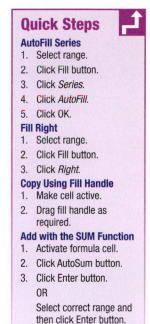

Quick Steps

AutoFill Series
1. Select range.
2. Click Fill button.
3. Click *Series*.
4. Click *AutoFill*.
5. Click OK.

Fill Right
1. Select range.
2. Click Fill button.
3. Click *Right*.

Copy Using Fill Handle
1. Make cell active.
2. Drag fill handle as required.

Add with the SUM Function
1. Activate formula cell.
2. Click AutoSum button.
3. Click Enter button.
   OR
   Select correct range and then click Enter button.

Beyond Basics   **More Examples of Using the Fill Command**

The Excel Fill command can detect patterns in values, dates, times, months, days, years, or other data. A pattern is detected based on the cells selected before dragging the fill handle. Following are some examples of series the fill handle can extend. In each example, you would select the range of cells in column A and column B, and then drag the fill handle to the right to extend the data.

Column A	Column B	Columns C, D, E, and so on
10	20	30, 40, 50, and so on
9:00	10:00	11:00, 12:00, 1:00, and so on
2018	2019	2020, 2021, 2022, and so on
Year 1	Year 2	Year 3, Year 4, Year 5, and so on

8.5 Inserting and Deleting Rows and Columns

Skills

Insert a new row

Insert a new column

Delete a row

Tutorials

Inserting Columns and Rows

Deleting Columns and Rows

New rows or columns are inserted or deleted using the Insert or Delete buttons in the Cells group on the Home tab. New rows are inserted above the row in which the active cell is positioned, and new columns are inserted to the left. Cell references within formulas and formula results are automatically updated when new rows or columns with data are added to or removed from a worksheet.

To insert a new row, activate any cell in the row below which a new row is required, and then choose *Insert Sheet Rows* from the **Insert button** drop-down list.

1 With the **SchoolBudget-YourName** workbook open, make any cell in row 4 active.

2 Click the Insert button arrow in the Cells group on the Home tab.

3 Click *Insert Sheet Rows* at the drop-down list.

A new blank row is inserted between *Housing* and *Food*.

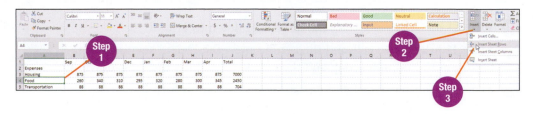

4 Type the following entries in the cells indicated:

A4	Utilities	F4	128
B4	110	G4	106
C4	115	H4	118
D4	132	I4	112
E4	147		

Oops!

Only one cell is inserted instead of an entire row? Clicking the top part of the Insert button inserts a cell instead of an entire new row. Use Undo and then try again, making sure to click the arrow on the bottom part of the Insert button to access the drop-down list.

5 Make J3 the active cell and then drag the fill handle down to J4 to copy the SUM formula to the new row.

	A	B	C	D	E	F	G	H	I	J
1		Sep	Oct	Nov	Dec	Jan	Feb	Mar	Apr	Total
2	Expenses									
3	Housing	875	875	875	875	875	875	875	875	7000
4	Utilities	110	115	132	147	128	106	118	112	968
5	Food	260	340	310	295	320	280	300	345	2450

Step 4　Step 5

6 Select A1:A2, click the Insert button arrow, and then click *Insert Sheet Rows*. Two rows are inserted above A1.

7 Make the new cell A1 the active cell and then type Proposed School Budget.

8 Make the new cell A2 the active cell and then type First Year of Program.

9 Use Merge & Center to format the range A1:J1.

10 Use Merge & Center to format the range A2:J2.

App Tip

To insert multiple rows, select row numbers along the left edge of the worksheet area, right-click, and then click *Insert*. Excel will insert the same number of blank rows as you selected and it will insert the blank rows above the first row you selected.

	A	B	C	D	E	F	G	H	I	J
1	Proposed School Budget									
2	First Year of Program									
3		Sep	Oct	Nov	Dec	Jan	Feb	Mar	Apr	Total
4	Expenses									
5	Housing	875	875	875	875	875	875	875	875	7000
6	Utilities	110	115	132	147	128	106	118	112	968

Step 9　Step 10

11 Make any cell in column J active.

12 Click the Insert button arrow and then click *Insert Sheet Columns* at the drop-down list.

A new column is inserted between *Apr* and *Total*. Notice also an **Insert Options button** () appears below and right of the active cell. Options from this button are used to format the new column the same as the column at its left or right, or to clear formatting in the new column.

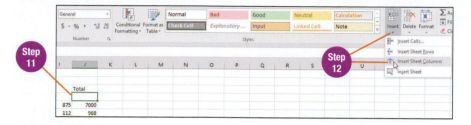

To delete a single row or column, position the active cell within that row or column, and then choose *Delete Sheet Rows* or *Delete Sheet Columns* from the **Delete button** drop-down list. Remove multiple rows or columns from the worksheet by first selecting the range of rows or columns to be deleted.

13 Make any cell in row 10 active.

14 Click the Delete button arrow and then click *Delete Sheet Rows* at the drop-down list.

Row 10 is removed from the worksheet, and existing rows below are shifted up to fill in the space.

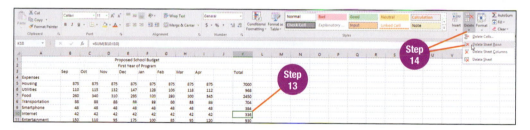

15 Save the workbook using the same name (**SchoolBudget-YourName**). Leave the workbook open for the next topic.

Beyond Basics   Inserting and Deleting Cells

The Insert and Delete buttons are also used to insert and delete cells within a worksheet. Select the range of cells you need to add and then choose *Insert Cells* at the Insert button drop-down list. At the Insert dialog box, choose whether to shift existing cells right or down.

Select the range of cells to delete, choose *Delete Cells* at the Delete button drop-down list, and then select whether to shift existing cells left or up to fill the space.

8.6 Sorting Data and Applying Cell Styles

Skills

Sort rows of data

Apply cell styles

Tutorials

Sorting Data

Applying Cell Styles

Applying and Modifying Themes

A range in Excel can be rearranged by sorting in either ascending or descending order on one or more columns. For example, you can sort a list of names and cities first by the city and then by the last name. To sort by more than one column, select the range and open the Sort dialog box from the **Sort & Filter button** drop-down list.

1. With the **SchoolBudget-YourName** workbook open, select A5:K10.

Notice you do not include the heading (in A4) or the totals (in A11:K11) in the sort range.

2. Click the Sort & Filter button in the Editing group on the Home tab.

3. Click *Sort A to Z* at the drop-down list.

4. Click any cell to deselect the range, and review the new order of the expenses.

5. Select A5:K10.

6. Click the Sort & Filter button and then click *Custom Sort* at the drop-down list.

7. At the Sort dialog box, click the *Sort by* list box arrow in the *Column* section, and then click *Column K*.

8. Click the *Order* list box arrow (currently displays *Smallest to Largest*) and then click *Largest to Smallest*.

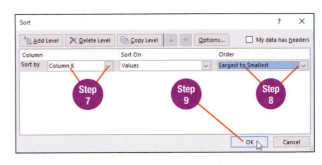

9. Click OK.

The range is rearranged in descending order from highest expense total to lowest.

10. Click any cell to deselect the range, and review the new order of expenses.

Similar to Word Styles feature, **Cell Styles** in Excel offers a set of predefined formatting options that can be applied to a single cell or a range. Using Cell Styles to format a worksheet is fast and promotes consistency. The Cell Styles gallery groups styles by the sections *Good, Bad and Neutral*; *Data and Model*; *Titles and Headings*; *Themed Cell Styles*; and *Number Format*.

11. Make A1 the active cell and then click the More button located at the bottom right of the Cell Styles gallery in the Styles group on the Home tab.

⑫ Click *Heading 1* in the *Titles and Headings* section of the Cell Styles gallery.

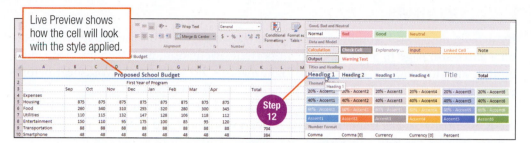

Live Preview shows how the cell will look with the style applied.

Step 12

⑬ Make A2 the active cell, click the More button in the Cell Styles gallery, and then click *Heading 4* in the *Titles and Headings* section.

⑭ Apply the Heading 4 cell style to A4, A13, and A15.

⑮ Select B3:K3, apply the Accent1 style in the *Themed Cell Styles* section of the Cell Styles gallery, and then center the content of those cells.

⑯ Select B5:K15 and apply the Comma [0] style in the *Number Format* section of the Cell Styles gallery.

⑰ Select B15:K15 and apply the Total style in the *Titles and Headings* section of the Cell Styles gallery.

⑱ Click any cell to deselect the range and compare your worksheet with the one shown in Figure 8.4.

⑲ Save the workbook using the same name (**SchoolBudget-YourName**). Leave the workbook open for the next topic.

App Tip

You can hold down the Ctrl key and click A4, A13, and A15 to format all three cells in one operation.

Quick Steps

Sort a Range by the First Column
1. Select range.
2. Click Sort & Filter button.
3. Click *Sort A to Z* or *Sort Z to A*.

Custom Sort
1. Select range.
2. Click Sort & Filter button.
3. Click *Custom Sort*.
4. Change options and/or add levels as needed.
5. Click OK.

Apply Cell Styles
1. Select cell or range.
2. Click desired cell style in Styles gallery.

◢	A	B	C	D	E	F	G	H	I	J	K
1					**Proposed School Budget**						
2					First Year of Program						
3			Sep	Oct	Nov	Dec	Jan	Feb	Mar	Apr	Total
4	Expenses										
5	Housing		875	875	875	875	875	875	875	875	7,000
6	Food		260	340	310	295	320	280	300	345	2,450
7	Utilities		110	115	132	147	128	106	118	112	968
8	Entertainment		150	110	95	175	100	85	95	120	930
9	Transportation		88	88	88	88	88	88	88	88	704
10	Smartphone		48	48	48	48	48	48	48	48	384
11	Total Expenses		1,531	1,576	1,548	1,628	1,559	1,482	1,524	1,588	12,436
12											
13	Income		1,750	1,750	1,750	1,750	1,750	1,750	1,750	1,750	14,000
14											
15	Cash left over		219	174	202	122	191	268	226	162	1,564

Figure 8.4
This is the sorted worksheet with cell styles applied.

Beyond Basics   **Workbook Themes**

Options in the *Titles and Headings* and *Themed Cell Styles* sections in the Cell Styles gallery change depending on the active theme (set of colors, fonts, and effects). Change the theme for a workbook using the Themes gallery, accessed by clicking the Themes button arrow in the Themes group on the Page Layout tab.

8.7 Changing Orientation and Scaling, and Displaying Cell Formulas

Skills

Preview a worksheet

Change orientation

Display cell formulas

Change scaling

Tutorials

Printing a Worksheet

Displaying Formulas

By default, new Excel workbooks have print options set to print the active worksheet on a letter-size page (8.5 x 11 inches), in portrait orientation, with 0.75-inch top and bottom margins, and 0.7-inch left and right margins. Preview a new worksheet before printing to determine whether these print options are appropriate.

Workbooks are often distributed as PDF files and circulated electronically. A PDF file is essentially an electronic view of a printed worksheet and has the same default settings. Always preview a worksheet and change print options as needed before exporting as a PDF.

1. With the **SchoolBudget-YourName** workbook open, click the File tab, click *Print*, and then compare your screen with the one shown in Figure 8.5.

2. Click the Next Page button at the bottom of the preview panel to view the second page of the worksheet.

3. Click the *Orientation* list box arrow in the *Settings* category and then click *Landscape Orientation* at the drop-down list.

Notice the worksheet now fits on one page. Landscape is a common layout used for wide worksheets.

4. Click the Back button to return to the worksheet and then save the revised workbook using the same name (**SchoolBudget-YourName**).

Quick Steps

Change the Orientation
1. Display Print backstage area.
2. Click *Orientation* list box arrow.
3. Click *Landscape Orientation*.

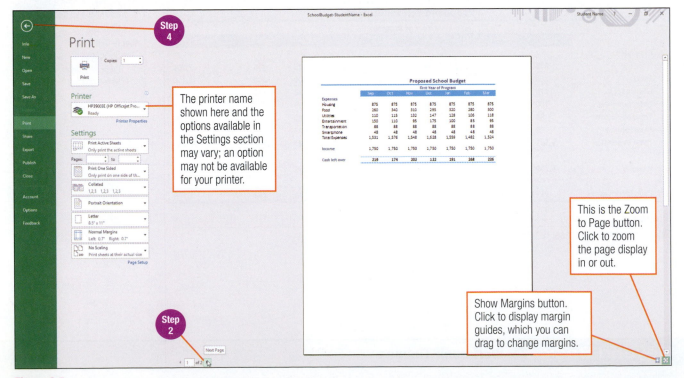

The printer name shown here and the options available in the Settings section may vary; an option may not be available for your printer.

This is the Zoom to Page button. Click to zoom the page display in or out.

Show Margins button. Click to display margin guides, which you can drag to change margins.

Figure 8.5

Print backstage area is shown with first page displayed for the **SchoolBudget-StudentName** worksheet.

On a printed copy of the worksheet, only the formula result is printed. You may want to print a second copy of the worksheet with the formulas displayed in the cell as a backup or documentation strategy for a complex or otherwise important worksheet.

5 Click the Formulas tab.

6 Click the Show Formulas button in the Formula Auditing group.

The **Show Formulas button** is a toggle button that switches the display between showing the formula in each cell, and showing the result in each cell.

7 Scroll right if necessary to review the worksheet with formulas displayed.

App Tip

Ctrl + ` (grave accent, usually located above the Tab key) is the keyboard command to turn on or off the display of formulas.

	A	B	C	D	E		F	G	H	I	J	K
1					Proposed School Budget							
2					First Year of Program							
3		Sep	Oct	Nov	Dec		Jan	Feb	Mar	Apr		Total
4	Expenses											
5	Housing	875	875	875	875		875	875	875	875		=SUM(B5:I5)
6	Food	260	340	310	295		320	280	300	345		=SUM(B6:I6)
7	Utilities	110	115	132	147		128	106	118	112		=SUM(B7:I7)
8	Entertainment	150	110	95	175		100	85	95	120		=SUM(B8:I8)
9	Transportation	88	88	88	88		88	88	88	88		=SUM(B9:I9)
10	Smartphone	48	48	48	48		48	48	48	48		=SUM(B10:I10)
11	Total Expenses	=SUM(B5:B10)	=SUM(C5:C10)	=SUM(D5:D10)	=SUM(E5:E10)		=SUM(F5:F10)	=SUM(G5:G10)	=SUM(H5:H10)	=SUM(I5:I10)		=SUM(B11:I11)
12												
13	Income	1750	1750	1750	1750		1750	1750	1750	1750		=SUM(B13:I13)
14												
15	Cash left over	=B13-B11	=C13-C11	=D13-D11	=E13-E11		=F13-F11	=G13-G11	=H13-H11	=I13-I11		=SUM(B15:I15)

8 Display the Print backstage area.

9 Click the *Scaling* list box arrow in the *Settings* category and then click *Fit Sheet on One Page*.

Fit Sheet on One Page shrinks the size of text on the printout to fit all columns and rows on one page.

10 Click the Back button to return to the worksheet.

11 Use Save As to save a copy of the worksheet with the formulas displayed, named **SchoolBudget–Formulas–YourName** in the Ch8 folder in CompletedTopicsByChapter.

12 Close the workbook.

App Tip

Other methods used to print wide worksheets are decreasing the margins, and changing the scaling percentage.

View

Model Answer

Compare your completed file with the model answer.

Quick Steps

Display Cell Formulas
1. Click Formulas tab.
2. Click Show Formulas button.

Scale a Worksheet
1. Display Print backstage area.
2. Click *Scaling* list box arrow.
3. Click required scaling option.

Alternative Method **Changing Print Options Using the Page Layout Tab**

You can change some print options on the Page Layout tab, using the Margins and Orientation buttons in the Page Setup group and the Width, Height, and Scale options in the Scale to Fit group.

Beyond Basics   **More Scaling Options**

In the Print backstage area, the scaling option *Fit All Columns on One Page* shrinks the size of text until all the columns fit the page width; more than one page may print if there are many rows. *Fit All Rows on One Page* shrinks the size of text until all the rows fit the page height; more than one page may print if there are many columns.

8.8 Inserting and Renaming a Worksheet, Copying Cells, and Indenting Cell Contents

Skills

Insert a new worksheet

Rename worksheets

Copy cells

Indent cells

Tutorials

Inserting and Renaming Worksheets

Copying and Pasting Cells

Moving Cells

Using Format Painter and the Repeat Command

Navigating and Scrolling

A workbook can contain more than one worksheet. Use multiple worksheets as a method to organize or group data into manageable units. For example, a homeowner might have one household finance workbook and keep track of bills and loans in one worksheet, savings and investments in a second worksheet, and a household budget in a third worksheet. Insert, rename, and navigate between worksheets using the sheet tabs in the **Sheet tab bar** near the bottom of the window. Use the **New sheet button** in the Sheet tab bar to add a new worksheet to the workbook.

1. Open the **SchoolBudget-YourName** workbook.

2. Click the New sheet button next to the Sheet1 tab on the Sheet tab bar.

3. Click the Sheet1 tab to make Sheet1 the active worksheet.

4. Right-click the Sheet1 tab and then click *Rename* at the shortcut menu.

5. Type First Year and then press Enter.

6. Right-click the Sheet2 tab, click *Rename*, type Second Year, and then press Enter.

7. Click the First Year tab to make First Year the active worksheet.

8. Select A4:A15 and then click the Copy button in the Clipboard group on the Home tab.

9. Make Second Year the active worksheet, make A4 the active cell, and then click the top portion of the Paste button (not the down-pointing arrow) in the Clipboard group.

10. Click the Paste Options button and then click *Keep Source Column Widths*.

11. Make First Year the active worksheet, select A1:K3, and then click the Copy button.

12. Make Second Year the active worksheet, make A1 the active cell, and then click the Paste button.

13. Edit A2 in the Second Year worksheet to change *First* to *Second* so that the title now reads Second Year of Program.

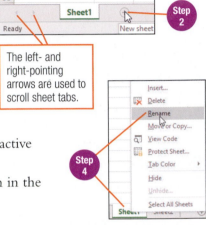

The left- and right-pointing arrows are used to scroll sheet tabs.

Step 2

Step 4

Step 6

Step 10

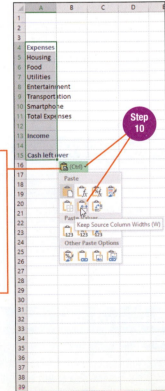

On a touch device, the Paste Options button does not appear here after text is pasted into the worksheet; instead, the options must be accessed by tapping the Paste button on the ribbon.

App Tip

Other options on the sheet tab shortcut menu are used to delete, move, copy, hide, or protect entire sheets, and to change the color of the background in the sheet tab.

Oops!

Pasted to the wrong starting cell? Drag the border of the selected range to the correct starting point, or use Cut and Paste to move cells.

14 Enter the data, complete the formulas for the cells in row 15 and column K, and format the cells in the Second Year worksheet, as shown in Figure 8.6.

	A	B	C	D	E	F	G	H	I	J	K
1					Proposed School Budget						
2					Second Year of Program						
3		Sep	Oct	Nov	Dec	Jan	Feb	Mar	Apr		Total
4	Expenses										
5	Housing	910	910	910	910	910	910	910	910		7,280
6	Food	245	330	298	285	308	275	295	355		2,391
7	Utilities	112	118	140	151	131	118	122	124		1,016
8	Entertainment	160	95	100	185	110	95	90	125		960
9	Transportation	90	90	90	90	90	90	90	90		720
10	Smartphone	50	50	50	50	50	50	50	50		400
11	Total Expenses	1,567	1,593	1,588	1,671	1,599	1,538	1,557	1,654		12,767
12											
13	Income	1,800	1,800	1,800	1,800	1,800	1,800	1,800	1,800		14,400
14											
15	Cash left over	233	207	212	129	201	262	243	146		1,633

Figure 8.6
The completed Second Year worksheet is shown here.

The Increase Indent button in the Alignment group on the Home tab moves an entry approximately one character width inward from the left edge of a cell each time the button is clicked. Use this feature to indent entries in a list below a subheading. The Decrease Indent button moves an entry approximately one character width closer to the left edge of the cell each time the button is clicked.

15 With Second Year still the active worksheet, select A5:A10 and then click the Increase Indent button in the Alignment group on the Home tab.

16 Make A11 the active cell and then click the Increase Indent button twice.

17 Change the orientation to landscape for the Second Year worksheet.

18 Make First Year the active worksheet, indent A5:A10 once, and indent A11 twice.

19 Save the workbook using the same name (**SchoolBudget–Your Name**) and then close the workbook.

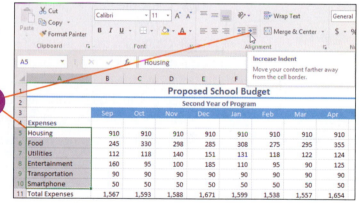

Step 16

Alternative Method **Renaming a Worksheet**

Another way to rename a worksheet is to double-click the sheet tab, type a new name, and then press Enter.

Beyond Basics   **Scrolling Sheet Tabs in the Sheet Tab Bar**

Use the left- and right-pointing arrows to the left of the first sheet tab to scroll to sheet tabs not currently visible, or right-click an arrow to open the Activate dialog box, click the sheet to make active, and then click OK.

8.9 Using Go To; Freezing Panes; and Shading, Wrapping, and Rotating Cell Entries

Skills

Use Go To

Freeze panes

Add fill color

Wrap text

Rotate text

Tutorials

Freezing and Unfreezing Panes

Adding Fill Color to Cells

In large worksheets where you cannot see all cells at once, the **Go To** and **Go To Special** commands accessed from the **Find & Select button** in the Editing group on the Home tab are used to move the active cell to a specific location in a worksheet. Column or row headings not visible when you scroll right or down beyond the viewing area can be fixed in place using the **Freeze Panes** option when working with large worksheets.

1. Open the **NSCSuppliesInventory** workbook from the Ch8 folder in Student_Data_Files.

2. Save the workbook as **NSCSuppliesInventory-YourName** in the Ch8 folder in CompletedTopicsByChapter.

3. Scroll down the worksheet area until the titles and column headings are no longer visible.

4. Click the Find & Select button in the Editing group on the Home tab and then click *Go To* at the drop-down list.

5. At the Go To dialog box, type a4 in the *Reference* text box and then press Enter or click OK.

6. Click the Find & Select button and then click *Go To Special*.

7. Click *Last cell* in the Go To Special dialog box and then click OK.

Use the *Last cell* option in the Go To Special dialog box in a large worksheet to move the active cell to the bottom right of the worksheet.

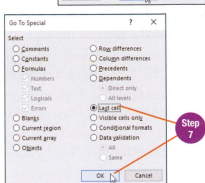

8. Use Go To to move the active cell back to A4.

9. Scroll up until you can see the first three rows, containing titles and column headings.

10. With A4 still the active cell, click the View tab and then click the Freeze Panes button in the Window group.

11. Click *Freeze Panes* at the drop-down list.

12. Scroll down past all data. Notice that rows 1 to 3 do not scroll out of the viewing area.

App Tip

Ctrl + G is the keyboard command for Go To.

App Tip

The active cell position determines which rows and columns are fixed—all rows above and all columns left of the active cell are frozen.

App Tip

Once cells have been frozen, *Freeze Panes* at the Freeze Panes button drop-down list changes to *Unfreeze Panes*.

Column headings can be formatted to stand out from the rest of the worksheet by shading the background of the cell using the **Fill Color button** in the Font group on the Home tab or by rotating the cell entries. Cells with long entries can be housed in narrower columns by formatting the text to automatically wrap within the width of the cell using the **Wrap Text button** in the Alignment group on the Home tab. In Steps 13–20 you will format the worksheet headings to match the appearance shown in Figure 8.7.

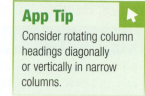

13 Scroll to the top of the worksheet and then select A1.

14 Click the Home tab, click the Fill Color button arrow in the Font group, and then click *Orange, Accent 6* (last color in first row of *Theme Colors* section) at the drop-down gallery.

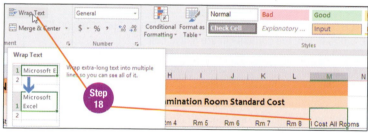

15 Select A2:M2 and apply the Orange, Accent 6, Lighter 60% fill color (last option in third row of *Theme Colors* section).

16 Select A3:M3 and apply the Orange, Accent 6, Lighter 80% fill color (last option in second row of *Theme Colors* section).

17 Make M3 the active cell, change the row height to 30, and then change the column width to 10.

18 With M3 still the active cell, click the Wrap Text button in the Alignment group.

The entire column heading is visible again, with the text wrapping within the cell.

19 Make D3 the active cell, change the column width to 9, and then click the Wrap Text button.

20 Select E3:L3, click the Orientation button in the Alignment group, and then click *Angle Counterclockwise* at the drop-down list.

Angle Counterclockwise rotates the text within the cell boundaries 45 degrees.

21 Click any cell to deselect the range and then display the Print backstage area.

22 Change settings to *Landscape Orientation* and *Fit All Columns on One Page* and then go back to the worksheet display.

23 Save the workbook using the same name (**NSCSuppliesInventory-YourName**) and then close the workbook.

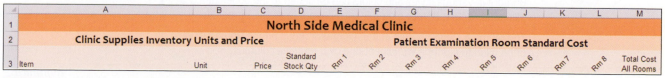

Figure 8.7
This worksheet shows the formatted headings.

Topics Review

Topic	Key Concepts	Key Terms
8.1 Creating and Editing a New Worksheet	A spreadsheet is an application in which data is created, analyzed, and presented in a grid-like structure of columns and rows.	spreadsheet application
	A workbook is an Excel file that consists of a collection of individual worksheets.	workbook
	A new workbook opens with a blank worksheet into which you add text, insert values, and create formulas.	worksheet
	The intersection of a column and a row is called a *cell*.	cell
	The active cell is indicated with a green border, and is the location into which the next data typed will be stored, or the next command will be acted upon.	active cell
	Create a worksheet by making a cell active and typing text, a value, or a formula.	formula
	A formula is used to perform mathematical operations on values.	Clear button
	Formula entries begin with an equals sign and are followed by cell references with operators between the references.	
	Edit a cell by typing new data to overwrite existing data, by double-clicking to open the cell for editing, or by inserting or deleting characters in the Formula bar.	
	Press the Delete key, or use the Clear button in the Editing group on the Home tab to delete the contents in the active cell.	
8.2 Formatting Cells	The Font group on the Home tab contains buttons to change the font, font size, font color, and font style of selected cells.	range
	Select cells with the mouse by positioning the cell pointer over the starting cell and dragging in the required direction.	Comma Style button
	Select cells using touch by tapping the starting cell, and then sliding your finger over a selection handle until the remaining cells are inside the selection rectangle.	Accounting Number Format button
	A rectangular-shaped group of cells is called a *range*, and is referenced with the starting cell address, a colon, and the ending cell address (e.g., A1:F15).	Decrease Decimal button
	By default, cells in a new worksheet have the General format, which has no specific formatting options.	Increase Decimal button
	The Comma Style button in the Number group on the Home tab adds a comma in thousands and two decimal places.	Merge & Center button
	The Accounting Number Format button in the Number group on the Home tab adds a dollar symbol, a comma in the thousands place, and displays two digits after the decimal point.	
	The Decrease Decimal button and Increase Decimal button in the Number group on the Home tab remove or add one decimal place each time the button is clicked.	
	Borders in various styles and colors can be added to the edges of a cell using the Bottom Border button in the Font group on the Home tab.	
	The Merge & Center button in the Alignment group on the Home tab is often used to center a worksheet title over multiple columns.	

continued…

Topic	Key Concepts	Key Terms
8.3 Adjusting Column Width and Row Height, and Changing Cell Alignment	A technique to add more space between cells to improve readability or emphasize a section is to widen a column or increase the height of a row. Open the Column Width dialog box from the Format button in the Cells group on the Home tab to enter a new value for the width of the column in which the active cell is positioned. Excel displays a series of pound symbols when a column width is too narrow to display all the cell contents. AutoFit changes the width of a column to fit the contents of the active cell. Open the Row Height dialog box from the Format button in the Cells group on the Home tab to enter a new value for the height of the row in which the active cell is positioned. Align a cell at the left, center, or right horizontally, or top, middle, or bottom vertically using buttons in the Alignment group. The Middle Align button in the Alignment group on the Home tab centers cell contents vertically.	AutoFit Middle Align
8.4 Entering or Copying Data with the Fill Command and Using AutoSum	Auto Fill can be used to automatically enter data in a series or pattern based upon an entry in an adjacent cell. When no pattern is detected the Fill feature copies an entry or formula to other cells. The Flash Fill feature automatically suggests entries when a pattern is detected. Select a range and use the Fill button to open the Series dialog box or choose Down, Right, Up, or Left to copy an entry. The small, green square at the bottom right of an active cell is the fill handle and can be used to copy data and formulas or to enter a series. Excel includes the AutoSum button in the Editing group on the Home tab that is used to enter a SUM function to add a column or row of numbers.	Auto Fill Fill Flash Fill Fill Right fill handle AutoSum button
8.5 Inserting and Deleting Rows and Columns	A new row is inserted above the active cell or selected range. A new column is inserted left of the active cell or selected range. Use the Insert button in the Cells group on the Home tab to insert new rows or columns. Options for formatting new rows or columns are available from the Insert Options button that appears when rows or columns are inserted. Delete rows or columns using the Delete button in the Cells group on the Home tab. The Insert and Delete buttons can also be used to insert or delete cells within the worksheet.	Insert button Insert Options button Delete button
8.6 Sorting Data and Applying Cell Styles	Select a range and choose the sort order option from the Sort & Filter button in the Editing group on the Home tab to arrange the rows by the entries in the first column. Open the Sort dialog box to sort by more than one column or to choose a different column in the range by which to sort. Cell Styles are a set of predefined formatting options applied to selected cells using options in the Cell Styles gallery in the Styles group on the Home tab. Use Cell Styles to format faster and/or promote consistency among worksheets.	Sort & Filter button Cell Styles

continued...

Topic	Key Concepts	Key Terms
8.7 Changing Orientation and Scaling, and Displaying Cell Formulas	New workbooks print on a letter-size page, in portrait orientation, with top and bottom margins of 0.75 inch, and left and right margins of 0.7 inch. Change print options even if you are only exporting the workbook as a PDF because PDFs are generated using the print settings. Change to landscape orientation using the Orientation list box at the Print backstage area. Landscape is a common layout used for wide worksheets. Turn on or turn off the display of formulas in cells instead of the results of formulas by clicking the Formulas tab, and then clicking the Show Formulas button in the Formula Auditing group. Fit Sheet on One Page scales text on a printout so that all columns and rows print on one page. Fit All Columns on One Page is a scaling option that shrinks text to fit all columns in one page width. Fit All Rows on One Page is a scaling option that shrinks text to fit all rows in one page height.	Show Formulas button Fit Sheet on One Page
8.8 Inserting and Renaming a Worksheet, Copying Cells, and Indenting Cell Contents	The Sheet tab bar near the bottom left of the window is used to insert, rename, and navigate among sheets in a workbook. Click the New sheet button on the Sheet tab bar to insert a new worksheet. Right-click a sheet tab and choose Rename to type a new name for a worksheet. Copy and paste cells between worksheets using the Copy and Paste buttons in the Clipboard group. *Keep Source Column Widths* from the Paste Options button lets you paste new cells with the same column width as the source cell.	Sheet tab bar New sheet button
8.9 Using Go To; Freezing Panes; and Shading, Wrapping, and Rotating Cell Entries	The Go To dialog box and Go To Special dialog box from the Find & Select button in the Editing group on the Home tab are used to move to a specific cell in a large worksheet. Freeze Panes fixes rows and/or columns in place for scrolling in large worksheets so that column and row headings do not scroll out of the viewing area. All rows above and all columns left of the active cell are frozen when Freeze Panes is turned on. Cells are shaded with color using the Fill Color button in the Font group on the Home tab. Long text entries in cells can be displayed in narrow columns by wrapping text within the cell column width using the Wrap Text button in the Alignment group on the Home tab. *Angle Counterclockwise* is an option from the Orientation button in the Alignment group on the Home tab that rotates text within a cell 45 degrees.	Go To Go To Special Find & Select button Freeze Panes Fill Color button Wrap Text button *Angle Counterclockwise*

 **Recheck**
Recheck your understanding of the topics covered in this chapter.

 **Workbook**
Chapter review and assessment resources are available in the *Workbook* ebook.

Chapter 9

Working with Functions, Charts, Tables, and Page Layout Options in Excel

Precheck

Check your understanding of the topics covered in this chapter.

Excel's function library is updated and expanded with each new software release. Several hundred preprogrammed formulas grouped by category in the function library are used to perform data analysis, decision making, and data modeling. Charts present data or analysis results in a visual snapshot. Data organized in a list format for use with the Table feature allows you to easily format, sort, and filter large blocks of data. Page Layout view is useful for previewing page layout and print options and for adding headers or footers. Collaborative tools, such as Comments, allow individuals to add notes or other feedback into a worksheet.

In this chapter, you continue working with formulas by learning the types of references used in formulas and how to use functions to perform statistical, date, financial, and logical analysis. Next, you explore strategies for presenting data using charts, comments, and tables, as well as formatting a worksheet using page layout options and print options.

Learning Objectives

9.1 Create formulas with absolute addresses and range names

9.2 Create formulas with AVERAGE, MAX, and MIN statistical functions

9.3 Enter and format dates and use the TODAY date function

9.4 Perform decision making using the logical IF function

9.5 Use the PMT function to calculate a loan payment

9.6 Create and edit a pie chart

9.7 Create and edit a column chart

9.8 Create and edit a line chart

9.9 Use Page Layout view, insert a header, change margins, and center a worksheet

9.10 Create and edit sparklines and insert comments into a worksheet

9.11 Sort and filter a range defined as a *table*

 If you are a SNAP user, go to your SNAP Assignments page to complete the Precheck, Tutorials, and Recheck.

Data Files

Before beginning this chapter, be sure you have copied the student data files for this course to your storage medium. Steps on downloading and extracting the data files are provided in Chapter 1, Topic 1.8, on pages 22–23.

9.1 Using Absolute Addresses and Range Names in Formulas

Skills

Create a formula with an absolute address

Create a range name

Create a formula with a range name

Tutorials

Absolute Addressing

Naming and Using a Range

Managing Range Names

The formulas in the previous chapter used cell references that are considered a **relative address**, where a column letter and row number change relative to the destination when a formula is copied. For example, the formula *=SUM(A4:A10)* becomes *=SUM(B4:B10)* when copied from a cell in column A to a cell in column B. Relative addressing is the most common addressing method.

Sometimes you need a formula in which one or more addresses should not change when the formula is copied. In these formulas, use cell references that are an **absolute address**. A dollar symbol precedes a column letter and/or row number in an absolute address, for example, *=$A$10*. Some formulas have both a relative and an absolute reference; these are referred to as a **mixed address**. See Table 9.1 for formula addressing examples.

Table 9.1

Cell Addressing and Copying Examples

Formula	Type of Reference	Action If Formula Is Copied
=B4*B2	relative	Both addresses will update.
=$B$4*$B$2	absolute	Neither address will update.
=B4*$B$2	mixed	The address B4 will update; the address $B$2 will not update.
=B4*$B2	mixed	The address B4 will update; the row number in the second address will update, but the column letter will not.
=B4*B$2	mixed	The address B4 will update; the column letter in the second address will update, but the row number will not.

1. Start Excel 2016 and open the workbook named *FinancialPlanner* from the Ch9 folder in Student_Data_Files.

2. Use Save As to save a copy of the workbook as **FinancialPlanner-YourName** in a new folder, Ch9, in the CompletedTopicsByChapter folder.

3. Review the worksheet noticing the three rates in row 2; these rates will be used to calculate gross pay, payroll deductions, and savings.

4. Make C4 the active cell, type =b4*b2, and then click the Enter button on the Formula bar, or press Ctrl + Enter.

5. Use the fill handle in C4 to copy the formula to C5:C30.

6. Make C5 the active cell and look at the formula in the Formula bar.

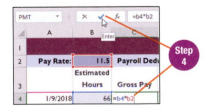

Step 4

The #VALUE! error occurs because B3 is a label and has no mathematical value.

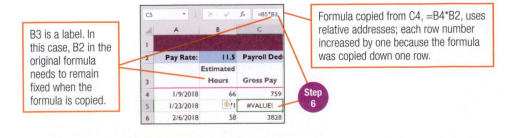

B3 is a label. In this case, B2 in the original formula needs to remain fixed when the formula is copied.

Formula copied from C4, =B4*B2, uses relative addresses; each row number increased by one because the formula was copied down one row.

Step 6

7 Select C5:C30 and press the Delete key or click the Clear button in the Editing group on the Home tab and then click *Clear All*.

8 Make C4 the active cell and edit the formula so that it reads =*B4*\$B\$2*.

9 Use the fill handle in C4 to copy the formula to C5:C30.

10 Make D4 the active cell and enter the formula =c4*\$e\$2.

11 Make E4 the active cell and enter the formula =c4-d4.

12 Select D4:E4 and use the fill handle to copy the formulas to D5:E30.

A cell or a range can be referenced by a descriptive label, which makes a formula easier to understand. For example, the formula =*Hours*PayRate* is readily understood. Names are also used when a formula needs an absolute reference because a cell or range name is automatically absolute. Cell or range names are assigned using the Name box at the left end of the Formula bar. Use the Name Manager button in the Formulas tab to manage cell names after they are created.

13 Make G2 the active cell, click in the Name box at the left end of the Formula bar, type SaveRate, and then press Enter.

A range name can use letters, numbers, and some symbols. Spaces are not valid in a range name, and the first character in a name must be a letter, an underscore, or a backward slash (\\).

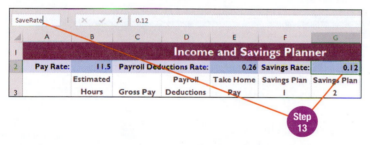

14 Make F4 the active cell, type =e4*SaveRate, and then press Ctrl + Enter or click the Enter button.

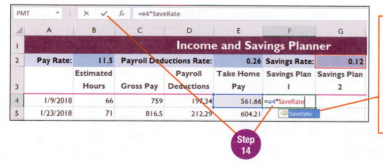

AutoComplete displays range names in the current workbook. Press Tab to accept the name in the list, or select the name (when necessary), and then press Tab.

15 Use the fill handle in F4 to copy the formula to F5:F30.

16 Save the revised workbook using the same name (**FinancialPlanner-YourName**). Leave the workbook open for the next topic.

App Tip

Press Function key F4 to make an address absolute. F4 cycles through variations of absolute and mixed addresses for the reference in which the insertion point is positioned.

Quick Steps

Make Cell Reference Absolute
Type dollar symbol before column letter and/or row number, or press F4 to cycle through variations of addressing.

Name a Cell or Range
1. Select cell or range.
2. Type name in Name box.
3. Press Enter.

App Tip

A range name can also be used in the Go To dialog box to move the active cell.

App Tip

Upper and lowercase letters separate words in a range name; however, range names are not case sensitive—*SaveRate* and *saverate* are considered the same name.

9.2 Entering Formulas Using Statistical Functions

The function library in Excel contains more than 400 preprogrammed formulas grouped into 13 categories. All formulas based on functions begin with the name of the function followed by the function's parameters in parentheses. The parameters for a function (referred to as an **argument**) will vary depending on the formula chosen and can include a value, a cell reference, a range, multiple ranges, or a combination of values with references.

1. With the **FinancialPlanner-YourName** workbook open, make I4 the active cell.

2. Click the AutoSum button arrow in the Editing group on the Home tab and then click *Average* at the drop-down list.

Excel enters *=AVERAGE(B4:H4)* in the cell with the range *B4:H4* selected. In this instance, Excel suggests the wrong range.

3. With the range B4:H4 highlighted in the formula cell, select B4:B30, and then press Enter or click the Enter button.

Excel returns the result *68.96296296* in the formula cell, which is the arithmetic mean of the hours in column B. If empty cells or cells containing text are included in the formula's argument, they are ignored.

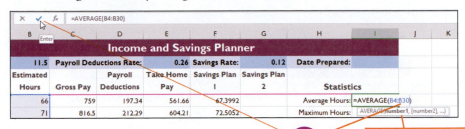

4. Make I5 the active cell, click the AutoSum button arrow, and then click *Max* at the drop-down list.

Excel provides the format for the function argument in a ScreenTip.

5. Type b4:b30 and then press Enter or click the Enter button.

Excel returns the value *80* in the formula cell. MAX returns the largest value found in the range included in the argument.

6. Make I6 the active cell, type =min(b4:b30), and then press Enter or click the Enter button.

The result *48* is shown in I6. MIN returns the smallest value within the range included in the argument.

Excel's Insert Function dialog box assists with finding and entering functions and their arguments into a formula cell. A variety of methods can be used to open the Insert Function dialog box, including clicking the Insert Function button on the Formula bar or clicking the Formulas tab and then clicking the Insert Function button in the Function Library group.

7 Make I8 the active cell and then click the Insert Function button on the Formula bar (button right of Enter button).

8 Click the *Or select a category* list arrow and then click *Statistical* at the drop-down list.

9 Click *AVERAGE* in the *Select a function* list box and then click OK.

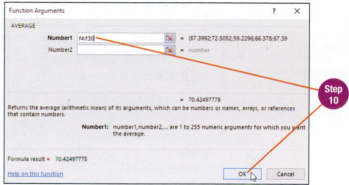

10 Type f4:f30 in the *Number1* text box at the Function Arguments dialog box and then press Enter or click OK.

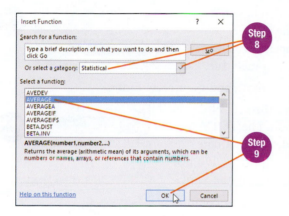

11 Make I9 the active cell, click the AutoSum button arrow, and then click *More Functions* at the drop-down list.

12 With the text already selected in the *Search for a function* text box, type max and then press Enter or click the Go button.

13 With MAX selected in the *Select a function* list box, click OK.

14 Type f4:f30 in the *Number1* text box at the Function Arguments dialog box, and then press Enter or click OK.

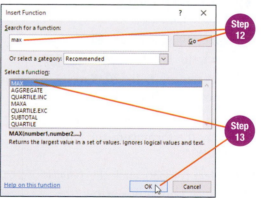

15 Make I10 the active cell, click the AutoSum button arrow, and then click *Min* at the drop-down list.

16 Type f4:f30 and then press Enter, or click the Enter button.

17 Save the revised workbook using the same name (**FinancialPlanner-YourName**). Leave the workbook open for the next topic.

Quick Steps

AVERAGE, MAX, or MIN Functions
1. Activate formula cell.
2. Click AutoSum button arrow.
3. Click required function.
4. Type or select argument range.
5. Press Enter.

Beyond Basics   COUNT Function

Count Numbers from the AutoSum button arrow inserts a COUNT function, which returns the number of cells in the argument range that have values. Use the function COUNTA if you want to count all the cells within the argument range including cells that contain text. Empty cells are ignored in both cases.

9.3 Entering, Formatting, and Calculating Dates

Skills

Enter a valid date

Enter the current date using a function

Create a formula using a date

Format dates

Tutorial

Using Date and Time Functions

Oops! !

General instead of *Date* appears? Excel did not recognize your entry as a valid date. Generally, this is because of a typing error. Try Step 2 again. Still *General*? You may need to check the Region in the Control Panel.

App Tip

The function =*NOW()* returns the current date and time in the active cell.

A date typed into a cell in normal date format, such as *May 1, 2018* or *5/1/2018*, is stored as a numerical value. A time is stored as a decimal value representing a fraction of a day. Because dates and times are stored as values, calculations can be performed using the cells, and various date and time formats can be applied to the results.

Consider using Excel to calculate elapsed time for scheduling, payroll, membership, or other purposes that involve analysis of date or time.

1. With the **FinancialPlanner-YourName** workbook open, make I2 the active cell.

2. Type 12/20/2018 and then press Enter.

3. Make I2 the active cell and notice that *Date* appears in the *Number Format* option box in the Number group on the Home tab. See Table 9.2 on page 219 for examples of cell entries that Excel will recognize as a valid date.

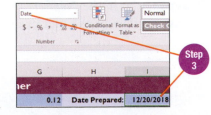

Step 3

4. Clear the contents of I2, type =today(), and then press Enter.

Excel enters the current date into the cell. No argument is required for this function. The TODAY function updates the cell entry to the current date whenever the worksheet is opened or printed. Do not use =*TODAY()* if you want the date to stay the same.

Step 4

5. Make B3 the active cell and insert a new column.

6. Type Pay Date in B3 and then click A3.

7. Type End Date in A3 and then click the Enter button.

8. Select A3:B3 and then click the Center button in the Alignment group on the Home tab.

9. Make B4 the active cell, type =a4+7, and then click the Enter button.

Excel returns *1/16/2018* in B4, which is seven days from January 9, 2018.

Step 9

10. Use the fill handle in B4 to copy the formula to the range B5:B30.

11. Select A4:B30, click the *Number Format* option box arrow (the arrow next to *Date*), and then click the *More Number Formats* option at the bottom of the drop-down list.

12. At the Format Cells dialog box with *Date* selected in the *Category* list box, click *14-Mar-12* in the *Type* list box and then click OK.

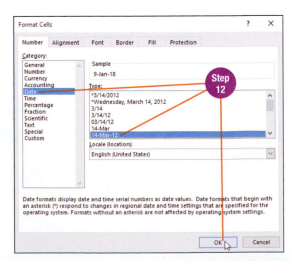

Step 12

⓭ Click any cell to deselect the range.

⓮ Save the revised workbook using
the same name (**FinancialPlanner–
YourName**). Leave the workbook open
for the next topic.

	A	B
3	**End Date**	**Pay Date**
4	9-Jan-18	16-Jan-18
5	23-Jan-18	30-Jan-18
6	6-Feb-18	13-Feb-18
7	20-Feb-18	27-Feb-18
8	6-Mar-18	13-Mar-18
9	20-Mar-18	27-Mar-18
10	3-Apr-18	10-Apr-18
11	17-Apr-18	24-Apr-18
12	1-May-18	8-May-18
13	15-May-18	22-May-18
14	29-May-18	5-Jun-18
15	5-Jun-18	12-Jun-18
16	19-Jun-18	26-Jun-18
17	3-Jul-18	10-Jul-18
18	17-Jul-18	24-Jul-18
19	31-Jul-18	7-Aug-18
20	7-Aug-18	14-Aug-18
21	21-Aug-18	28-Aug-18
22	4-Sep-18	11-Sep-18
23	18-Sep-18	25-Sep-18
24	2-Oct-18	9-Oct-18
25	16-Oct-18	23-Oct-18
26	30-Oct-18	6-Nov-18
27	6-Nov-18	13-Nov-18
28	20-Nov-18	27-Nov-18
29	4-Dec-18	11-Dec-18
30	18-Dec-18	25-Dec-18

Formatted
dates in A4:B30

Table 9.2

Entries Excel Recognizes as Valid Dates or Times

Dates	Times
12/20/18; 12-20-18	4:45 (stored as 4:45:00 AM)
Dec 20, 2018	4:45 PM (stored as 4:45:00 PM)
20-Dec-18 or 20 Dec 18 or 20/Dec/18	16:45 (stored as 4:45:00 PM)

Note: *The year can be entered as two digits or four digits and the month as three characters or spelled in full.
Times are generally entered as hh:mm, but in situations that require a higher level of accuracy, they are entered
as hh:mm:ss.*

Alternative Method   **Entering a Date Using a Date Function Formula**

You can also enter dates into cells as DATE functions. A DATE function is typed as
=DATE(Year,Month,Day). For example *=DATE(2018,12,20)*. Use the Date & Time button in the
Function Library group on the Formulas tab to browse other Date or Time functions.

Beyond Basics   **Region Setting and Dates**

The Region setting in the Control Panel affects the format that Excel 2016 will recognize as a valid date.
Following are examples of date format by region:

English (United States)	*m/d/yy*
English (Canada)	*yy/m/d* (Windows 10) or *d/m/yy* (Windows 7)
English (United Kingdom)	*d/m/yy*

To change the Region, open the Control Panel from the desktop and select the Region icon or the Clock,
Language, and Region category.

9.4 **Using the IF Function**

Logical functions are used when you need a formula to perform a calculation based on a condition or comparison of a cell, with a value or the contents of another cell. For example, in column G of the Income and Savings Planner worksheet, you calculated a savings value based on the take-home pay amounts in column F. Suppose you decide that you cannot afford to contribute to your savings plan unless your take-home pay is more than $500. The formula in column G does not accommodate this scenario; however, an IF formula can analyze the take-home pay and calculate the savings for those values that are over your minimum.

1 With the **FinancialPlanner-YourName** workbook open, make H4 the active cell.

2 Click the Formulas tab.

The category drop-down lists in the Function Library group are another way that you can find an Excel function to insert into a cell.

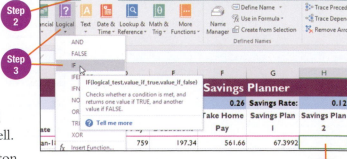

3 Click the Logical button in the Function Library group and then click *IF* at the drop-down list.

4 With the insertion point positioned in the *Logical_test* text box at the Function Arguments dialog box, type f4>500 and then press Tab or click in the *Value_if_true* text box.

A logical test is a statement to evaluate a comparison so that one of two actions can be performed. In this case, the statement *f4>500* tells Excel to determine if the value that resides in F4 is greater than 500. All logical tests result in either a true or a false response—either the value is greater than 500 (true) or the value is not greater than 500 (false). See Table 9.3 on page 221 for more examples of logical tests.

5 Type f4*SaveRate and then press Tab or click in the *Value_if_false* text box.

The statement in the *Value_if_true* text box is the formula you want Excel to calculate when the logical test proves true. In other words, if the value in F4 is greater than 500, you want Excel to multiply the value in F4 times the value in the cell named SaveRate (.12).

6 Type 0 and then click OK.

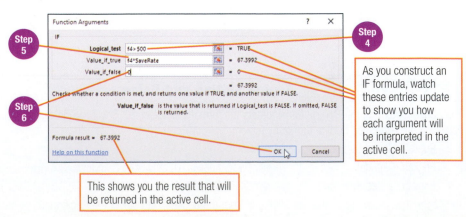

This shows you the result that will be returned in the active cell.

As you construct an IF formula, watch these entries update to show you how each argument will be interpreted in the active cell.

The *Value_if_false* statement is the formula you want Excel to calculate when the logical test proves false. In other words, if the value in F4 is 500 or less, you want zero placed in the cell because you have decided that you cannot afford to contribute to your savings plan.

7 Look in the Formula bar at the IF statement entered into the active cell =IF(F4>500,F4*SaveRate,0).

Using the Function Arguments dialog box to build an IF statement is a good idea because the commas and parentheses are inserted automatically in the correct positions within the formula.

8 Use the fill handle in H4 to copy the formula to the range H5:H30.

9 Review the results in H5:H30. Notice the cells that have 0 appear in a row where the take-home pay value in column F is 500 or less.

10 Select C4:H30, and then click the Quick Analysis button that appears below the selection.

11 Click the Totals tab, and then click the Sum button in the *Totals* gallery (first button).

Excel creates SUM functions for each column in the selected range and enters them in row 31.

12 Select D4:H31, click the Home tab, and then click the Comma Style button.

13 Apply the Comma Style format to C2 and J4:J10.

14 Apply the Percent Style format to F2 and H2.

15 Select C31:H31 and then add a Top and Double Bottom Border.

16 Save the revised workbook using the same name (**FinancialPlanner-YourName**). Leave the workbook open for the next topic.

Quick Steps

IF Function
1. Activate formula cell.
2. Click Formulas tab.
3. Click Logical button.
4. Click *IF*.
5. Type *Logical_test* statement.
6. Type value or formula in *Value_if_true* text box.
7. Type value or formula in *Value_if_false* text box.
8. Click OK.

App Tip

Percent Style multiplies the value in the cell by 100 and adds a percent symbol (%) to the cell.

First 10 rows of worksheet with formatting applied at Steps 12 to 14.

Table 9.3

IF Statement Logical Test Examples

Logical Test	Condition Evaluated	IF Statement Example
F4>=500	Is the value in F4 greater than or equal to 500?	=IF(F4>=500,F4*SaveRate,0)
F4<500	Is the value in F4 less than 500?	=IF(F4<500,0,F4*SaveRate)
F4<=500	Is the value in F4 less than or equal to 500?	=IF(F4<=500,0,F4*SaveRate)
F4=$K$2	Is the value in F4 equal to the value in K2? Assume value in K2 is the take-home pay value for which you will set aside savings.	=IF(F4=$K$2,F4*SaveRate,0)
Hours<>0	Is the value in the cell named Hours not equal to 0?	=IF(Hours<>0,Hours*PayRate,0) Calculates Gross Pay when hours have been logged

Skills

Enter PMT function

Tutorial

Using Financial
Functions

Financial functions in Excel can be used for a variety of tasks that involve saving or borrowing money, such as calculating the future value of an investment, calculating the present value of an investment, or calculating borrowing criteria, such as interest rates, terms, or payments. If you are considering a loan or mortgage, use Excel's PMT function to determine an estimated loan payment. The PMT function uses a specified interest rate, number of payments, and loan amount to calculate a regular payment. Once a payment is shown, you can manipulate the interest rate, term, or loan amount to find a payment with which you are comfortable.

1. With the **FinancialPlanner-YourName** workbook open, click the LoanPlanner sheet tab.

2. Make B7 the active cell.

3. Click the Formulas tab, click the Financial button in the Function Library group, scroll down the drop-down list, and then click *PMT*.

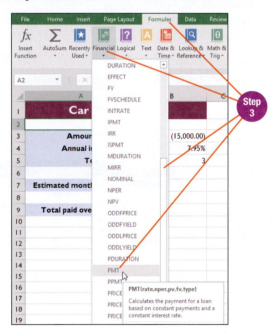

Oops!

Can't see B4? Drag the Function Arguments title bar right until the dialog box is no longer obscuring your view of columns A and B.

4. With the insertion point positioned in the *Rate* text box at the Function Arguments dialog box, click B4 and then type /12.

The interest rate in B4 is expressed as the interest rate per year. Typing /12 after B4 causes Excel to divide the interest rate in B4 by 12 to calculate the monthly interest rate. To use the PMT function correctly, you need to ensure that the time periods are all the same. In other words, if you want to find a monthly payment, you need to make sure the rate and terms are also in monthly units. Most lending institutions express interest with the annual rate (not monthly) but compound the interest monthly.

5. Click in the *Nper* text box, click B5, and then type *12.

The value in B5 is the number of years you will take to pay back the loan. Multiplying the value times 12 will convert the value to the number of months to repay the loan. Most lending institutions express the repayment term in years (not months).

6 Click in the *Pv* text box and then click B3.

Pv stands for *present value* and represents the amount you want to borrow (referred to as the *principal*). Notice the amount borrowed is entered as a negative value in this worksheet. By default, Excel considers payments as negative values because money is subtracted from your bank balance when you make a loan payment. By entering a negative number for the amount borrowed, the PMT formula will return a positive value for the calculated loan payment. Whether you prefer to show a negative value for the amount borrowed or for the estimated monthly loan payment is a matter of personal preference; both options are acceptable.

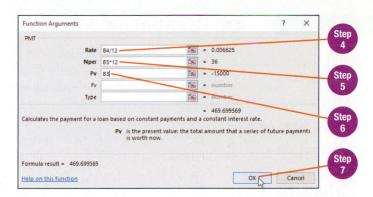

7 Click OK.

Excel returns the payment *$469.70* in B7.

8 Look in the Formula bar at the PMT statement entered into the active cell =PMT(B4/12,B5*12,B3).

9 Make B9 the active cell and enter the formula =b7*b5*12.

Excel calculates the total cost for the loan to be $16,909.18.

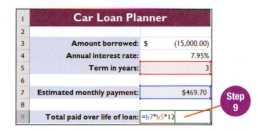

10 Change the value in B5 from *3* to *4*.

Notice that increasing the term one more year reduces your monthly payment; however, the total cost of the loan increases because you are making more payments.

11 Save the revised workbook using the same name (**FinancialPlanner-YourName**) and then close the workbook.

Beyond Basics   **Using FV to Calculate the Future Value of an Investment**

Another useful financial function is FV, which is used to calculate the future value of a series of regular payments that earn a constant interest rate. For example, if you deposit $100 each month for 10 years into an investment account that earns 9.75% per year (compounded monthly), the FV function *=FV(9.75%/12,10*12,100)* calculates the value of the account after the 10-year period to be $20,193.76.

9.6 Creating and Modifying a Pie Chart

Skills

Create a pie chart

Add data labels

Modify the legend position

Edit the chart title

Tutorials

Creating Charts

Formatting with Chart Buttons

Inserting a Shape

Formatting a Shape

Changing Chart Design

Changing Chart Formatting

Charts provide a visual snapshot of data. Charts can illustrate trends, proportions, and comparisons more distinctly than numbers alone. Excel provides 15 categories of charts with multiple styles in each category. The Quick Analysis button recommends charts based on the type of data you are analyzing and allows you to preview the chart style live with your data. For example, a **pie chart** is a circular graph with each data point (pie slice) sized to show its proportion to a total. Governments often use a pie chart to illustrate how tax dollars are allocated across various programs and services.

1. Open the workbook named *SocialMediaStats*.

2. Use Save As to save a copy of the workbook as **SocialMediaStats-YourName** in the Ch9 folder in CompletedTopicsByChapter.

3. Select A5:B10.

Before you can insert a chart you first need to select the data that you want Excel to represent in a chart.

4. Click the Quick Analysis button that appears below the selection area and then click the Charts tab.

5. Click *Pie* at the *Charts* gallery.

Excel graphs the data in a pie chart and places the chart overlapping the cells within a chart object window. Notice also that the chart object is selected with selection handles, three chart editing buttons, and the Chart Tools Design and Format tabs in the ribbon.

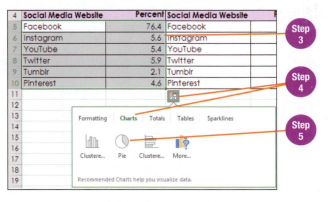

6. With the chart selected, drag the chart with the mouse positioned over any white unused area inside the chart borders, to the approximate location shown below.

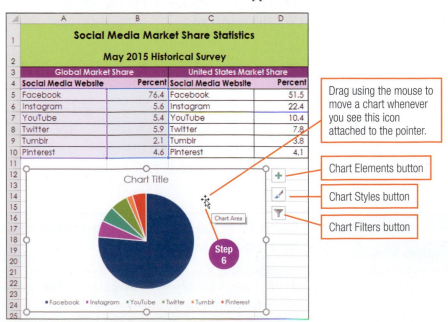

7 Click the Chart Elements button (displays as a plus symbol next to the chart).

8 Click the *Data Labels* check box to insert a check mark, click the right-pointing arrow that appears at the right end of the *Data Labels* option, and then click *Outside End*.

Although pie charts show proportions well, adding data labels allows a reader to include context with the size of each pie slice.

Quick Steps

Create a Pie Chart
1. Select range.
2. Click Quick Analysis button.
3. Click Charts.
4. Click *Pie*.
5. Move and/or modify chart elements as required.

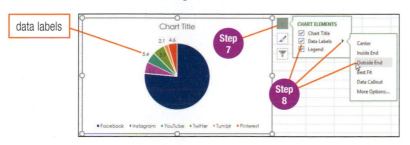

9 With Chart Elements still displayed, click at the right end of *Legend* when the right-pointing arrow appears and then click *Right*.

10 Click to select the *Chart Title* object inside the chart window, drag to select *Chart Title*, type Global Market Share, and then click in any white, unused area within the chart to deselect the title.

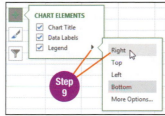

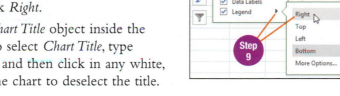

11 Select C5:D10 and insert a second pie chart, as shown in Figure 9.1, by completing steps similar to those in Steps 4 to 10.

12 Save the revised workbook using the same name (**SocialMediaStats-YourName**). Leave the workbook open for the next topic.

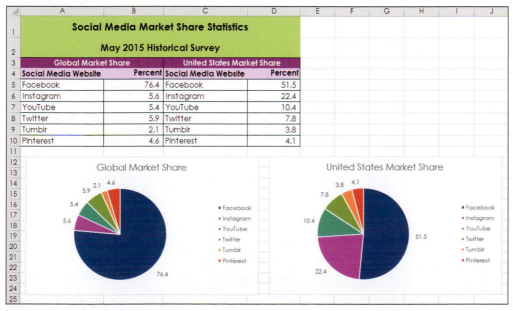

App Tip

A popular technique to emphasize a pie slice is to move the slice away from the rest of the pie (called a *point explosion*). To do this, click to select the pie slices, then click to isolate the individual slice and drag the slice away from the pie.

Figure 9.1

The revised workbook shows side-by-side pie charts for Social Media Market Share Statistics.

9.7 Creating and Modifying a Column Chart

Skills

Create a column chart

Change the chart style

Change the chart color scheme

Add axis titles

In a **column chart**, each data point is a colored bar that extends up from the **category axis** (horizontal axis, also called *x-axis*) with the bar height representing the value of the data points on the **value axis** (vertical axis, also called *y-* or *z-axis*). Use a column chart to compare one or more series of data side by side. Column charts are often used to identify trends or illustrate comparisons over time or categories.

1. With the **SocialMediaStats-YourName** workbook open, click the Facebook sheet tab.

2. Select A7:B14.

3. Click the Quick Analysis button, click Charts, and then click *Clustered Column* (first button).

4. Drag the chart until the top left corner is positioned in row 1 under column letter C (see Figure 9.2 on page 227).

5. Drag the bottom right selection handle down and right to resize the chart until the bottom right corner is at approximately the bottom right border of J16 (see Figure 9.2).

6. With the chart selected, click the Chart Styles button (displays as a paintbrush next to the chart).

7. Scroll down to the bottom of the Style list and then click the last option in the gallery (*Style 16*).

8. With the *Chart Styles* gallery still open, click the Color tab and then click the third row in the *Colorful* section of the color gallery (*Color 3*).

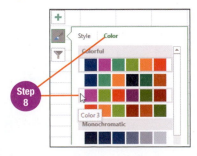

9. Click the Chart Elements button.

   Additional chart elements options are available for column charts that are not possible with a pie chart.

10 Click the *Axis Titles* check box to insert a check mark.

Excel adds an Axis Title object to the vertical axis and to the horizontal axis.

11 With the Axis Title object along the vertical axis already selected, select the title text and type North American Users.

12 Click to select the *Axis Title* object along the horizontal axis and press Delete to remove the object.

13 Edit the *Chart Title* to Facebook Audience by Age Group.

14 Compare your chart to the one shown in Figure 9.2. If necessary, redo an action in Steps 4 to 13.

15 Save the revised workbook using the same name (**SocialMediaStats-YourName**). Leave the workbook open for the next topic.

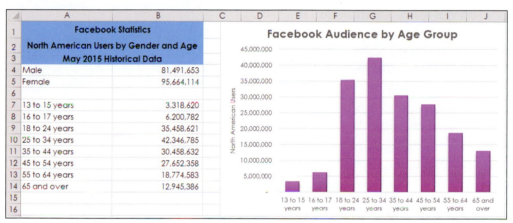

Figure 9.2

Shown above is a column chart for Facebook North American Audience Statistics by Age Group.

Alternative Method | **Inserting and Modifying a Chart Using the Ribbon**

Buttons in the Charts group on the Insert tab can also be used to create a chart. Select the data range, click the Insert tab, and then click the button for the desired chart type in the Charts group. Once a chart has been inserted, the Chart Tools Design and Format tabs contain the same options to modify the chart as those found in the *Chart Elements* and *Chart Styles* galleries.

App Tip

Microsoft added six new chart types to Excel 2016: Treemap, Sunburst, Histogram, Box & Whisker, Funnel, and Waterfall.

Beyond Basics   **Recommended Charts**

Not sure which chart type best represents your data? Select the data you want to graph and Excel shows a series of customized charts that best suit the selection when you click the Charts tab from the Quick Analysis button. Use the *More Charts* option to view other chart types. Alternatively, click the Insert tab and then click the Recommended Charts button in the Charts group to view recommended charts in the Insert Chart dialog box.

9.8 Creating and Modifying a Line Chart

Skills

Create a line chart

Move a chart to a new sheet

Format an axis

Format data labels

Line charts are best suited for data where you want to illustrate trends and changes in values over a period of time. With a **line chart**, a reader can easily spot a trend, or identify growth spurts, dips, or unusual points in the series. Line charts are also often used to help predict future values based on the direction of the line.

1 With the **SocialMediaStats–YourName** workbook open, click the FacebookUserTimeline sheet tab.

2 Select A4:B15.

3 Click the Quick Analysis button, click Charts, and then click *Line*.

4 With the chart selected, click the Move Chart button in the Location group on the Chart Tools Design tab.

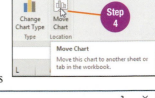

5 Click *New sheet*, type FBTimelineChart, and then press Enter or click OK.

Click the **Move Chart** button to move a chart to its own chart sheet, which automatically scales the chart to fit a letter-size page in landscape orientation.

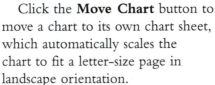

6 With the Chart Tools Design tab active, click the second option in the *Chart Styles* gallery (*Style 2*).

7 Click the Change Colors button in the Chart Styles group and then click the third row in the *Colorful* section of the color gallery (*Color 3*).

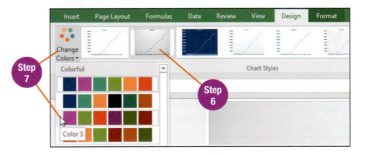

Oops!

No border around dates? Click the axis labels a second time. Sometimes the chart is selected the first time you click.

8 Edit the Chart Title to Facebook Active Users Historical Timeline.

In the next step you will correct the axis labels. The dates in column A were incorrectly converted, changing *Dec* to *Jan*.

9 Click to select the dates in the category axis along the bottom of the chart. Make sure you see a border and selection handles around the axis labels.

10 Double-click inside the selected axis labels to open the Format Axis task pane at the right side of the window.

11 Click *Text axis* in the *Axis Options* section of the task pane.

Notice the axis labels change to show the December dates as they appeared in the worksheet.

12 Close the Format Axis task pane.

13 Click any data value on a data point in the line chart to select the entire series of data labels.

14 Right-click any of the selected data values and then click *Format Data Labels* at the shortcut menu.

This opens the Format Data Labels task pane.

15 Click *Above* in the *Label Position* section of the Format Data Labels task pane with the Label Options tab active.

16 Click Text Options and then click *Text Fill* to expand the options list.

17 Click the Color button and then click *Black, Text 1* (second color in first row).

18 Close the Format Data Labels task pane and then click in the window outside the chart to deselect the data labels.

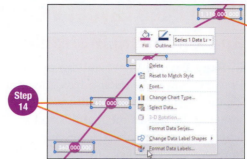

Click any data value in the line chart at Step 13 to select the data labels.

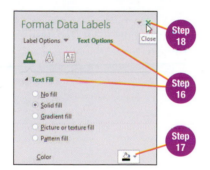

19 Compare your chart with the chart shown in Figure 9.3 and make corrections if necessary.

20 Save the revised workbook using the same name (**SocialMediaStats-YourName**). Leave the workbook open for the next topic.

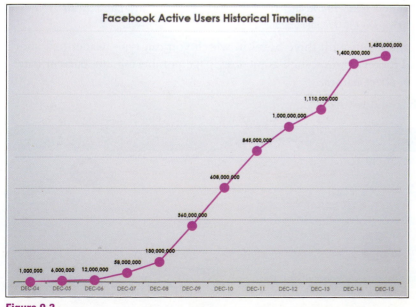

Figure 9.3

Line chart for Facebook Active Users Historical Timeline created in new chart sheet

9.9 Using Page Layout View, Adding a Header, and Changing Margins

In **Page Layout view**, you can preview page layout options similarly to Print Preview; however, you also have the advantage of being able to edit the worksheet. The worksheet is divided into pages with white space around the edges of each page showing the size of the margins and a ruler along the top and left of the column letters and row numbers. Pages and cells outside the active worksheet are grayed out; however, you can click any page or cell to add new data.

1. With the **SocialMediaStats-YourName** workbook open, click the SocialMediaWebsites sheet tab and then click E1.

2. Click the View tab and then click the Page Layout button in the Workbook Views group.

   Notice in Page Layout view you can see that the right pie chart is split over two pages.

3. Click the Page Layout tab, click the Orientation button in the Page Setup group, and then click *Landscape* at the drop-down list.

4. Click the *Width* list arrow in the Scale to Fit group (displays *Automatic*) and then click *1 page* at the drop-down list.

Inserting a Header or Footer

A header prints at the top of each page and a footer prints at the bottom of each page. Headers and footers are divided into three sections, with the left section left-aligned, the center section centered, and the right section right-aligned by default.

5. Click the dimmed text *Add header* near the top center of the page. **Hint:** *You may need to scroll up to see the Header pane.*

   The Header pane opens with three boxes in which you can type header text and/or add header and footer options, such as pictures, page numbering, the current date or time, and file or sheet names.

6. Type your first and last names.

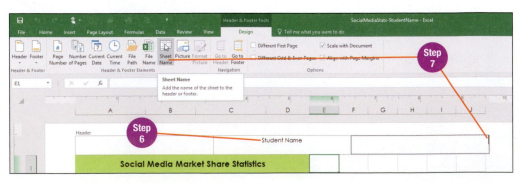

7 Click at the right of the center section text box in the Header pane to open the right section text box and then click the Sheet Name button in the Header & Footer Elements group on the Header & Footer Tools Design tab.

Excel inserts the code *&[Tab]*, which is replaced with the sheet tab name when you click outside the *Header* section.

8 Click at the left of the Header pane to open the left section text box and then click the File Name button in the Header & Footer Elements group.

Excel inserts the code *&[File]*, which is replaced with the file name when you click in the worksheet area.

9 Click any cell in the worksheet area.

Changing Margins

Worksheet margins are 0.75 inch top and bottom and 0.7 inch left and right with the header or footer printing 0.3 inch from the top or bottom of the page. Adjust margins to add more space around the edges of a page, or between the header and footer text and the worksheet. Center a smaller worksheet horizontally and/or vertically to improve the page appearance.

10 Click the Page Layout tab, click the Margins button in the Page Setup group, and then click *Wide*.

The *Wide* preset margin option changes the top, bottom, left, and right margins to 1 inch and the header and footer margins to 0.5 inch.

11 Click the Margins button and click *Custom Margins*.

12 Click the *Horizontally* and the *Vertically* check boxes in the *Center on page* section to insert a check mark in each box and then click OK.

13 Click the Facebook sheet tab, click A6, change to Page Layout view, and modify print options by completing steps similar to those in Steps 3 to 12 to improve the appearance of the printed worksheet.

14 Save the revised workbook using the same name (**SocialMediaStats-YourName**) and then close the workbook.

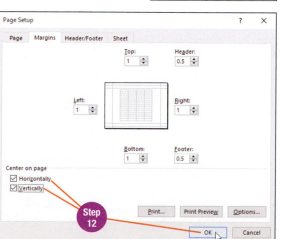

App Tip

Page Layout view is not available for chart sheets; however, you can add a header or change margins in Print Preview by using the Margins button or Page Setup hyperlink.

View
Model Answer
Compare your completed file with the model answer.

9.10 Creating and Modifying Sparklines and Inserting Comments

A **sparkline chart** is a miniature chart inserted into an individual cell within a worksheet. Sparkline charts are used to draw attention to trends or variations in data on a smaller scale than a column or line chart. Excel offers three types of sparkline charts: Line, Column, or Win/Loss.

A comment attached to a cell pops up when the reader points or clicks the cell. Comments are used to add explanatory information, pose questions, or provide other feedback to readers when a workbook is shared.

1. Open the workbook named *SchoolBudget*.

2. Use Save As to save a copy of the workbook as **SchoolBudget–YourName** in the Ch9 folder in CompletedTopicsByChapter.

3. Make K3 the active cell.

4. Click the Insert tab and then click the Column button in the Sparklines group.

5. Type b3:i3 in the *Data Range* text box at the Create Sparklines dialog box and then press Enter or click OK.

Excel embeds a column chart within the cell.

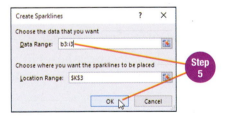

6. Use the fill handle to copy the sparklines column chart from K3 to K4:K11.

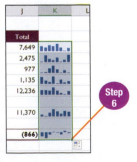

7 Click the *High Point* check box in the Show group on the Sparkline Tools Design tab to insert a check mark.

Excel highlights the bar in the column chart with the highest value by coloring it red. Other options in the tab are used to change the type or style, emphasize other points, show markers, or edit the data source.

Step 7

8 Make K2 the active cell, type Trend, and then click E9.

9 Click the Review tab and then click the New Comment button in the Comments group.

10 Type Assuming extra hours during Christmas break. and then click I3.

Step 10

Excel inserts a red triangle in the upper right corner of a cell to indicate a comment exists for the cell.

11 Click the New Comment button, type May be able to use last month's rent., and then click any cell.

12 Point to I3 with the mouse to display the comment in a pop-up box.

13 Click I3 to activate the cell.

14 Click the Edit Comment button in the Comments group and edit the comment text to *May be able to use last month's rent, which lowers this value to 55.*

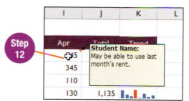

Step 12

15 Click any cell to finish editing the Comment box.

16 Click the Show All Comments button in the Comments group to display both comment boxes in the worksheet.

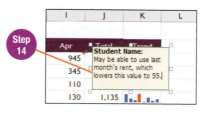

Step 14

17 Click the Show All Comments button to turn off the display of comment boxes.

18 Save the revised workbook using the same name (**SchoolBudget–YourName**) and then close the workbook.

View

Model Answer
Compare your completed file with the model answer.

Beyond Basics   **Printing Comments**

By default, comments do not print with a worksheet. You can choose to print a list of comments on a separate page after the worksheet prints or you can turn on the display of the comment boxes and print the worksheet with the comments as shown in the worksheet. Use the *Comments* option in the *Print* section of the Page Setup dialog box with the Sheet tab active to specify the print option.

9.11   **Working with Tables**

Skills

Format a range as a table

Sort a table

Filter a table

Tutorials

Formatting Data as a Table

Adding Rows to a Table

Sorting a Table

Filtering a Table

Filtering Data Using Conditional Formatting or Cell Attributes

Applying Conditional Formatting Using Quick Analysis

App Tip

Recall from Chapter 7 that different formatting applied to every other row is referred to as *banded rows* and is used to improve readability. The fill color, border style, and other options vary by table style.

Format a range of cells as a table to analyze, sort, and filter data as an independent unit. A worksheet can have more than one table, which means you can isolate and analyze data in groups. A table also allows you to choose from a variety of preformatted table styles, which is faster than manually formatting a range. Use tables for any block of data organized in a list format.

A **filter** temporarily hides any data that does not meet a criterion. Use filters to look at subsets of data without deleting rows in the table.

1. Open the workbook named *CalorieActivityTable*.

2. Use Save As to save a copy of the workbook as **CalorieActivityTable-YourName** in the Ch9 folder in CompletedTopicsByChapter.

3. Select A3:D23.

4. Click the Quick Analysis button, click the Tables tab, and then click the *Table* option.

5. Select A1:A2 and apply *White, Background 1, Darker 5%* fill color (first color in second row).

6. Make A4 the active cell.

7. Click the Table Tools Design tab and then click the *Table Style Medium 1* option in the *Table Styles* gallery (option to the left of the active style).

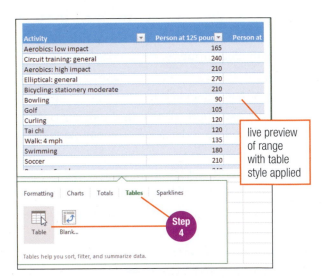

live preview of range with table style applied

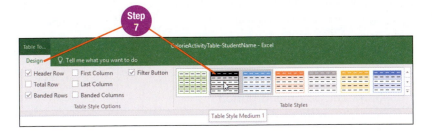

8. Click the filter arrow button at the top of the *Activity* column in the table (displays as down-pointing arrow).

9. Click *Sort A to Z* at the drop-down list.

The table rows are sorted in ascending order by the activity descriptions.

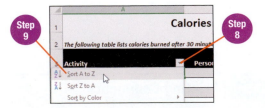

10. Click the filter arrow button at the top of the *Person at 155 pounds* column.

A check box is included for each unique value within the column. Filter a table by clearing check boxes for values or items you do not want to see in the filtered list or use the *Filter by Color* and *Number Filters* options to specify a filter condition.

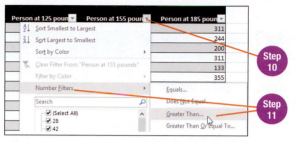

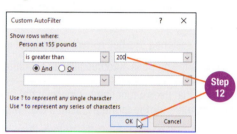

11 Point to *Number Filters* and then click *Greater Than*.

12 Type 200 at the Custom AutoFilter dialog box with the insertion point positioned in the text box at the right of *is greater than* and then click OK.

Excel filters the table and displays only those activities in which the calories burned are more than 200 for a person at 155 pounds.

13 Click the filter arrow button at the top of the *Person at 155 pounds* column and click *Clear Filter From "Person at 155 pounds"*.

Clearing a filter redisplays the entire table.

14 Change the orientation to landscape and center the worksheet horizontally.

15 Save the revised workbook using the same name (**CalorieActivityTable–YourName**) and then close the workbook.

Alternative Method   **Filtering a Table Using the Filter Slicer Pane**

A Slicer pane contains all the unique values for a column within the table. Click a value within the pane to filter the table. Use the Insert Slicer button in the Tools group on the Table Tools Design tab and insert a check mark for each column heading for which a Slicer pane is needed.

Beyond Basics   Conditional Formatting

Another tool that is used to highlight or review cells is conditional formatting. Conditional formatting applies formatting options only to cells that meet a specified criterion. Select a range, click the Quick Analysis button, and then click the desired conditional formatting button on the Formatting tab. More conditional formatting options are available from the Conditional Formatting button in the Styles group on the Home tab.

Topics Review

Topic	Key Concepts	Key Terms
9.1 Using Absolute Addresses and Range Names in Formulas	By default, a cell address in a formula is a relative address, which means the column letter or row number will update as the formula is copied relative to the destination column and row. A dollar symbol in front of a column letter or row number makes an address absolute, and means the reference will not update when the formula is copied. A formula that has both relative and absolute addresses is referred to as a *mixed address*. A descriptive label can be assigned to a cell or range and used in a formula. A name is assigned to a cell by typing the label in the Name box. A cell or range name is automatically an absolute address.	relative address absolute address mixed address
9.2 Entering Formulas Using Statistical Functions	A formula that uses a function begins with the function name followed by the parameters for the formula (called the *argument*) within parentheses. The AVERAGE function returns the arithmetic mean from the range used in the formula. The MAX function returns the largest value from the range. The MIN function returns the smallest value from the range. The Insert Function dialog box accessed from the Insert Function button provides tools to find and enter a function and argument. The *Count Numbers* option from the AutoSum drop-down list returns the number of cells with values in the range, while COUNTA returns a count of cells with values or text.	argument
9.3 Entering, Formatting, and Calculating Dates	A valid date or time entered into a cell is stored as a numerical value and can be used in formulas. The TODAY function enters the current date into the cell and updates the date whenever the worksheet is opened or printed. Date and time cells can be formatted to a variety of month, day, and year combinations at the Format Cells dialog box with the *Date* category selected. The format in which Excel expects a date to be entered is dependent on the Region setting in the Control Panel. In the United States, the date is expected to be in the format m/d/y.	
9.4 Using the IF Function	The IF function performs a comparison of a cell with a value of another cell and performs one of two calculations depending on whether the comparison proves true or false. Use the Insert Function dialog box to assist with entering an IF statement's arguments. The *logical_test* argument is the statement you want Excel to evaluate to determine which calculation to perform. The *value_if_true* argument is the value or formula if the logical test proves true. The *value_if_false* argument is the value or formula if the logical test proves false.	

continued…

Topic	Key Concepts	Key Terms
9.5 Using the PMT Function	Financial functions can be used for a variety of calculations that involve saving or borrowing money. The PMT function calculates a regular loan payment from a specified interest rate, term, and amount borrowed. Make sure the interest rate and terms are in the same units as the payment you want calculated. For example, divide the interest rate by 12 and/or multiply the term times 12 to calculate a monthly payment from an annual rate or terms. In the PMT argument, *Rate* means the interest rate, *Nper* means the term, and *Pv* means the amount borrowed. The FV function calculates the future value of a regular series of payments that earn a constant interest rate.	
9.6 Creating and Modifying a Pie Chart	Charts are often used to portray a visual snapshot of data. A pie chart shows each data point as a pie slice. The size of each slice in the pie chart represents the value of the data point in proportion to the total of all the values. Use the Charts tab in the *Quick Analysis* gallery to create a pie chart from a selected range. The Chart Elements button is used to add or modify a chart title, data labels, or legend.	pie chart
9.7 Creating and Modifying a Column Chart	A column chart shows one bar for each data point extending upward from a horizontal axis with the height of the bar representing its value. The horizontal axis in a column chart is the category axis, also called the *x-axis*, and shows the labels for each bar. The vertical axis in a column chart is called the *value axis*, also known as the *y-* or *z-axis*, and is scaled to the values of the bars graphed. A column chart is often used to illustrate trends or comparisons over time or by category. The Chart Styles button is used to choose a preformatted style for a column chart, or to change the color scheme. The *Axis Titles* option from the Chart Elements button is used to add titles to each axis in a column chart. The Recommended Charts feature provides a set of customized charts recommended for the data you have selected.	column chart category axis value axis
9.8 Creating and Modifying a Line Chart	A line chart helps a reader identify trends, growth spurts, dips, or unusual points in a data series. Use the Move Chart button in the Location group on the Chart Tools Design tab to move a chart from the worksheet into a chart sheet. A chart in a chart sheet is automatically scaled to fill a letter-sized page in landscape orientation. Change axis options in the Format Axis task pane, or data label options in the Format Data Labels task pane.	line chart Move Chart button

continued…

Topic	Key Concepts	Key Terms
9.9 Using Page Layout View, Adding a Header, and Changing Margins	In Page Layout view the worksheet is divided into pages with white space depicting the size of the margins, and a ruler along the top and left edges. You can see page layout and print options in Page Layout view while viewing and editing the worksheet. Add a header in Page Layout view by clicking the dimmed text *Add header*. Use buttons in the Header & Footer Tools Design tab to add options to a header or footer such as a picture, page numbering, date or time, or file or sheet names. Change to a preset set of margins from the Margins button in the Page Layout tab, or choose *Custom Margins* to enter your own margin settings. Open the Page Setup dialog box with the Margins tab active to center a worksheet horizontally and/or vertically.	Page Layout view
9.10 Creating and Modifying Sparklines and Inserting Comments	A miniature chart embedded into a cell is called a *sparkline chart*. Sparkline charts emphasize trends or variations in data on a smaller scale. Activate a cell and choose a Line, Column, or Win/Loss Sparkline chart from the Sparklines group on the Insert tab. Once created, add or modify sparkline options using buttons on the Sparkline Tools Design tab. A comment appears in a pop-up box when you point or click a cell with an attached comment. Excel displays a red triangle in a cell containing a comment. Use the New Comment button on the Review tab to add a comment in the active cell. To change the text in an existing comment, activate the cell containing the comment and use the Edit Comment button.	sparkline chart
9.11 Working with Tables	A block of data set up in list format can be formatted as a table for formatting, analyzing, sorting, or filtering purposes. A filter temporarily hides data that does not meet a criterion. Use a filter to review subsets of data without deleting rows. Click the filter arrow button at the top of a column to sort or filter a table. Click the filter arrow button at the top of a filtered column and then use the *Clear Filter From (column title)* option to redisplay the hidden rows in the table. Conditional formatting applies formatting options to cells within a range that meet a criterion.	filter

Recheck

Recheck your understanding of the topics covered in this chapter.

Workbook

Chapter review and assessment resources are available in the *Workbook* ebook.

Creating, Editing, and Formatting a PowerPoint Presentation

Precheck

Check your understanding of the topics covered in this chapter.

Presentations occur in meetings, seminars, and classrooms for a variety of purposes. Some presentations are informational, while others are designed to persuade you to buy a product or service. Some people use presentations at weddings, anniversaries, or family reunions to entertain an audience. In some organizations, presentations are used to provide information at a kiosk where a slide show is set up to run continuously for individuals to view as they walk by or enter a booth. For example, at a trade show, a company might provide a slide show with information about a product. At school you may have used a presentation as a study guide to prepare for an exam.

A presentation is made up of a collection of slides (referred to as a **slide deck**) containing text and multimedia. PowerPoint is the **presentation application** in the Microsoft Office suite. The program is widely used to create a slide deck for presentations. In this chapter, you will learn how to create, edit, and format a presentation. You will create a presentation with a variety of text-based slide layouts; edit content and placeholders; move, duplicate, and delete slides; format slides using a variety of techniques; add notes and comments; and preview the presentation as a slide show. Finally, you will preview options for audience and speaker handouts.

Note: If you are using a tablet, consider connecting it to a USB or wireless keyboard because you will be typing text in several slides for new presentations in this chapter.

Learning Objectives

10.1 Create a new presentation based on a theme, insert slides, and add content to slides

10.2 Change the design theme and theme variant, and insert a table

10.3 Format text using font and paragraph options

10.4 Create a slide with the comparison layout and select, resize, align, and move slide placeholders

10.5 Use Slide Sorter view and duplicate, move, and delete slides

10.6 Modify the slide master

10.7 Add notes and comments to slides

10.8 Run a presentation in Slide Show view and Presenter view

10.9 Prepare slides for audience handouts or speaker notes

If you are a SNAP user, go to your SNAP Assignments page to complete the Precheck, Tutorials, and Recheck.

Data Files

Before beginning this chapter, be sure you have copied the student data files for this course to your storage medium. Steps on downloading and extracting the data files are provided in Chapter 1, Topic 1.8, on pages 22–23.

10.1 Creating a New Presentation and Inserting Slides

Skills

Create a new presentation

Choose a theme and
theme variant

Insert a slide

Edit text on a slide

Tutorials

Opening a Presentation
Based on a Template

Exploring the
PowerPoint Screen

Navigating to Slides

Choosing a Slide
Layout

Changing Views

Changing Slide Layout

Inserting and Deleting
Text in Slides

Changing the Display
of a Slide in the Slide
Pane

Inserting a New Slide

Selecting Text

Checking Spelling

App Tip

Double-click a theme to
start a new presentation
using the theme default
style and color scheme.

Begin creating a new presentation at the PowerPoint Start screen by choosing a template, a theme and variant on a theme, or by starting with a blank presentation. The first slide in a presentation is a **title slide** with a text **placeholder** for a title and a subtitle. A placeholder is a rectangular container on a slide that can hold text or other content. Each placeholder on a slide can be manipulated independently.

PowerPoint starts a new presentation with a title slide displayed in Normal view. In Normal view, the current slide displays in widescreen format in the **slide pane**. Numbered slide thumbnails display in the **slide thumbnail pane** at the left of the current slide. A notes pane at the bottom of the slide pane and a Comments pane at the right of the slide pane can be opened as needed.

1 Start PowerPoint 2016.

2 At the PowerPoint Start screen, click the *Ion* theme.

PowerPoint 2016 starts with a gallery of design themes. Click to preview a theme along with the theme variants. A **variant** is a different style and color scheme included in the theme family.

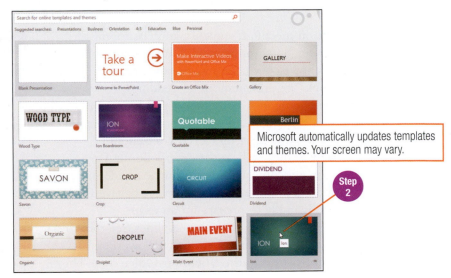

Microsoft automatically updates templates and themes. Your screen may vary.

3 Click the last variant (orange color scheme), and then click the right-pointing arrow below the preview slide next to *More Images*.

When previewing a variant, browse through the *More Images* slides to view the color scheme with a variety of content. This allows you to get a better perspective of the theme or variant style and colors before making your selection.

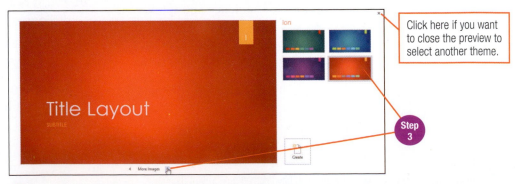

Click here if you want to close the preview to select another theme.

4 Click the second variant (blue color scheme), and then click the right-pointing arrow below the preview slide to view the blue color scheme with a Title and Content layout depicting a chart.

5 Click the right-pointing arrow below the preview slide two more times to view other types of content with the blue color scheme.

6 With the Photo Layout preview displayed, click the Create button.

Quick Steps

Start a New Presentation
1. Start PowerPoint 2016.
2. Click theme.
3. Click variant.
4. Click Create button.

Insert a Slide
Click New Slide button in Slides group.
OR
1. Click down-pointing arrow on New Slide button.
2. Click required slide layout.

Edit Text
1. Activate slide.
2. Select text or click in placeholder and move insertion point as needed.
3. Type new text or change text as required.

7 Compare your screen with the one shown in Figure 10.1.

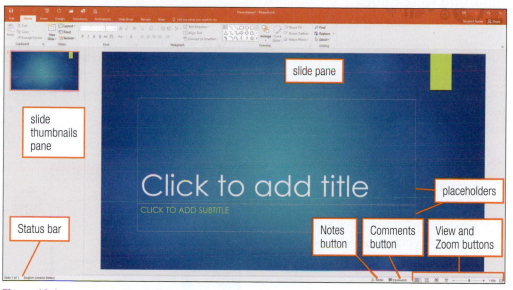

Figure 10.1

A new PowerPoint presentation with Ion theme and blue color variant in the default Normal view is shown above. See Table 10.1 for a description of screen elements.

Table 10.1

PowerPoint Features

Feature	Description
Comments button	Button to turn on or turn off Comments pane at the right side of the slide pane.
Notes button	Button to turn on or turn off the notes pane at the bottom of the slide pane.
placeholders	Containers in which you type or edit text, or insert other content such as an image or audio clip.
slide pane	Pane that displays the active slide. Add or edit content on a slide in this area.
slide thumbnails pane	Pane that displays numbered thumbnails of the slides in the presentation. Navigate to, insert, delete, or duplicate a slide in this pane.
Status bar	Bar that displays active slide number with total number of slides in the presentation and displays a message about an action in progress.
View and Zoom buttons	Buttons to change the display of the PowerPoint window. View buttons in order are: Normal, Slide Sorter, Reading View, and Slide Show. Zoom buttons enlarge or shrink the display of the active slide.

8 Click anywhere in *Click to add title* in the title slide on the slide pane and then type Car Maintenance.

9 Click anywhere in *CLICK TO ADD SUBTITLE* in the title slide on the slide pane and then type Tips for all seasons.

The subtitle text displays in all capital letters regardless of the case used when you type the text because the Ion theme uses the All Caps font effect for the subtitle text.

10 Click in an unused area of the slide to deactivate the subtitle placeholder.

Inserting New Slides

The **New Slide button** in the Slides group on the Home tab is used to insert a new slide following the active slide. The button has two parts. Clicking the top part of the button adds a new slide with the Title and Content layout, which is the layout used most frequently. The content placeholder in this layout provides options to add text or a table, chart, SmartArt graphic, picture, or video to the slide. Clicking the bottom of the New Slide button (down-pointing arrow) provides a drop-down list of slide layouts and other new slide options. A **slide layout** is an arrangement of placeholders that determine the number, position, and type of content placeholders included in a slide.

11 Click the top part of the New Slide button in the Slides group on the Home tab.

12 Click anywhere in *Click to add title* in the title placeholder and then type Why maintain a car?.

13 Click anywhere in *Click to add text* in the content placeholder, type Preserve vehicle value, and then press Enter.

Typing text in the content placeholder automatically creates a bulleted list. In the Ion theme, the bullet character is a green, right-pointing arrow.

14 Type the remaining bulleted list items, pressing Enter after each item except the last one.

Prolong vehicle life

Improve driver safety

Spend less for repairs

Lower operating costs

Improve vehicle appearance

Reduce likelihood of breakdowns

15. Click the top part of the New Slide button in the Slides group.

16. Type the text in the third slide as shown in the image below.

Step 16

Editing Text on Slides

Activate the slide you want to edit by clicking the slide in the slide thumbnails pane. Select the text you want to change or delete or click in the placeholder to place an insertion point at the location where you want to edit text and then type new text, change text, or delete text as needed.

17. Click to select Slide 1 in the slide thumbnails pane.

18. Select ALL SEASONS in the subtitle text placeholder and then type car owners so that the subtitle text now reads TIPS FOR CAR OWNERS.

Step 17

Step 18

19. Click to select Slide 2 in the slide thumbnails pane.

20. Click at the beginning of the text in the third bulleted list item, delete *Improve*, and then type Sustain.

Step 20

21. Click to place an insertion point within the title placeholder and edit the title text so that *m* in *maintain* and *c* in *car* are capital letters. The title should now read Why Maintain a Car?

22. Click in an unused area of the slide to deactivate the title placeholder.

23. Save the presentation as **CarMaintenance-YourName** in a new folder named *Ch10* in the CompletedTopicsByChapter folder on your storage medium. Leave the presentation open for the next topic.

10.2 Changing the Theme and Inserting and Modifying a Table

Skills

Change theme

Change variant

Insert a table on a slide

Modify table layout

Tutorial
Changing and
Modifying Design
Themes

The presentation theme and/or variant can be changed after a presentation has been created. To do this, click the Design tab and browse the themes and theme variants in the Themes and Variants galleries.

1. With the **CarMaintenance-YourName** presentation open, click Slide 1 in the slide thumbnails pane.

2. Click the Design tab.

3. Click the *Facet* theme in the Themes gallery (third option).

When deciding upon a theme, roll the mouse over the various theme options to view the active slide with a live preview of the theme. Changing the theme after slides have been created may cause some changes in capitalization in placeholders. For example, a theme that uses the All Caps font in a title or subtitle may mean that you have to do some corrections after the theme is changed if the new theme does not use the All Caps font.

4. Click the More button (⟱) located at the bottom right of the Variants gallery.

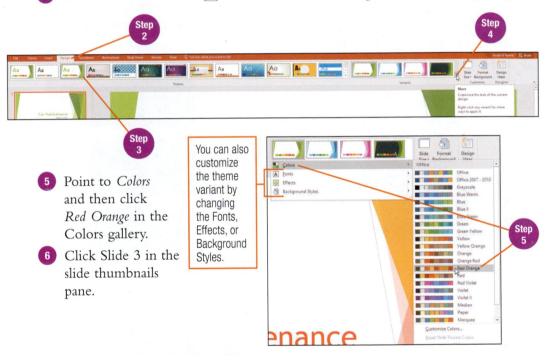

5. Point to *Colors* and then click *Red Orange* in the Colors gallery.

6. Click Slide 3 in the slide thumbnails pane.

You can also customize the theme variant by changing the Fonts, Effects, or Background Styles.

Inserting a Table on a Slide

PowerPoint includes a Table feature for organizing text on a slide in columns and rows similar to the Table feature in Word. To insert a table on a slide, click the Insert Table button in the content placeholder.

7. Click the Home tab and then click the top part of the New Slide button to insert a new slide with the Title and Content layout.

8. Click anywhere in *Click to add title* in the title placeholder and then type Typical Annual Maintenance Costs.

9. Click the Insert Table button in the content placeholder.

10. Select *5* in the *Number of columns* text box at the Insert Table dialog box and then type 2.

11 Select *2* in the *Number of rows* text box, type 6 and then click OK.

PowerPoint inserts a table in the slide with the colors in the theme variant.

Insert Table ? ✕

Number of columns: 2

Number of rows: 6

OK   Cancel

Step 10

Step 11

12 With the insertion point positioned in the first cell in the table, type Type of Car and then press Tab or click in the second cell.

13 Type Cost.

14 Type the remaining entries in the table by pressing Tab to move to the next cell or by clicking in the next cell and then typing the text as follows:

Small size	$600
Medium size	$675
Large family sedan	$750
Minivan	$775
SUV	$825

Modifying a Table

The Table Tools Design and Layout tabs provide options for modifying and customizing a table with the same tools you learned about in Word. Use the sizing handles to enlarge or shrink the table size. Drag the border of a table to move the table to a new position on the slide.

15 Drag the right middle sizing handle to the left until the right border of the table ends approximately below the first *e* in *Maintenance* in the title text as shown in the image below.

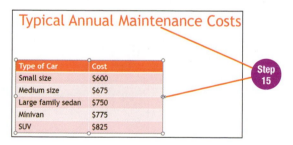

Typical Annual Maintenance Costs

Type of Car	Cost
Small size	$600
Medium size	$675
Large family sedan	$750
Minivan	$775
SUV	$825

Step 15

16 Click in any cell in the second column of the table.

17 Click the Table Tools Layout tab, click the Select button in the Table group, and then click *Select Column* at the drop-down list.

18 Click the Center button in the Alignment group.

19 Click in any cell in the first column of the table, select the current entry in the *Width* text box in the Cell Size group, type 3.5, and then press Enter.

20 Drag the top border of the table to move the table until it is positioned at the approximate location shown in the image at right.

21 Save the revised presentation using the same name (**CarMaintenance-YourName**). Leave the presentation open for the next topic.

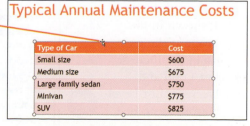

Step 20

Typical Annual Maintenance Costs

Type of Car	Cost
Small size	$600
Medium size	$675
Large family sedan	$750
Minivan	$775
SUV	$825

Quick Steps

Change Theme
1. Click Design tab.
2. Click desired option in Themes gallery.
3. If desired, click variant option.

Insert a Table on a Slide
1. Insert new slide.
2. Click Insert Table button in content placeholder.
3. Enter number of columns.
4. Enter number of rows.
5. Click OK.

Oops! !

Pressed Tab after last entry and added a new row by mistake? Click the Undo button to remove the extra blank row.

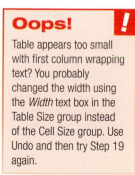

Oops! !

Table appears too small with first column wrapping text? You probably changed the width using the *Width* text box in the Table Size group instead of the Cell Size group. Use Undo and then try Step 19 again.

10.3 Formatting Text with Font and Paragraph Options

Skills

Create a multilevel
bulleted list

Change the font color

Center text in a
placeholder

Tutorial
Increasing and
Decreasing Indent

Font and paragraph formatting options in PowerPoint are the same as those in Word and Excel. Select text within a placeholder and apply a formatting option to only the selected text or select a placeholder and apply a formatting option to all the text in the placeholder.

A multilevel bulleted list is created using the **Increase List Level button** and the **Decrease List Level button** in the Paragraph group on the Home tab. Each time you click the Increase List Level button, the insertion point or text is indented to the next tab and the bullet character changes to indicate the text is being demoted to the next list level. Use the Decrease List Level button to move the insertion point or text back to the previous tab and promote the text to the previous list level.

1. With the **CarMaintenance-YourName** presentation open, click Slide 3 in the slide thumbnails pane.

2. Insert a new slide with the Title and Content layout.

New slides are inserted after the active slide. The new slide should be positioned between the Fall and Winter Maintenance slide and the Typical Annual Maintenance Costs slide.

3. With the new Slide 4 the active slide, type Spring and Summer Maintenance as the slide title.

4. Type Thoroughly clean vehicle as the first bulleted list item in the content placeholder and then press Enter.

5. With the insertion point positioned at the beginning of the second bulleted list item, click the Increase List Level button in the Paragraph group on the Home tab.

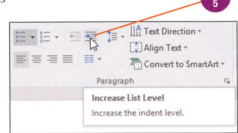

Step 5

6. Type Prevent rust by removing sand and salt accumulated from winter driving and then press Enter.

7. With the insertion point positioned at the beginning of the third bulleted list item, click the Decrease List Level button in the Paragraph group to move the bullet back to the previous level, type Check cooling system, and then press Enter.

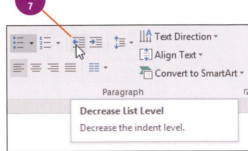

Step 7

8. Type the remaining text on the slide as shown in the image at the right using the Increase List Level and Decrease List Level buttons as needed.

Steps 4–8

App Tip

A bulleted list can have up to eight levels.

App Tip

You can also press Tab to increase the list level and Shift + Tab to decrease the list level.

Good to Know

When preparing a bulleted list, consider the size of room and number of people attending the presentation to make sure the slide will be readable by everyone in the room. If necessary, divide the content over two slides rather than cramming too much text on one slide.

Spring and Summer Maintenance

- Thoroughly clean vehicle
    - Prevent rust by removing sand and salt accumulated from winter driving
- Check cooling system
- Make sure all lights and turn signals are working properly
    - Check the small bulb above the license plate
- Check brakes
    - Salt from winter driving can damage brakes
    - Brakes should be checked every 12,000 miles (19,000 kilometers)
- Switch to summer tires
    - Check tire condition

9   Select *12,000* in the content placeholder, click the Font Color button arrow in the Mini toolbar or in the Font group on the Home tab and then click *Red, Accent 1* (fifth option in first row of *Theme Colors* section).

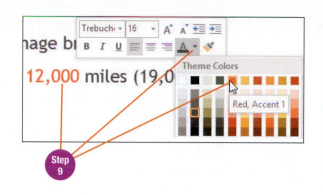

10   Click the Bold button in the Font group.

11   Click in the title text to activate the title placeholder.

12   Click the Center button in the Paragraph group.

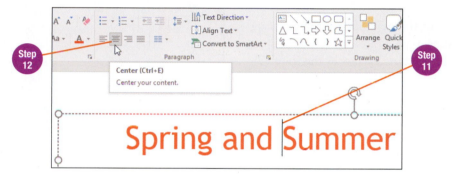

13   Click the Align Left button in the Paragraph group to return the title placeholder to the default paragraph alignment.

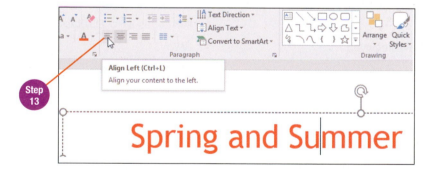

14   Save the revised presentation using the same name (**CarMaintenance-YourName**). Leave the presentation open for the next topic.

Beyond Basics   Changing the Bullet Symbol

The bullet symbols vary with each theme; however, you can change the bullet character to another symbol from the Bullets button arrow in the Paragraph group on the Home tab for an individual list item or for an entire list. To access more bullet symbol options, click *Bullets and Numbering* at the Bullets drop-down list to open the Bullets and Numbering dialog box. Choose the Picture button in the dialog box to select a bullet image or the Customize button to open a Symbol dialog box from which you can select a bullet character. You can also change the size and color of the bullet character in the Bullets and Numbering dialog box.

10.4 Selecting, Resizing, Aligning, and Moving Placeholders

Skills

Insert a slide with the comparison layout

Change a bulleted list to a numbered list

Resize a placeholder

Align a placeholder

Move a placeholder

Tutorial
Modifying Placeholders

The active placeholder displays with a border and selection handles, which are used to resize or move the placeholder. Paragraph or font options apply to the text in which the insertion point is positioned or to selected text. To apply a font or paragraph change to all the text in a placeholder, click the placeholder border to select the placeholder, which also removes the insertion point or deselects text.

1. With the **CarMaintenance-YourName** presentation open, click Slide 5 in the slide thumbnails pane.

2. Click the down-pointing arrow on the New Slide button and then click *Comparison* at the drop-down list.

3. With the new Slide 6 the active slide, type Top 5 Cars Rated by Maintenance Costs as the slide title.

4. Type the title and bulleted list text in the left and right content placeholders as shown in the image below.

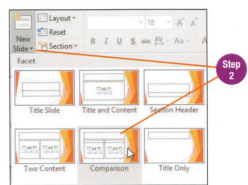

Least Expensive	Most Expensive
▶ Honda Fit	▶ Nissan GT-R
▶ Toyota Corolla	▶ Chevrolet Corvette
▶ Toyota Yaris	▶ Mercedes-Benz SL-Class
▶ Chevrolet Aveo	▶ BMW Z4
▶ Ford Focus	▶ Chevrolet Camaro

5. Click anywhere in the bulleted list below the heading *Least Expensive* to activate the placeholder.

6. Click anywhere along the border of the active placeholder to remove the insertion point, selecting the entire placeholder.

The placeholder border changes to a solid line from a dashed line when the entire placeholder is selected. The next action will affect all the text within the placeholder.

7. Click the Numbering button in the Paragraph group to change the bulleted list to a numbered list. (Do *not* click the down-pointing arrow on the button.)

Oops! !

Only one bullet changed to a number? This occurs if you have an insertion point within the placeholder; only the item in the list at which the insertion point was positioned is changed. Go back to Step 6 and try again.

8 Click anywhere in the bulleted list below *Most Expensive*, click along the border of the active placeholder to select the entire placeholder, and then click the Numbering button.

9 Select the numbered list placeholder below the title *Least Expensive*.

10 Drag the right middle sizing handle to the left until the right border of the placeholder is at the approximate location shown in the image below.

11 Select the *Least Expensive* title placeholder and then drag the right middle sizing handle to the left to resize the placeholder until the smart guide appears, indicating the title placeholder is the same width as the content placeholder below it.

Smart guides, also called *alignment guides*, appear automatically when moving or resizing objects. A **smart guide** is a colored horizontal and/or vertical guideline that helps you align, space, or size placeholders or objects evenly.

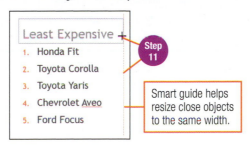

Smart guide helps resize close objects to the same width.

12 With the *Least Expensive* placeholder still selected, drag the border of the placeholder right when the pointer displays with the move icon attached to move the placeholder until the smart guides appear as shown in the image at right.

13 Select the numbered list placeholder below *Least Expensive* and drag right until left, right, top, and bottom smart guides appear, indicating the placeholder is aligned evenly with the placeholders above and right.

Smart guides help align and evenly space close objects.

14 Save the revised presentation using the same name (**CarMaintenance-YourName**). Leave the presentation open for the next topic.

Quick Steps

Resize a Placeholder
1. Select placeholder.
2. Drag sizing handle as needed.

Move a Placeholder
1. Select placeholder.
2. Drag placeholder border as needed.

App Tip

The AutoFit feature, which is on by default, automatically scales the font size and adjusts spacing between points to fit text within a placeholder.

Oops!

No right middle sizing handle? On a smaller-screened tablet, the right middle sizing handle may not appear. Use the top right or bottom right sizing handle instead. Remove your finger when you see two smart guides—one at the bottom and one at the right.

10.5 Using Slide Sorter View and Moving, Duplicating, and Deleting Slides

Slide Sorter view displays all the slides in a presentation as slide thumbnails. Change to Slide Sorter view to perform slide management tasks. For example, you can easily rearrange the order of the slides by dragging a slide thumbnail to a new position within the slide deck. Select a slide in Slide Sorter view or Normal view to duplicate or delete the slide.

1. With the **CarMaintenance-YourName** presentation open, click the View tab and then click the Slide Sorter button in the Presentation Views group.

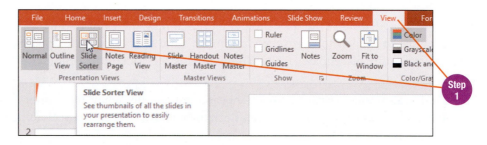

2. Click Slide 5 to select the slide.

3. Drag Slide 5 to place the slide to the right of Slide 2, and then release the mouse.

As you drag to move a slide in Slide Sorter view, the existing slides rearrange around the slide and are automatically renumbered.

The pointer displays with a dimmed box attached when you drag to move a slide.

Duplicating a Slide

When you need to create a new slide with the same layout as an existing slide and with the placeholders sized, aligned, and positioned the same, make a duplicate copy of the existing slide. Once the slide is duplicated, all you have to do is change the text inside the placeholders. A duplicated slide is inserted in the presentation immediately after the slide selected to be duplicated.

4. Click to select Slide 6.

5. Right-click Slide 6 to display the shortcut menu.

6. Click *Duplicate Slide* at the shortcut menu.

7. Double-click Slide 7 to return to Normal view.

Deleting a Slide

Slides can be deleted in Slide Sorter view or Normal view by selecting the slide, displaying the shortcut menu, and then choosing *Delete Slide* or by pressing the Delete key on the keyboard. Multiple slides can be deleted all at once. To do this, begin by holding down the Ctrl key while clicking each slide you want to remove. When all slides to be deleted have been selected, right-click any selected slide and then choose *Delete Slide* at the shortcut menu.

8 Right-click Slide 7 in the slide thumbnails pane to display the shortcut menu.

9 Click *Delete Slide* at the shortcut menu.

10 Save the revised presentation using the same name (**CarMaintenance–YourName**). Leave the presentation open for the next topic.

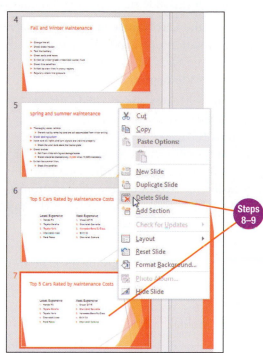

Quick Steps

Slide Sorter View
Click Slide Sorter button in Status bar. OR
1. Click View tab.
2. Click Slide Sorter button.

Move a Slide
Drag slide in Slide Sorter view to required location.

Duplicate a Slide
1. Select slide.
2. Display shortcut menu.
3. Click *Duplicate Slide*.

Delete a Slide
1. Select slide.
2. Display shortcut menu.
3. Click *Delete Slide*.

Alternative Method	**Moving or Duplicating Slides**
Move slide	In Normal view, drag slide up or down slide thumbnails pane.
Duplicate slide	Select slide, click down-pointing arrow on New Slide button in Slides group on Home tab, and then click *Duplicate Selected Slides*.

Beyond Basics   **Hiding a Slide**

You may have a slide in a presentation that you want to hide during a particular slide show because the slide does not apply to the current audience or provides more detail than you have time to explain. In Slide Sorter view or Normal view, right-click the slide to be hidden and choose *Hide Slide* from the shortcut menu. A hidden slide is dimmed and has a diagonal line drawn through the slide number in the slide thumbnails pane or in Slide Sorter view. Hide Slide is an on or off feature—redisplay a hidden slide by choosing the Hide Slide option again.

10.6 Modifying the Slide Master

Each presentation that you create includes a slide master. A **slide master** determines the default formatting and paragraph options for placeholders when you insert new slides. If you want to make a change to a font or paragraph option for the entire presentation, making the change in the slide master will apply the change automatically to all slides in the presentation. For example, if you want a different font color for all the slide titles, change the color on the slide master.

1 With the **CarMaintenance-YourName** presentation open and the View tab active, click the Slide Master button in the Master Views group.

In **Slide Master view**, a slide master at the top of the hierarchy in the slide thumbnails pane controls the font, colors, paragraph options, and background for the entire presentation. Below the slide master is a variety of layouts for the presentation. Changes made to the slide master at the top of the hierarchy affect all the slide layouts below it except the title slide.

2 Scroll up the slide thumbnails pane to the first slide at the top of the hierarchy.

3 Click to select Slide 1.

4 Click the border of the Master title placeholder on the slide master to select the placeholder.

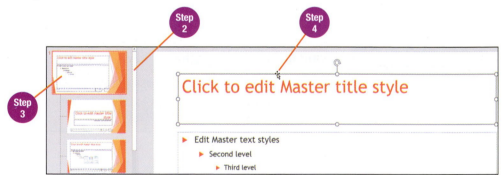

5 Click the Home tab, click the Font Color button arrow, and then click *Dark Red, Accent 6* (last color in first row of *Theme Colors* section).

6 Click the border of the content placeholder to select the placeholder and then click the Bullets button arrow in the Paragraph group.

7 Click *Bullets and Numbering* at the drop-down list.

8 Click the Color button in the Bullets and Numbering dialog box and then click *Dark Red, Accent 6*.

9 Click the *Hollow Square Bullets* option (first option in second row).

10 Click OK.

Quick Steps
Slide Master View
1. Click View tab.
2. Click Slide Master button.

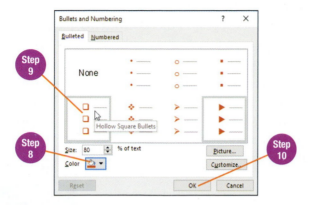

11 Click the Slide Master tab.

12 Click the Close Master View button in the Close group.

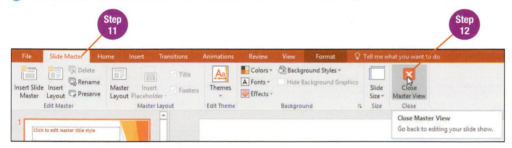

13 Click the Previous Slide or Next Slide buttons at the bottom of the vertical scroll bar or click each slide in the slide thumbnails pane to scroll through and view each slide in the presentation.

Notice that the font color for the title text and the bullet character are changed on each slide *after* the title slide. A title slide has its own slide master and is the first layout below slide 1 in the slide master hierarchy.

App Tip

Page Up and Page Down also display the previous or next slide in the presentation.

14 Save the revised presentation using the same name (**CarMaintenance-YourName**). Leave the presentation open for the next topic.

Beyond Basics   **Adding Text to the Bottom of Each Slide**

Add text to the bottom of each slide in a footer placeholder by selecting the *Footer* check box and typing footer text at the Header and Footer dialog box with the Slides tab active (click Insert tab, then click Header & Footer button in the Text group). Use the slide master to format the footer text to a different font, font size, or color as required. For example, many speakers use the Footer placeholder to add a company name or presentation title at the bottom of each slide.

10.7 Adding Notes and Comments

Skills

Display the notes and Comments panes

Add speaker notes

Add comments

**Tutorials**

Customizing the Handout Master

Customizing the Notes Master

Notes, generally referred to as *speaker notes*, are text typed in the **notes pane** below the slide pane in Normal view. Use notes to type reminders for the presenter or use this pane to add more details about the slide content for the person giving the presentation. In a presentation designed to be used as a self-study aid, notes are used to provide more explanation to the learner.

Comments added to slides appear in the **Comments pane** at the right side of the slide pane. If you are creating a presentation with a group of people, use comments to provide feedback or pose questions to others in the group.

1 With the **CarMaintenance–YourName** presentation open, display Slide 1 in the slide pane.

2 Click the Notes button in the Status bar to turn on the display of the notes pane at the bottom of the slide pane. Skip this step if the notes pane is already visible.

3 Click anywhere in *Click to add notes* in the notes pane and then type Begin this slide with the statistic that approximately 5.2% of motor vehicle accidents are caused by vehicle neglect..

Begin this slide with the statistic that approximately 5.2% of motor vehicle accidents are caused by vehicle neglect.

App Tip

Text in the notes pane is visible to the presenter but not the audience during a slide show.

4 Display Slide 3 in the slide pane.

5 Drag the top border of the notes pane upward to increase the height of the pane by approximately one-half inch.

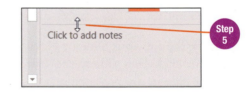

6 Click anywhere in *Click to add notes* in the notes pane and then type Mention that these costs are estimated for a driving distance of 12,000 miles (19,000 kilometers) per year..

7 Press Enter twice and then type Ask the audience if anyone wants to share the total amount paid each year to maintain his or her vehicle..

Mention that these costs are estimated for a driving distance of 12,000 miles (19,000 kilometers) per year.

Ask the audience if anyone wants to share the total amount paid each year to maintain his or her vehicle.

8 Click the Notes button to turn off the notes pane.

9 Click the Comments button in the Status bar to turn on the Comments pane at the right side of the slide pane.

10 Click the New button near the top of the Comments pane.

PowerPoint opens a comment box in the Comments pane with your account name associated with the comment.

11 Type Consider adding the source of these statistics to the slide..

12 Click Slide 4 in the slide thumbnails pane, click the New button in the Comments pane, and then type Add more information for any of these points?.

13 Click in the Comments pane outside the Comment box to close the comment.

14 Close the Comments pane.

A comment balloon appears in the top left corner of a slide for which a comment has been added.

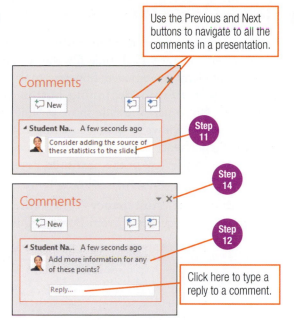

Use the Previous and Next buttons to navigate to all the comments in a presentation.

Step 11

Step 14

Step 12

Click here to type a reply to a comment.

Quick Steps

Add Notes
1. Activate required slide.
2. Click Notes button.
3. Type note text in notes pane.

Add Comments
1. Activate required slide.
2. Click Comments button.
3. Click New button.
4. Type comment text.

App Tip

A prompt appears in the Status bar when you open a presentation that has comments, informing you of their existence in the file.

Comment balloon displays on slides with comments. Click the balloon to open the Comments pane, and view the comments and replies.

Fall and Winter Maintenance

15 Save the revised presentation using the same name (**CarMaintenance-YourName**). Leave the presentation open for the next topic.

Alternative Method   **Adding a Comment to Selected Text on a Slide**

A comment can also be added to selected text on a slide. The comment balloon displays at the end of the selected text. To do this, select the text and reveal the Comments pane or click the New Comment button in the Comments group on the Review tab.

Beyond Basics   **Deleting or Hiding Comments**

Pointing to a comment in the Comments pane displays a Delete icon (black ⊠) at the top right of the comment. Click the Delete icon when it appears to remove a comment. To hide comment balloons, click the Show Comments button in the Comments group on the Review tab and then click *Show Markup* at the drop-down list to remove the check mark.

10.8 Displaying a Slide Show

Skills

Display a presentation
in Slide Show view

Display a presentation
in Presenter view

Tutorials

Running a Slide Show

Changing the Display
when Running a Slide
Show

Using the Pen Tool
during a Slide Show

Display the presentation in **Slide Show view** to preview the slides as they will appear to an audience. Each slide fills the screen with the ribbon and other PowerPoint elements removed; however, tools to navigate and annotate slides are available. Use the **From Beginning button** in the Start Slide Show group on the Slide Show tab to start the slide show at Slide 1.

In **Presenter view**, the slide show displays full screen on one monitor (the monitor the audience will see) and in Presenter view on a second monitor. Presenter view displays a preview of the next slide, notes from the notes pane, a timer, and a slide show toolbar along with other options.

1 With the **CarMaintenance-YourName** presentation open, click the Slide Show tab, and then click the From Beginning button in the Start Slide Show group.

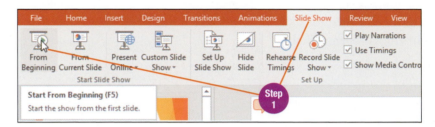

App Tip

Press F5 to start a slide show from Slide 1.

Oops!

Using Touch? Tap the slide to display the Slide Show toolbar.

2 Click the right-pointing arrow that appears in the Slide Show toolbar near the bottom left corner of the screen to move to Slide 2. Move the mouse to display the toolbar if the Slide Show toolbar is not visible.

You can also click anywhere on a slide or press the Page Down key to move to the next slide. The buttons on the Slide Show toolbar are shown in Figure 10.2 and described in Table 10.2 below.

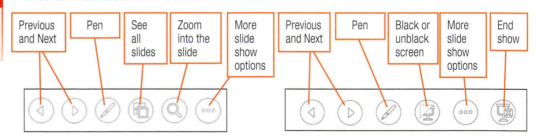

Figure 10.2

The Slide Show toolbar for a mouse-enabled device (left), and for a touch-enabled device (right) is shown above. See Table 10.2 for a description of each button.

Table 10.2

Slide Show Toolbar Buttons

Button	Description
Previous and Next	Displays the previous or next slide in the presentation.
Pen	Displays a pop-up list of options for using the laser pointer, pen, or highlighter when running the presentation.
See all slides	View all slides in the presentation similarly to Slide Sorter view. Use this option to jump to a slide out of sequence during a presentation.
Zoom into the slide	Use this button to click on a portion of a slide that you want to enlarge to temporarily fill the screen for a closer look. Right-click or press the Esc key to restore the slide.
More slide show options	Displays a pop-up list of options for customizing the presentation. On mouse-enabled devices, use this button to end the show or black/unblack the screen during a presentation.
Black or unblack screen	On a touch-enabled device, this button blacks the screen or unblacks the screen.
End show	On a touch-enabled device, use this button to end the show.

3 Continue clicking the Next Slide arrow to navigate through the remaining slides in the presentation until the black screen appears.

After the last slide is viewed, a black screen is shown with the message *End of slide show, click to exit.* Many presenters leave the screen black when their presentation is ended until the audience has left because clicking to exit displays the presentation in Normal view on the screen.

4 At the black screen that appears, click anywhere on the screen to return to Normal view.

5 Display Slide 1 in the slide pane and then click the Slide Show button in the Status bar.

The **Slide Show button** in the Status bar starts the slide show at the active slide.

Step 5

6 Click the More slide show options button (button with three dots) in the Slide Show toolbar and then click *Show Presenter View* at the pop-up list.

You can use Presenter view on a system with only one monitor to preview or rehearse a presentation. At a presentation venue, PowerPoint automatically detects the computer setup and chooses the correct monitor on which to show Presenter view.

7 Click the Next Slide button in the slide navigator near the bottom of Presenter view until you have navigated to Slide 3 (see Figure 10.3).

8 Compare your screen with the one shown in Figure 10.3.

9 Continue clicking the Next Slide button until you reach Slide 6 and then at the top of the screen click END SLIDE SHOW at the top of Presenter view.

10 Leave the presentation open for the next topic.

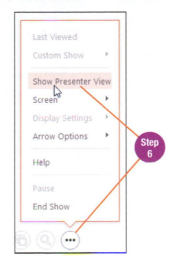
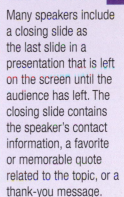

Last Viewed
Custom Show ▶
Show Presenter View
Screen ▶
Display Settings ▶
Arrow Options ▶
Help
Pause
End Show

Step 6

Good to Know

Many speakers include a closing slide as the last slide in a presentation that is left on the screen until the audience has left. The closing slide contains the speaker's contact information, a favorite or memorable quote related to the topic, or a thank-you message.

App Tip

You can also preview a slide show in Reading view where each slide fills the screen. A Title bar, Status bar, and Taskbar remain visible with buttons to navigate slides in the Status bar.

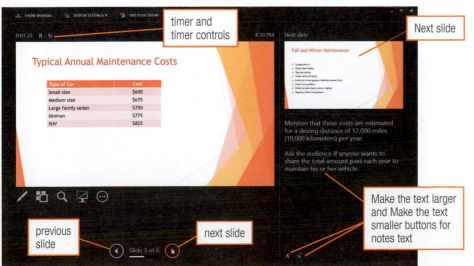

timer and timer controls

Next slide

previous slide

next slide

Make the text larger and Make the text smaller buttons for notes text

Figure 10.3
Shown above is Slide 3 of the Car Maintenance-YourName presentation in Presenter View

10.9 Preparing Audience Handouts and Speaker Notes

Skills

Preview slides as handouts

Hide comments on printouts

Preview notes pages

Add header and footer text

Tutorial
Previewing Slides and Printing

Some speakers provide audience members with a printout of their slides in a format that allows an individual to add his or her own handwritten notes during the presentation. PowerPoint provides several options for printing slides as handouts. Speakers who do not use Presenter view during a presentation may also print a copy of the slides with the notes included for reference during the presentation.

1. With the **CarMaintenance-YourName** presentation open, click the File tab and then click *Print*.

2. At the Print backstage area, click the Full Page Slides list arrow in the *Settings* category.

3. Click *3 Slides* in the *Handouts* section of the drop-down list.

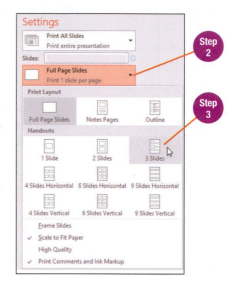

The option to print three slides per page provides horizontal lines next to each slide for writing notes.

4. Click the 3 Slides list arrow and then click *6 Slides Horizontal* at the drop-down list.

Notice that the printout requires two pages even though only six slides are in the presentation. By default, comments print with the presentation; the second page is for printing the comments.

App Tip

Preview various handout options before making a selection to make sure printouts will be legible for most people when slides have a lot of detailed content.

5. Click the 6 Slides Horizontal list arrow and then click *Print Comments and Ink Markup* at the drop-down list to remove the check mark. The printout is now only one page.

6. Click the 6 Slides Horizontal list arrow, and click *Notes Pages* in the *Print Layout* section of the drop-down or pop-up list.

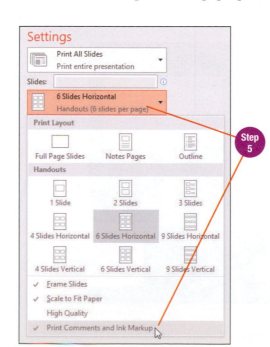

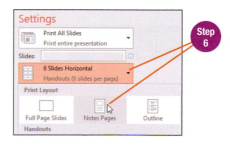

Notes Pages prints one slide per page with the slide at the top half of the page and notes or blank space in the bottom half.

7. Click the Next Page button to display Slide 2 in the *Preview* section.

8. Click the Next Page button to display Slide 3.

   Notice the notes text is displayed below the slide.

9. Click the <u>Edit Header & Footer</u> hyperlink at the bottom of the *Settings* section.

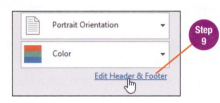

10. At the Header and Footer dialog box with the Notes and Handouts tab active, click the *Header* check box to insert a check mark, click in the *Header* text box, and then type your first and last names.

11. Click the *Footer* check box to insert a check mark, click in the *Footer* text box, and then type your school name.

12. Click the Apply to All button.

Click here to print the date at the top right of each page. By default, the date updates to the date the slides are printed; choose *Fixed* to enter a specific date.

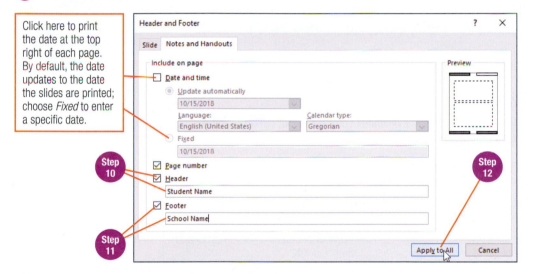

13. Preview the header and footer text by scrolling through the remaining slides.

14. Click the Back button to exit the Print backstage area.

15. Save the revised presentation using the same name (**CarMaintenance-YourName**) and then close the presentation.

Good to Know

To conserve paper and ink, some presenters publish their presentations to a web service such as <u>SlideShare</u> instead of printing handouts.

Beyond Basics   **Slide Size and Orientation for Printing**

By default, slides print in landscape orientation when printed as slides or in portrait orientation when printed as notes or handouts. Change the orientation with the orientation list arrow in the *Settings* category of the Print backstage area or at the *Custom Slide Size* option from the Slide Size button in the Customize group on the Design tab.

Topics Review

Topic	Key Concepts	Key Terms
10.1 Creating a New Presentation and Inserting Slides	A slide deck is a collection of slides in a presentation. A presentation application is a software program used to create a slide deck. PowerPoint is the presentation application in the Microsoft Office suite. A new presentation is created from the PowerPoint Start screen by choosing a template, a theme, or the blank presentation option. PowerPoint starts a new presentation with a title slide in widescreen format in Normal view. A title slide is the first slide in a presentation. A placeholder is a rectangular container in which you type text or insert other content. Normal view includes the slide pane, which displays the active slide on the right side of the screen, and the slide thumbnails pane, which displays numbered thumbnails of all the slides in a single column along the left side of the screen. A variant of a theme family is based upon the same theme, but with different colors, styles, and effects. Add or edit text on a slide by clicking inside the placeholder and then typing or editing text. Use the New Slide button to add a slide to the presentation. A variety of slide layouts is available. Each slide layout option sets the number, placement, and type of placeholders on a slide.	slide deck presentation application title slide placeholder slide pane slide thumbnail pane variant New Slide button slide layout
10.2 Changing the Theme and Inserting and Modifying a Table	Change the theme and/or variant for a presentation after the presentation has been started using options on the Design tab. Insert a table on a slide using the Insert Table button located within the content placeholder of a new slide. At the Insert Table dialog box, type the number of columns for the table and the number of rows for the table, and then click OK. Use the same methods and tools for entering text and modifying the table layout as the methods and tools that you learned in Word.	
10.3 Formatting Text with Font and Paragraph Options	Select text within a placeholder or select the entire placeholder to apply formatting changes using the options in the Font and Paragraph groups on the Home tab. Create a multilevel bulleted list using the Increase List Level button and the Decrease List Level button in the Paragraph group. Each time you click the Increase List Level button, the insertion point or text is indented to the next tab, and the level changes to the next level. Click the Decrease List Level button to move the insertion point or text back to the previous tab and level. The bullet symbols vary for each theme. You can change the bullet symbol character using the Bullets button arrow.	Increase List Level button Decrease List Level button

continued…

Topic	Key Concepts	Key Terms
10.4 Selecting, Resizing, Aligning, and Moving Placeholders	The active placeholder displays with sizing handles and a border with which you can resize or move the placeholder.	smart guide
	Click the border of a placeholder to remove the insertion point and select the entire placeholder to apply a formatting change.	
	A smart guide is a colored line that appears on the slide as you resize or move a placeholder to assist in aligning the placeholder or evenly spacing the placeholder with other close objects.	
10.5 Using Slide Sorter View and Moving, Duplicating, and Deleting Slides	Slide Sorter view displays all slides as slide thumbnails and is used to rearrange the order of slides or otherwise manage slides in the presentation.	Slide Sorter view
	Slide or drag a slide in Slide Sorter view to move the slide to a new position within the presentation.	
	Duplicating a slide makes a copy of an existing slide with the placeholders sized, aligned, and positioned the same as the original slide.	
	Delete a slide you no longer need in a presentation by selecting the slide or slides and using the Delete Slide option from the shortcut menu or by pressing the Delete key.	
	A slide can be hidden in the presentation if you do not want the slide to display in a slide show.	
10.6 Modifying the Slide Master	Each presentation has a slide master that determines the formatting and paragraph options for placeholders in slides.	slide master
	Display the slide master to make formatting changes that you want to apply to all slides in the presentation automatically.	Slide Master view
	Change to Slide Master view from the View tab to modify the slide master.	
	In Slide Master view, the slide thumbnails pane displays the slide master at the top of the hierarchy.	
	Below the slide master, individual slide layouts are included for formatting separately from the slide master.	
	Add text to the bottom of each slide in a footer placeholder by typing footer text in the Header and Footer dialog box (click Insert tab, then click Header & Footer button in Text group).	
	Use the footer placeholder on the slide master to apply formatting options to the footer text.	
10.7 Adding Notes and Comments	The notes pane appears along the bottom of the slide pane and is used to type speaker notes, reminders for the presenter, or more detailed information about the slide content for a reader.	notes pane
	Reveal or hide the notes pane with the Notes button in the Status bar.	Comments pane
	A comment is added to the active slide by displaying the Comments pane, clicking the New button, and then typing the comment text.	
	Reveal or hide the Comments pane with the Comments button in the Status bar.	
	Delete a comment using the Delete icon that appears when you point to the comment text in the Comments pane.	
	Hide comments by removing the check mark next to *Show Markup* at the Show Comments button drop-down list on the Review tab.	

continued…

Topic	Key Concepts	Key Terms
10.8 Displaying a Slide Show	Slide Show view previews each slide as the audience will see the slide with a full screen.	Slide Show view
	Display a slide show starting at Slide 1 by clicking the From Beginning button in the Start Slide Show group on the Slide Show tab.	From Beginning button
		Presenter view
	The Slide Show toolbar provides buttons to navigate slides, annotate slides, zoom into a slide, or black/unblack the screen during the presentation.	Slide Show button
	After the last slide is shown, a black screen displays indicating the end of the slide show.	
	The Slide Show button in the Status bar starts the slide show from the active slide in the slide pane.	
	Display a slide show and click the More slide show options button to switch the view to Presenter view.	
	Presenter view works with two monitors, where one monitor displays the slide show as the audience will see it, and the second monitor displays the slide show in Presenter view.	
	Presenter view can also be seen on a computer with only one monitor so that you can rehearse a presentation.	
	In Presenter view, the speaker's monitor displays a timer and timer controls, a preview of the next slide, notes, and a slide show toolbar along with other options.	
10.9 Preparing Audience Handouts and Speaker Notes	Preview slides formatted as handouts at the Print backstage area.	Notes Pages
	The *3 Slides* Handouts option in the *Settings* category at the Print backstage area provides lines next to each slide for writing notes.	
	Various other horizontal or vertical options are available in the *Settings* category of the Print backstage area for printing slide thumbnails as a handout.	
	By default, comments print on a separate page after the slides; to prevent comments from printing, remove the check mark next to *Print Comments and Ink Markup* from the slides option list at the Print backstage area.	
	Choose the *Notes Pages* option at the Print backstage area to print one slide per page with the notes from the notes pane.	
	Add header and/or footer text to a printout using the Edit Header & Footer hyperlink at the bottom of the *Settings* category at the Print backstage area.	
	By default, slides printed as slides print in landscape orientation, while slides printed as handouts or notes pages print in portrait orientation.	

 Recheck

Recheck your understanding of the topics covered in this chapter.

 **Workbook**

Chapter review and assessment resources are available in the *Workbook* ebook.

Chapter 11

Enhancing a Presentation with Multimedia and Animation Effects

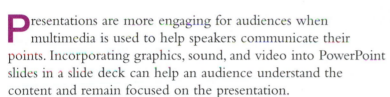

 Precheck

Check your understanding of the topics covered in this chapter.

P resentations are more engaging for audiences when multimedia is used to help speakers communicate their points. Incorporating graphics, sound, and video into PowerPoint slides in a slide deck can help an audience understand the content and remain focused on the presentation.

In this chapter, you will learn how to add graphics to slides using clip art, pictures, SmartArt, WordArt, charts, and drawn shapes. You will also learn to add text in a text box; add sound and video; and complete a slide show presentation by adding transitions and animation effects. Lastly, you will learn how to set up a slide show that advances through the slide deck automatically.

Learning Objectives

11.1 Insert and resize pictures and clip art on a slide

11.2 Insert and modify a SmartArt graphic on a slide

11.3 Convert existing text to a SmartArt graphic and insert and modify a WordArt object on a slide

11.4 Create and modify a chart on a slide

11.5 Draw and modify shapes and text boxes on a slide

11.6 Insert a video clip into a presentation

11.7 Insert a sound clip into a presentation

11.8 Add transition and animation effects into a slide show

11.9 Set up a self-running presentation

 If you are a SNAP user, go to your SNAP Assignments page to complete the Precheck, Tutorials, and Recheck.

 Data Files

Before beginning this chapter, be sure you have copied the student data files for this course to your storage medium. Steps on downloading and extracting the data files are provided in Chapter 1, Topic 1.8, on pages 22–23.

11.1 Inserting Graphic Images from Picture Collections

Adding a picture, illustration, diagram, or chart on a slide emphasizes content, adds visual interest to a slide, and helps an audience understand and make connections with the information more easily than with text alone. As you did in Chapter 7, you can insert pictures from a file on your PC or from an online resource.

Inserting Pictures from a File on Your Computer

To add a picture to an existing slide, use the Pictures button in the Images group on the Insert tab to add an image stored as a file in a folder on your computer, or a computer to which you are connected. Once inserted, move, resize, and/or modify the picture using buttons in the Picture Tools Format tab. On a new slide, use the Pictures icon in the content placeholder to add a picture to a slide.

1 Start PowerPoint 2016 and open the presentation **PaintedBunting** from the Ch11 folder in Student_Data_Files.

2 Use Save As to save a copy of the presentation as **PaintedBunting-YourName** in a new folder *Ch11* in the CompletedTopicsByChapter folder.

3 Browse through the presentation and read the slides.

4 Make Slide 2 the active slide in the slide pane.

5 Click the Insert tab and then click the Pictures button in the Images group.

6 At the Insert Picture dialog box, navigate to the Ch11 folder within Student_Data_Files and then double-click the file *PaintedBunting_NPS*.

7 Using one of the four corner selection handles, resize the image smaller to the approximate size shown in the image below.

8 Drag the image to move the picture to the right side of the slide and then align it with the horizontal and vertical smart guides that appear when the picture is even with the top of the text and the right margin on the slide.

Step 5

Align picture with horizontal and vertical smart guides.

Steps 7-8

Painted Bunting Facts

- *Passerina ciris*
- Medium-sized finch
  - Male sports a colorful plumage
  - Female has distinctive green plumage
- Eastern and western populations
  - North and South Carolina, Georgia, Florida
  - Kansas and Missouri south to Texas

9 Insert the picture *PaintedBunting_Female* near the bottom right of the slide as shown on the next page by completing steps similar to Steps 5 to 8.

- *Passerina ciris*
- Medium-sized finch
  - Male sports a colorful plumage
  - Female has distinctive green plumage
- Eastern and western populations
  - North and South Carolina, Georgia, Florida
  - Kansas and Missouri south to Texas
- Habitat
  - Prefer thick vegetation next to open, grassy areas
- Winter migration to Cuba, Mexico, and South America

Step 9

Quick Steps

Insert an Image from the Computer
1. Activate slide.
2. Click Insert tab.
3. Click Pictures button.
4. Navigate to drive and/or folder.
5. Double-click desired image.
6. Resize and/or move as required.

Insert Clip Art or Online Photo
1. Activate slide.
2. Click Insert tab.
3. Click Online Pictures button.
4. Type search keyword.
5. Press Enter.
6. Double-click desired image.
7. Resize and/or move as required.

Inserting a Picture from a Web Resource

The Online Pictures button is in the Images group on the Insert tab or in a content placeholder. Use the Online Pictures button to find a suitable image from a website.

10 Make Slide 6 the active slide in the slide pane.

11 Click the Insert tab and then click the Online Pictures button in the Images group.

12 With the insertion point positioned in the search text box to the right of the *Bing Image Search* option, type team and then press Enter or click the Search button (displays as a magnifying glass).

13 Double-click the picture shown in the image below. If you cannot locate the image shown in the search results list, close the Insert Pictures dialog box and insert the picture from a file in the Ch11 folder in Student_Data_Files using the file *Team*.

14 Resize and move the picture to the left side of the bullet list and align it with the horizontal smart guide that appears when the center of the image is evenly positioned with the bullet list.

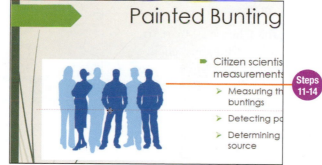

Painted Bunting

- Citizen scientis[t] measurements
  - Measuring th[e] buntings
  - Detecting po[...]
  - Determining source

Steps 11-14

15 Save the revised presentation using the same name (**PaintedBunting-YourName**). Leave the presentation open for the next topic.

Beyond Basics   **Editing Images**

Edit an image with buttons on the Picture Tools Format tab using techniques similar to those you learned in Chapter 7. For example, you can apply a picture style or artistic effect; adjust the brightness, contrast, or sharpness; or change the color properties. Use buttons in the Arrange and Size groups to layer the image with other objects, specify the position of the image on the slide, crop unwanted portions of the picture, or specify measurements for height and width.

11.2 Inserting a SmartArt Graphic

Skills

Add a SmartArt graphic

Modify a SmartArt graphic

**Tutorials**

Inserting, Sizing, and Moving SmartArt

Formatting SmartArt

SmartArt is a graphic object that visually communicates a relationship in a list, process, cycle, hierarchy, or some other diagram. Begin creating a SmartArt graphic by choosing a predesigned layout and then adding text in the Text pane or by typing text directly in the text placeholders within the shapes. You can add and delete shapes to the graphic as needed and choose from a variety of color schemes and styles. See Table 11.1 for a description of layout category diagrams created using SmartArt.

Table 11.1

SmartArt Graphic Layout Categories

Layout Category	Description
List	Nonsequential tasks, processes, or other list items
Process	Illustrate a sequential series of steps to complete a process or task
Cycle	Show a sequence of steps or tasks in a circular or looped process
Hierarchy	Show an organizational chart or decision tree
Relationship	Show how parts or elements are related to one another
Matrix	Depict how individual parts or ideas relate to a whole or central idea
Pyramid	Show proportional or hierarchical relationships that build upward
Picture	Add pictures inside shapes with small amounts of text to show ideas, a process, or a relationship

App Tip

On a new slide with no other content, use the Insert a SmartArt Graphic icon in the content placeholder to create a SmartArt object on a slide.

App Tip

If you are unsure which SmartArt graphic to use, click a layout in the center pane so you can read a description of it in the right pane including suggested usage.

1. With the **PaintedBunting-YourName** presentation open and Slide 6 the active slide in the slide pane, click the Insert tab if the Insert tab is not the active tab.

2. Click the SmartArt button in the Illustrations group.

3. At the Choose a SmartArt Graphic dialog box, click *Process* in the Category pane at the left, click *Basic Chevron Process* in the layout pane in the center (second option in fifth row), and then click OK.

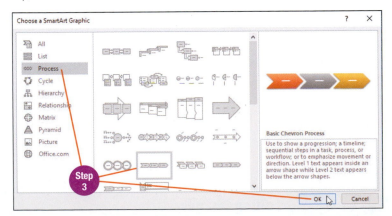

PowerPoint places the SmartArt graphic in the center of the slide. Three shapes are automatically included in the *Basic Chevron Process* layout.

4 Click the left-pointing arrow along the left border of the graphic if the Text pane is not visible and then click in the pane next to the first bullet; if the Text pane is already visible, proceed to Step 5.

5 With the insertion point in the Text pane next to the first bullet, type Band; click next to the second bullet and type Observe; and click next to the third bullet and then type Analyze.

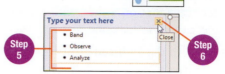

The SmartArt graphic updates as each word is typed in the Text pane to show the text in the shape. You can also add text to the shapes by typing text directly within the text placeholders inside each shape.

6 Click the Close button to close the Text pane.

7 If necessary, click the SmartArt Tools Design tab.

8 Click the More button at the bottom right of the SmartArt Styles gallery.

9 Click *Polished* at the drop-down gallery (first option in *3-D* section).

10 Click the Change Colors button in the SmartArt Styles group and then click *Colorful – Accent Colors* at the drop-down gallery (first option in the *Colorful* section).

11 Drag the border of the SmartArt graphic until the diagram is positioned near the bottom center of the slide, as shown in Figure 11.1.

12 Click in an unused area of the slide to deselect the graphic.

13 Save the revised presentation using the same name (**PaintedBunting-YourName**). Leave the presentation open for the next topic.

App Tip

Press the Down Arrow key to move to the next bullet in the Text pane.

Quick Steps

Insert a SmartArt Graphic
1. Activate slide.
2. Click Insert tab.
3. Click SmartArt button.
4. Select category in left pane.
5. Select layout in center pane.
6. Click OK.
7. Add text in Text pane or in shapes.
8. Format and/or move graphic object as required.

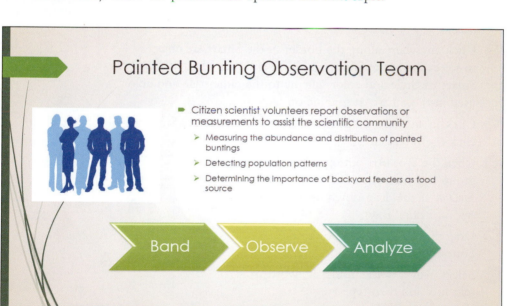

Figure 11.1
The completed Slide 6 with SmartArt graphic is shown here.

Beyond Basics | **Modifying a SmartArt Graphic**

Use buttons in the Create Graphic group on the SmartArt Tools Design tab to add shapes, change the direction of the layout (switch between *Right to Left* and *Left to Right*), and move shapes up or down the layout. Each shape in the layout can also be selected and moved or resized individually.

11.3 Converting Text to SmartArt and Inserting WordArt

An existing bullet list on a slide can be converted to a SmartArt graphic using the **Convert to SmartArt button** in the Paragraph group on the Home tab. **WordArt** is text that is created and formatted as a graphic object. With WordArt you can create decorative text on a slide with a variety of WordArt Styles and text effects. A WordArt object can also have the text formed around a variety of shapes.

1. With the **PaintedBunting–YourName** presentation open, make Slide 5 the active slide in the slide pane.

2. Click in the bullet list to activate the placeholder.

3. Click the Convert to SmartArt button in the Paragraph group on the Home tab.

4. Click *Hierarchy List* at the drop-down gallery (second option in second row).

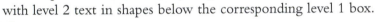
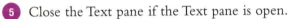

PowerPoint converts the text in the bullet list into the selected SmartArt layout. Level 1 text from the bullet list is placed inside shapes at the top level in the hierarchy diagram, with level 2 text in shapes below the corresponding level 1 box.

5. Close the Text pane if the Text pane is open.

6. Select and delete *Need to* in the second shape in the top level of the hierarchy, and capitalize *m* so that the text inside the shape reads *Manage and preserve natural habitat*.

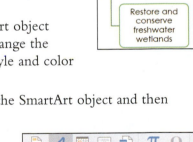

7. Click anywhere along the border of the SmartArt object to select the entire SmartArt object and then change the SmartArt Style and color scheme to the same style and color used in the SmartArt graphic on Slide 6.

8. Click in an unused area of the slide to deselect the SmartArt object and then click the Insert tab.

9. Click the WordArt button in the Text group and then click *Fill – Green, Accent 1, Shadow* at the drop-down list (second option in first row).

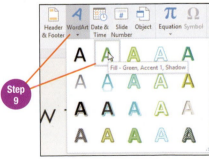

10. Drag the WordArt text box to the bottom center of the slide.

11. Select *Your text here* inside the WordArt text box and then type Help Save the Painted Bunting!.

12 Click the border of the WordArt text box to remove the insertion point and select the entire placeholder.

13 Click the Text Effects button in the WordArt Styles group on the Drawing Tools Format tab, point to *Glow*, and then click *Lime, 5 pt glow, Accent color 3* (third option in first row of *Glow Variations* section).

Glow: 5 points lime, Accent color 3

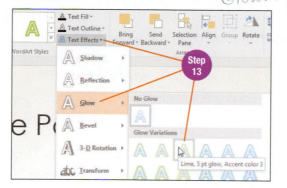

14 Drag the border of the WordArt text box until the smart guide appears, indicating the object is aligned with the center of the object above, as shown in Figure 11.2.

15 Save the revised presentation using the same name (**PaintedBunting-YourName**). Leave the presentation open for the next topic.

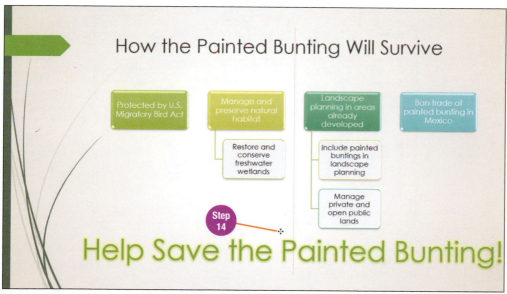

Figure 11.2
Smart guides aid in the alignment of the WordArt object with the center of the SmartArt object.

Use the *Transform* option from the Text Effects drop-down list to choose a shape around which WordArt text is formed. Text can be shaped to follow a circular or semicircular path, be slanted, or otherwise be altered to create a distinctive effect. Experiment with options in the Shape Styles gallery to add a rectangular box around the WordArt.

11.4 Creating a Chart on a Slide

Skills

Create a chart on a slide

Modify the chart style and color

**Tutorials**

Creating a Chart

Formatting with Chart Buttons

Changing Chart Design

Changing Chart Format

Copying and Pasting between Programs

Charts similar to the ones you created with Excel in Chapter 9 can be added to a PowerPoint slide. Add a chart using the Insert Chart icon in a content placeholder or with the Chart button in the Illustrations group on the Insert tab. Charts are commonly used in presentations to show an audience dollar figures, targets, budgets, comparisons, patterns, trends, or variations in numerical data.

1. With the **PaintedBunting–YourName** presentation open, make Slide 3 the active slide in the slide pane.

2. Click the Insert Chart icon in the content placeholder.

3. At the Insert Chart dialog box with *Column* selected in the All Charts category list and *Clustered Column* selected as the chart type, click OK.

PowerPoint creates a sample chart on the slide and opens a chart data grid into which the data to be graphed is typed. As you enter labels and values into the chart data grid, the chart on the slide updates.

4. With B1 in the chart data grid the active cell, type 2016.

5. Click C1 and then type 2017.

6. Type the remaining data in the cells in the chart data grid as shown in the image below.

7. Close the chart data grid.

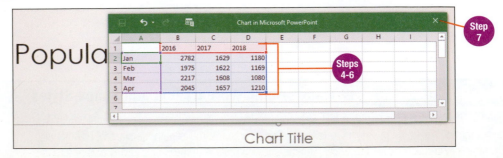

8 Select the text *Chart Title* inside the chart object and then type Monthly Sightings January to April.

9 Click the border of the chart to select the entire chart.

10 Click *Style 4* (fourth option) in the Chart Styles gallery on the Chart Tools Design tab.

11 Click *Color 2* (second row in *Colorful* section) in the Change Colors drop-down list.   → colorful Palette 2

12 Click outside the chart to deselect the chart object and then compare your slide with the one shown in Figure 11.3.

13 Save the revised presentation using the same name (**PaintedBunting-YourName**). Leave the presentation open for the next topic.

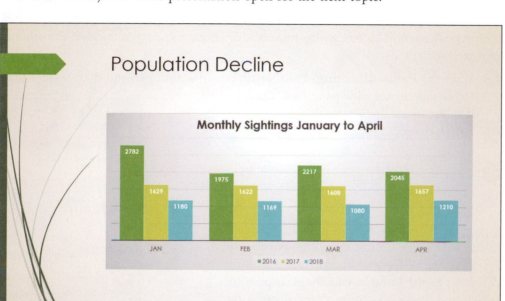

Figure 11.3
Your slide should match the completed Slide 3 with Clustered Column Chart.

> **Quick Steps**
> **Insert a Chart**
> 1. Activate slide.
> 2. Click Insert Chart icon in content placeholder.
> 3. Choose category and chart type.
> 4. Click OK.
> 5. Add data in chart data grid.
> 6. Close data grid.
> 7. Format chart as required.

Alternative Method **Copying and Pasting a Chart from Excel to PowerPoint**

Sometimes the data needed to create a chart resides in an Excel worksheet. In that case, create the chart in Excel and paste a copy of the chart onto the PowerPoint slide. You will practice this method in Chapter 14.

Beyond Basics   **Formatting Charts and Editing Data**

Use the same tools you learned in Excel to modify and format charts in PowerPoint, such as the Chart Elements and Chart Styles buttons at the top right of the chart or with buttons on the Chart Tools Design and Chart Tools Format tabs.

To make a change to the source data for the chart, redisplay the chart data grid by activating the chart and then clicking the top part of the Edit Data button in the Data group on the Chart Tools Design tab.

11.5   Drawing a Shape and Adding a Text Box

Skills

Draw and modify
a shape on a slide

Add text inside a shape

Format a shape

Create a text box on a
slide

Tutorials

Formatting Shapes

Displaying Rulers,
Gridlines, and Guides;
Copying and Rotating
Shapes

Inserting and
Formatting Text Boxes

Inserting Action Buttons

App Tip

When drawing other
shapes, hold down the
Shift key while dragging
the mouse to create a
perfect square, circle, or
straight line.

App Tip

Use the yellow handles
that appear for a selected
shape to change the
appearance of the shape.
For example, the yellow
handle at the top of the
arrow can be used to
change the length of the
arrowhead.

A graphic can be added to a slide by drawing a line, rectangle, circle, arrow, star, banner, or other shape. Once the shape is drawn, text can be added inside the shape, and the shape can be formatted by changing the outline color or fill color or by adding a visual effect. Draw a shape by clicking the **Shapes button** in the Illustrations group on the Insert tab and then selecting the type of shape from the drop-down list. Click the slide where you want the shape to appear to insert a default sized shape or drag the crosshairs to create the desired shape.

A text box is text inside a rectangular object that can be manipulated independently from other objects on a slide. Add a text box using the Text Box button in the Text group on the Insert tab or from the Insert Shapes group on the Drawing Tools Format tab. Drag the downward pointing arrow the approximate length and width desired or click the slide where you want the text box to begin and then type the text. The box width automatically expands to accommodate the amount of text typed.

1. With the **PaintedBunting-YourName** presentation open and with Slide 3 the active slide in the slide pane, click the Insert tab if Insert is not the active tab.

2. Click the Shapes button in the Illustrations group.

3. Click the *Striped Right Arrow* shape (fifth option in second row of the *Block Arrows* section). ~~Arrow's Striped Right~~

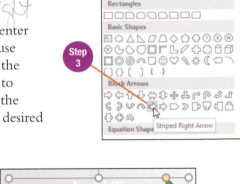

For touch users, a shape is placed in the center of the slide at the default shape size. For mouse users, crosshairs display. A mouse user moves the crosshairs to the location where the shape is to appear and then clicks to insert the shape at the default shape size, or drags the crosshairs the desired height and width.

4. Click inside the chart near the JAN bar for 2017 (with the value *1629*); if you are using touch, proceed to Step 5.

5. With the shape selected, type A 41% decline!

6. Drag the middle right sizing handle to the right until the text fits on one line inside the arrow shape.

7. Drag the rotation handle (circled arrow above top center sizing handle) in an upward diagonal direction toward the left until the shape is at the approximate angle shown in the image at the right.

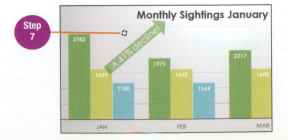

8 Drag the border of the arrow shape down to the bottom left of the chart so that it points to the 2017 January bar, as shown in the image at right.

9 With the arrow shape still selected and the Drawing Tools Format tab active, click the More button at the bottom right of the Shape Styles gallery, and then click *Intense Effect – Turquoise, Accent 6* (last option in last row of *Theme Styles* section in the drop-down gallery).

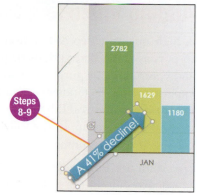

10 Click the Insert tab and then click the Text Box button in the Text group.

11 Click anywhere at the left side of the slide below the chart to insert a text box with an insertion point; if you are using touch, proceed to Step 12.

12 Type Source: Painted Bunting Observer Team, University of North Carolina, Wilmington.

13 Click the border of the text box to remove the insertion point and select the entire placeholder, click the Home tab if the Home tab is not active, and then click the Italic button in the Font group.

14 Move and/or resize the text box, aligning the text box with the bottom of the chart as shown in the image below.

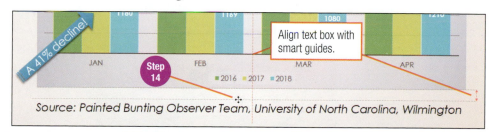

15 Save the revised presentation using the same name (**PaintedBunting-YourName**). Leave the presentation open for the next topic.

App Tip

Click the View tab and then click the *Gridlines* check box in the Show group to insert a check mark and display evenly spaced horizontal and vertical dotted lines on the slide to assist with placing objects at precise locations.

Beyond Basics   **Adding an Action Button to a Slide**

A category of shapes called *Action Buttons* contains a series of buttons with actions assigned that are used to create a navigation interface or launch other items during a slide show. For example, draw the *Action Button: Home* button (shown at the right) on a slide to move to the first slide in the slide deck when the button is clicked during a slide show.

Home Action Button

11.6 Adding Video to a Presentation

Skills

Insert a video from a file

Trim the video

Set video playback options

Tutorial
Inserting and Modifying Video Files

A high-quality video can demonstrate a process or task that is otherwise difficult to portray using descriptions or pictures. Video is widely used for instructional and entertainment purposes. Appropriately used, video provides a more enjoyable experience for the audience. You can play a video from a file stored on your PC or link to a video at YouTube or another online source. Use the **Trim Video button** in the Editing group on the Video Tools Playback tab to crop a portion of the video playing at the beginning or end of the video clip.

1. With the **PaintedBunting–YourName** presentation open, make Slide 6 the active slide in the slide pane.

2. Insert a new slide with the Title and Content layout and then type A Beautiful Bird as the slide title.

3. Click the Insert Video icon in the content placeholder.

4. At the Insert Video dialog box, click the <u>Browse From a file</u> hyperlink in the *From a file* section.

5. Navigate to the Ch11 folder in Student_Data_Files and then double-click the file *PaintedBuntingVideo*.

6. Click the Play/Pause button below the video to preview the video clip.

    The video plays for approximately 52 seconds.

App Tip

PowerPoint recognizes most video file formats, such as QuickTime movies, MP4 videos, MPEG movie files, Windows Media Video (wmv) files, and Adobe Flash Media.

7. Click the Video Tools Playback tab.

8. Click the Trim Video button in the Editing group.

Trimming a video allows you to show only a portion of a video file if the video is too long or if you do not wish to show parts at the beginning or end. Drag the green or red slider to start playing at a later starting point and/or end before the video is finished, or enter the start and end times at the Trim Video dialog box.

9 At the Trim Video dialog box, select the current entry in the *Start Time* text box and then type 00:10.

10 Select the current entry in the *End Time* text box and then type 00:30.

11 Click the Play button to preview the shorter video clip.

12 Click OK.

13 Click the *Start* list box arrow (displays *On Click*) in the Video Options group and then click *Automatically* at the drop-down list.

The *Automatically* option means the video will begin playing as soon as the slide is displayed in the slide show.

14 Drag the video object left until the smart guide appears at the left, indicating the object is aligned with the slide title.

15 Click the Video Tools Format tab and then click the *Soft Edge Rectangle* option in the Video Styles group (third option).

16 Save the revised image using the same name (**PaintedBunting-YourName**). Leave the presentation open for the next topic.

Alternative Method	**Adding Video to an Existing Slide and Playing a Video from YouTube**

Add video to the active slide by clicking the Insert tab, clicking the Video button in the Media group, and then choosing *Online Video* or *Video on my PC* at the drop-down list.

Link to a video at YouTube by displaying the Insert Video dialog box, typing the name of the video and pressing Enter in the *YouTube* text box, and then double-clicking the desired video in the search results list.

Quick Steps

Add Video from a File on a PC
1. Click Insert Video icon in content placeholder.
2. Click Browse From a file.
3. Navigate to drive and/or folder.
4. Double-click video file.
5. Edit and/or format video clip object as required.

Beyond Basics   **Other Video Playback Options**

The video can be set to display full screen, to loop continuously so that the video repeats until the slide show has ended, and to fade in or out. Configure these settings in the Editing and Video Options groups on the Video Tools Playback tab.

11.7 Adding Sound to a Presentation

Skills

Insert sound from a file

Set audio playback options

Tutorials

Inserting and Modifying Audio Files

Adding Sound to Slide Transitions

Adding music or other sound during a slide show is another way to interest and entertain your audience. For example, you might time introductory music to play while the title slide displays and your audience gathers, with the end of the music cueing the audience that the presentation is about to begin. Music can also be timed to play during a segment of a presentation to accompany a series of images. To add a sound clip or music to a presentation, activate the slide at which the sound should begin, click the **Audio button** in the Media group on the Insert tab, choose *Audio on My PC* at the drop-down list, and then select the sound or music file at the Insert Audio dialog box.

Note: *You will need headphones or earbuds if you are completing this topic in a computer lab at school where sound through the speakers is disabled.*

1. With the **PaintedBunting-YourName** presentation open, make Slide 1 the active slide in the slide pane and then click the Insert tab if the Insert tab is not the active tab.

2. Click the Audio button in the Media group and then click *Audio on My PC* at the drop-down list.

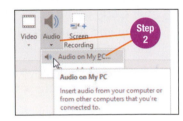

3. If necessary, navigate to the Ch11 folder in Student_Data_Files at the Insert Audio dialog box, and then double-click the file ***Allemande***.

4. Drag the sound icon to position the icon and playback tools near the bottom right of the slide.

5. Click the Play/Pause button and listen to the recording for a few seconds.

The entire music clip plays for approximately two and a half minutes.

6. With the Audio Tools Playback tab active, click the *Hide During Show* check box in the Audio Options group to insert a check mark.

7. Click the *Start* list box arrow (displays *On Click*) and then click *Automatically* at the drop-down list.

This option will start the music as soon as the slide is displayed in the slide show.

8. Click the *Loop until Stopped* check box to insert a check mark.

This option will cause the music to replay continuously until the slide is advanced during the slide show.

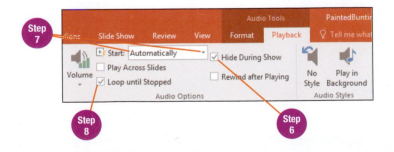

9. Make Slide 7 the active slide in the slide pane.

App Tip

Choose *Record Audio* to record a new sound clip using the Record Sound dialog box. Type a name for the sound clip and then use the record, stop, and play buttons.

App Tip

PowerPoint recognizes most audio file formats including MIDI files, MP3 and MP4 audio files, Windows audio files (.wav), and Windows Media Audio files (.wma).

10 Insert the audio file *PaintedBunting_Song* in the slide by completing steps similar to those in Steps 2 to 8.

The audio recording of the Painted Bunting bird song is slightly less than two seconds in length.

11 Insert a new slide after Slide 7 with the Title and Content layout and then type Photo, Video, and Audio Credits as the slide title.

12 Insert a table, type the information shown in Figure 11.4, and then adjust the layout using your best judgment for column widths and position on the slide.

Always credit the source of images, audio, and video used in a presentation if you did not create the multimedia yourself. In this instance, the music from Slide 1 is not credited because the recording is in the public domain.

13 Save the revised presentation using the same name (**PaintedBunting-YourName**). Leave the presentation open for the next topic.

Good to Know

Many websites offer copyright-free or public domain multimedia. Include the keywords *copyright free* or *public domain* in a search for a picture, audio, or video file. Check the terms of use at each site to be sure you credit sources appropriately because copyright-free allows free usage but usually requires attribution to the creator.

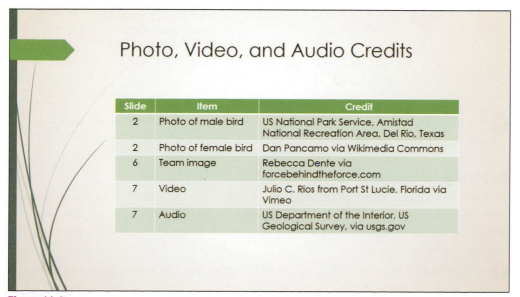

Photo, Video, and Audio Credits

Slide	Item	Credit
2	Photo of male bird	US National Park Service, Amistad National Recreation Area, Del Rio, Texas
2	Photo of female bird	Dan Pancamo via Wikimedia Commons
6	Team image	Rebecca Dente via forcebehindtheforce.com
7	Video	Julio C. Rios from Port St Lucie, Florida via Vimeo
7	Audio	US Department of the Interior, US Geological Survey, via usgs.gov

Figure 11.4
This table for Slide 8 shows multimedia credits.

11.8 Adding Transitions and Animation Effects to Slides for a Slide Show

Skills

Add a transition to all slides

Add an animation effect to an object on the slide master

Add an animation effect to an individual object

Tutorials

Adding Transitions

Applying and Removing Animations

Modifying Animations

Applying an Action to an Object

A **transition** is a special effect that appears as one slide is removed from the screen and the next slide appears. **Animation** adds a special effect to an object on a slide that causes the object to move or change. Animation is a powerful way to enliven a presentation by focusing the viewer's attention on specific text or objects, but don't overuse transition and animation effects as too much movement can become a distraction.

1. With the **PaintedBunting–YourName** presentation open, make Slide 1 the active slide in the slide pane.

2. Click the Transitions tab and then click the More button at the bottom right of the Transition to This Slide gallery.

3. Click the *Blinds* option in the *Exciting* section of the gallery.

PowerPoint previews the effect with the current slide so that you can experiment with various transitions and effects before making your final selection.

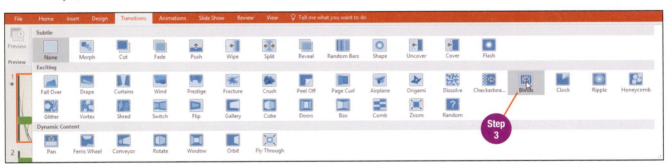

4. Click the Apply To All button in the Timing group.

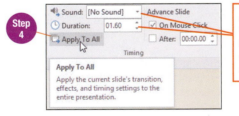

Add a sound effect that plays with the transition and/or speed up or lengthen the time of transition using the two options shown here.

Oops! !

Trouble finding *Blinds*? On small-screen devices such as tablets, the Transitions gallery displays fewer buttons per row. Look for *Blinds* further down.

5. Display the slide show, advance through the first three slides to view the transition effect, and then end the show to return to Normal view.

Applying Animation Effects Using the Slide Master

To apply the same animation effect to all the titles and/or bullet lists in a presentation, apply the effect using the slide master. Generally, animation effects for similar objects should be consistent throughout a presentation; a different effect on each slide might become a distraction.

6. Make Slide 2 the active slide in the slide pane, click the View tab, and then click the Slide Master button.

7. Click to select the border of the title placeholder on the slide master.

App Tip

Use the Effect Options button in the Transition to This Slide group to choose a variation for the selected transition (such as the direction the blinds move).

8 Click the Animations tab.

9 Click the *Split* option in the Animation gallery.

See Table 11.2 for a description of Animation categories.

10 Click to select the border of the content placeholder and then click the More button in the Animation gallery.

11 Click the *Zoom* option in the *Entrance* section.

12 Click the *Start* list box arrow (displays *On Click*) in the Timing group and then click *After Previous*.

13 Select the current entry in the *Duration* text box, type 1.5 and then press Enter.

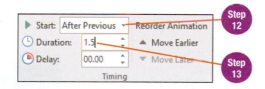

14 Click the Slide Master tab and then click the Close Master View button.

15 Make Slide 1 the active slide in the slide pane, run through the presentation in a slide show to view the transition and animation effects, and then return to Normal view.

Table 11.2

Animation Categories

Category	Description
Entrance	Most common animation effect in which the object animates as it appears on the slide.
Emphasis	Animates text or object already in place by causing the object to move or to change in appearance; includes effects such as darkening, changing color, bolding, or underlining, to name a few.
Exit	Animates the text or object after it has been revealed, such as by fading or flying off the slide.
Motion Paths	An object moves along a linear path, an arc, or some other shape.

Applying Animation Effects to Individual Objects

As you previewed the slide show, you probably noticed that images and other objects, such as shapes or text boxes, appeared on the slide before the title. You may want these items to remain hidden until the title and text have been revealed. To apply animation to an individual object, display the slide, select the object to be animated, and then apply the desired animation option.

To copy an animation effect from one object to another, use the **Animation Painter button** in the Advanced Animation group on the Animations tab, which operates similarly to the Format Painter button that you learned to use in Chapter 3.

16 Make Slide 2 the active slide in the slide pane.

17 Click to select the male bird picture at the top right of the slide.

18 Click the Animations tab and then click the *Wipe* option in the Animation gallery.

19 Click the *Start* list box arrow and then click *After Previous*.

20 Click to select the male bird picture and then click the Animation Painter button in the Advanced Animation group.

21 Click to select the female bird picture at the bottom right of the slide.

22 Make Slide 3 the active slide in the slide pane.

23 Click to select the arrow shape, apply the *Fly In* animation option, and then change the *Start* option to *After Previous*.

24 Copy the arrow shape animation options to the text box object below the chart by completing steps similar to those in Steps 20 and 21.

25 Make Slide 5 the active slide in the slide pane.

26 Click to select the WordArt object at the bottom of the slide, apply the *Float In* animation option, and then change the *Start* option to *After Previous*.

27 Make Slide 6 the active slide in the slide pane.

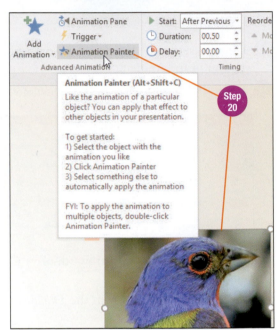

Step 20

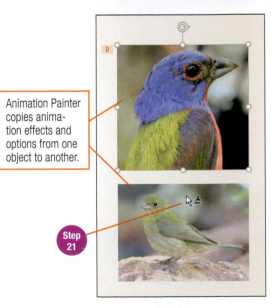

Animation Painter copies animation effects and options from one object to another.

Step 21

28 Click to select the picture at the left side of the slide, apply the *Shape* animation option, and then change the *Start* option to *After Previous*.

29 Click to select the SmartArt graphic at the bottom of the slide. Make sure the entire graphic is selected and not an individual shape within the graphic.

30 Apply the *Fly-in* animation option and then change the *Start* option to *After Previous*.

31 Click the Effect Options button and then click the *One by One* option in the *Sequence* section of the drop-down list.

This option will cause each chevron in the graphic to animate on the slide one at a time, starting with the leftmost shape first.

32 Run through the presentation in a slide show from the beginning to view the revised animation effects and then return to Normal view when the slide show ends.

33 Save the revised presentation using the same name (**PaintedBunting-YourName**). Leave the presentation open for the next topic.

Beyond Basics   **Changing the Animation Sequence**

To adjust the order in which objects are animated, display the required slide and open the Animation pane at the right side of the window. To do this, click the Animation Pane button in the Advanced Animation group on the Animations tab. Select the object that you want to move in the Animation pane and then use the Move Earlier or Move Later buttons in the Reorder Animation group.

11.9 Setting Up a Self-Running Presentation

Skills

Add timings to slides

Change the show type to kiosk

Tutorials

Advancing Slides Automatically

Setting Timings for a Slide Show

Looping a Slide Show Continuously

Some presentations are designed to be self-running, meaning that the slides are intended to be shown continuously at a kiosk or viewed at a PC by an individual. To create a presentation that advances through slides automatically, you need to set up a time for each slide to display and ensure that each slide animation is set to start automatically for each object. When the animation options and timing settings are complete, open the Set Up Show dialog box to change the show type to *Browsed at a kiosk* by clicking the **Set Up Slide Show button** in the Set Up group on the Slide Show tab.

1. With the **PaintedBunting-YourName** presentation open, use Save As to save a copy of the presentation in the current folder naming it **PaintingBuntingSelfRunning-YourName**.

2. Make Slide 2 the active slide in the slide pane and then display the slide master.

3. Select the border of the title placeholder.

4. Click the Animations tab and then change the *Start* option in the *Timing* group to *After Previous*.

5. Close Slide Master view.

6. Make Slide 1 the active slide in the slide pane and then click to select the sound icon.

7. Click the Audio Tools Playback tab and then click the *Play Across Slides* check box in the Audio Options group to insert a check mark.

This option will cause the music that starts at Slide 1 to continue playing through the remaining slides.

8. Click the Volume button and then click *Low* at the drop-down list.

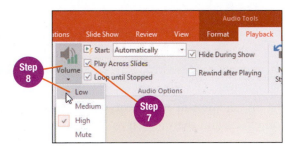

9. Make Slide 7 the active slide in the slide pane and then select and delete the sound icon to remove the audio.

10. Make Slide 8 the active slide in the slide pane and then select and delete the last row in the table.

11. Click the Transitions tab.

12. Click the *After* check box in the *Timing* section to insert a check mark, select the current entry in the *After* text box, type 0:25 and then press Enter.

13. Click the Apply To All button.

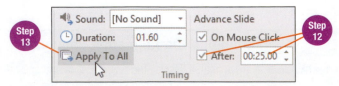

All the slides will advance automatically after the same 25-second duration. To set individual times for slides, activate a slide and enter a different time in the *After* text box.

14 Make Slide 3 the active slide in the slide pane, select the entry in the *After* text box, type 0:10 and then press Enter.

15 Change the *After* time for Slide 5 and Slide 6 to 0:15 and for Slide 8 to 0:10.

16 Click the Slide Show tab and then click the Set Up Slide Show button in the Set Up group.

17 At the Set Up Show dialog box, click the *Browsed at a kiosk (full screen)* option in the *Show type* section.

18 Click OK.

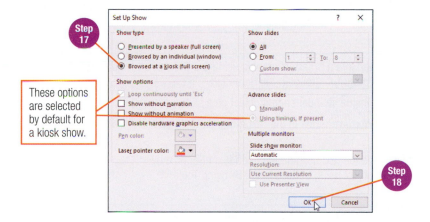

These options are selected by default for a kiosk show.

19 Start the slide show from the beginning and watch the presentation as it advances through all the slides automatically. End the show when the presentation starts at Slide 1 again by pressing the Esc key.

20 Save the revised presentation using the same name (**PaintedBuntingSelf Running–YourName**) and then close the presentation.

App Tip

You can create an MPEG 4 (MP4) movie file that you can burn to a disc or upload to a website using the *Create a Video* panel at the Export backstage area.

View

Model Answer
Compare your completed file with the model answer.

Beyond Basics   **Using Rehearse Timings**

The Rehearse Timings feature on the Slide Show tab lets you assign a time to each slide as you run through a slide show with a timer active and a Recording toolbar. Use the Next button on the Recording toolbar to advance each slide, and PowerPoint will enter the times for each slide transition. This method lets you time each slide for a typical audience member after a suitable time has elapsed in the slide show.

Topics Review

Topic	Key Concepts	Key Terms
11.1 Inserting Graphic Images from Picture Collections	A picture from your PC can be added to a slide using the Pictures button in the Images group on the Insert tab or with the Pictures icon in the content placeholder on a new slide. Resize, move, or edit a picture by selecting the image and using the selection handles and/or buttons in the Picture Tools Format tab. Find a picture or other image from a website using the Online Pictures button in the Images group on the Insert tab.	
11.2 Inserting a SmartArt Graphic	SmartArt graphics use shapes with text to illustrate information in lists, processes, cycles, hierarchies, or other diagrams. Add a SmartArt graphic to a slide using the SmartArt button in the Illustrations group on the Insert tab or the Insert a SmartArt Graphic icon in a content placeholder. Choose a SmartArt category and layout at the Choose a SmartArt Graphic dialog box and then click OK to enter the desired text. Text can be added to shapes in the Text pane or by typing directly inside a shape. Modify SmartArt styles or colors or edit the graphic using buttons in the SmartArt Tools Design tab.	SmartArt
11.3 Converting Text to SmartArt and Inserting WordArt	A bullet list can be converted into a SmartArt graphic using the Convert to SmartArt button in the Paragraph group on the Home tab. WordArt is decorative text inside an independent object on a slide. Create WordArt using the WordArt button in the Text group on the Insert tab. Type the WordArt text inside the text box and then add text effects, move, and or otherwise edit the object using buttons in the Drawing Tools Format tab.	Convert to SmartArt button WordArt
11.4 Creating a Chart on a Slide	Insert a chart using the Insert Chart icon in the content placeholder or the Chart button in the Illustrations group on the Insert tab. Choose the chart category and chart type at the Insert Chart dialog box and then click OK. Type the data to be graphed in the chart data grid, which is a small Excel worksheet object on top of the slide placeholders. Modify the chart using the buttons in the Chart Tools Design and Chart Tools Format tabs.	
11.5 Drawing a Shape and Adding a Text Box	Draw your own graphics on a slide using the Shapes button on the Insert tab. Type text inside a selected shape and then resize, move, or otherwise modify the shape using buttons in the Drawing Tools Format tab. A text box is a rectangular object in which you can type text and that can be moved, resized, or formatted independently. Create a text box using the Text Box button on the Insert tab or the Drawing Tools Format tab.	Shapes button

continued…

Topic	Key Concepts	Key Terms
11.6 Adding Video to a Presentation	Add a video clip to a slide using the Insert Video icon in a content placeholder or with the Video button in the Media group on the Insert tab.	Trim Video button
	You can select a video clip from a file on your PC or by finding a video clip at YouTube or another website.	
	Use buttons in the Video Tools Playback tab to edit a video or change the video options.	
	Use the Trim Video button to change the starting and/or ending position of the video if you do not want to play the entire clip.	
	Change the *Start* option if you want the video to start automatically when the slide is displayed in a slide show.	
	Options in the Video Tools Format tab are used to format the video object.	
11.7 Adding Sound to a Presentation	Add audio to a slide using the Audio button in the Media group on the Insert tab.	Audio button
	Use buttons in the Audio Options group on the Audio Tools Playback tab to hide the sound icon during a slide show, start the audio automatically, play the sound in the background across all slides, or loop the audio continuously until the slide is advanced.	
	Always credit the sources of all multimedia used in a presentation that you did not create yourself.	
11.8 Adding Transitions and Animation Effects to Slides for a Slide Show	A transition is a special effect that appears as one slide is removed from the screen and another is revealed during a slide show.	transition
	Animation causes an object to move or transform in some way.	animation
	Select a transition at the Transition to This Slide gallery on the Transitions tab.	Animation Painter button
	The Apply To All button in the Timing group on the Transitions tab sets the same transition effect to all slides.	
	Add an animation effect to a placeholder on the slide master to apply the effect to all slides in the presentation.	
	Animation effects are selected in the Animation gallery on the Animations tab.	
	Specify how the animation will start and the animation duration using options in the Timing group.	
	The Animation Painter button copies the animation effect and effect options from one object to another.	
	Animate an individual object on a slide by selecting the object and then adding an animation effect from the Animation gallery.	
	Animation effects are grouped into four categories: *Entrance*, *Emphasis*, *Exit*, and *Motion Paths*.	
	To change the sequence in which objects are animated, display the Animation pane, select the object to be reordered, and then use the Move Earlier or Move Later buttons in the Reorder Animation group.	

continued…

Topic	Key Concepts	Key Terms
11.9 Setting Up a Self-Running Presentation	A self-running presentation is set up to run a slide show continuously. To create a self-running presentation, each slide needs to have a time entered in the *After* text box in the Timing group on the Transitions tab, and each animated object needs to be set to start automatically. Open the Set Up Show dialog box from the Set Up Slide Show button on the Slide Show tab, and then choose *Browsed at a kiosk (full screen)* to instruct PowerPoint to play the slide show continuously until stopped. As an alternative to manually entering each slide time, you can use the Rehearse Timings feature from the Slide Show tab to set a time for each slide to display while watching a slide show with a timer and Recording toolbar active.	Set Up Slide Show button

 Recheck

Recheck your understanding of the topics covered in this chapter.

 **Workbook**

Chapter review and assessment resources are available in the *Workbook* ebook.

Chapter 12

Using and Querying an Access Database

Precheck
Check your understanding of the topics covered in this chapter.

Organizations and individuals rely on data to complete transactions, make decisions, and otherwise store and track information. Data that is stored in an organized manner to provide information to meet a variety of needs is called a **database**. Microsoft Access is a software program designed to organize, store, and maintain data in an application referred to as a **database management system (DBMS)**. You interact with a DBMS several times a day as you complete your daily activities. Examples of the types of transactions that involve a DBMS include withdrawing cash from your bank account, completing a purchase, looking up a telephone number, or programming your GPS to find a route to an address.

In this chapter you will learn database terminology and how to navigate a DBMS, including how to open and close objects; add and maintain records using a datasheet and form; find and replace data; sort and filter data; and use queries to look up information and perform a calculation on a numeric field.

Learning Objectives

12.1 Identify a database table, query, report, and form; and define *field*, *field value*, and *record*

12.2 Add a record using a datasheet

12.3 Edit and delete records using a datasheet

12.4 Add, edit, and delete records using a form

12.5 Find and replace data, and adjust column widths in a datasheet

12.6 Sort and filter records in a datasheet and form

12.7 Create a query using a wizard

12.8 Create a query using Design view

12.9 Select records in a query using criteria

12.10 Select records in a query using AND and OR criteria, and sort query results

12.11 Modify a query to insert and remove a field, add a calculated field to a query, and preview a database object

SNAP If you are a SNAP user, go to your SNAP Assignments page to complete the Precheck, Tutorials, and Recheck.

Data Files

Before beginning this chapter, be sure you have copied the student data files for this course to your storage medium. Steps on downloading and extracting the data files are provided in Chapter 1, Topic 8, on pages 22–23.

12.1 Understanding Database Objects and Terminology

Skills

Open and close
a database

Identify objects
in a database

Open and close objects

▷ Tutorials

Opening an Existing
Database

Closing a Database
and Closing Access

Opening and Closing
an Object

An Access database is structured and organized specifically to keep track of a large amount of similar data. For example, a library database is organized so that information such as the title, author, and publisher is maintained for each book in the library. Making sure that the data is entered and updated in the same manner for each item is important so that information is complete and accurate. For this reason, a database is created with a structure that defines the data that will be collected for each item. Working with objects and data in an existing database will help you understand the terms that are used and how data is organized before you create your own database.

Identifying Database Objects

An Access database is a collection of related objects in which data is recorded, edited, and viewed. Access opens with a Navigation pane along the left side of the window that is used to open an object. Objects are grouped in the Navigation pane by type. The four most common types are tables, queries, forms, and reports. See Table 12.1 for a description of each type.

Table 12.1

Access Objects

Object	Description
Table	Data is organized into tables, each of which opens in a datasheet that displays data in columns and rows. A table stores data about one topic or subject only. For example, in the LibraryFines database, one table contains data about each student and another table contains data about each fine.
Query	A query is used to extract information from one or more tables in a single datasheet that shows all the data or a subset of data that meets a specific condition. For example, a query could show all library fines that have been assessed or only those fines that remain unpaid.
Form	A form provides a user-friendly interface to enter or update data where one record is viewed at a time. The layout of a form can be customized to match closely an existing paper form used in a business.
Report	Reports are used for viewing or printing data from a table or query. Reports can include summary totals and a customized layout.

Oops! !

A Connect to dialog box appears asking for your user name and password and then a Save a Local Copy dialog box appears? If you open data files from your OneDrive account, you may be prompted to enter your OneDrive account name and password and then save a copy of the database to your PC. It will be easier for you to copy the student data files for Chapter 12 and 13 to a folder on a USB flash drive or other hard drive. Close the dialog box, exit Access, and then copy the files to another folder. Start Step 1 again, opening the database from the new folder.

① Start Access and open the database named *LibraryFines* from the Ch12 folder in Student_Data_Files.

② Click the Enable Content button in the SECURITY WARNING message bar that appears below the ribbon. If a SAVE CHANGES message bar appears with a Save to SharePoint Site button, click the Close this message button at the right end of the bar.

The Security Warning message bar appears each time you open a database unless the settings for Access on your PC have been changed. Microsoft disables some content as a way to protect your PC from potentially harmful files that may be embedded in the database without your knowledge (such as a virus). The data files provided with this textbook are safe, so you can click the Enable Content button.

③ Compare your screen with the one shown in Figure 12.1 on page 289.

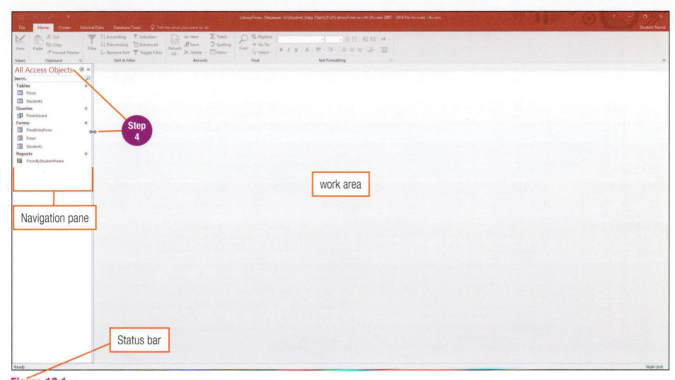

Figure 12.1

The LibraryFines database is shown opened in the Access window.

④ If necessary, drag the gray border along the right side of the Navigation pane to the right to expand the width of the pane until the title *All Access Objects* is entirely visible.

Opening and Closing Objects

Data in a database is organized by topic or subject about a person, place, event, item, or other category grouping in an object called a **table**. A database table is the first object that is created. The number of tables varies for each database depending on the information that needs to be stored. Tables are the building blocks for creating other objects, such as a query, form, or report. In other words, you cannot create a query, form, or report without first creating a table.

⑤ Double-click *Fines* in the Tables group in the Navigation pane.

The table opens in Datasheet view within a tab in the work area as shown in Figure 12.2 on page 290. A datasheet resembles a spreadsheet with the data organized in columns and rows. The information about the subject or topic of a table (such as library fines) is divided into columns, each of which is called a **field**. A field should store only one unit of information about a person, place, event, or item. For example, a mailing address is split into at least four fields so the street address, city, state or province, and zip or postal code are separated. This allows the database to be sorted, filtered, or searched by any piece of information.

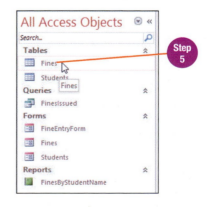

Oops! !

Trouble opening a table? Another way to open the table is to right-click the object name and then choose *Open* at the Shortcut menu.

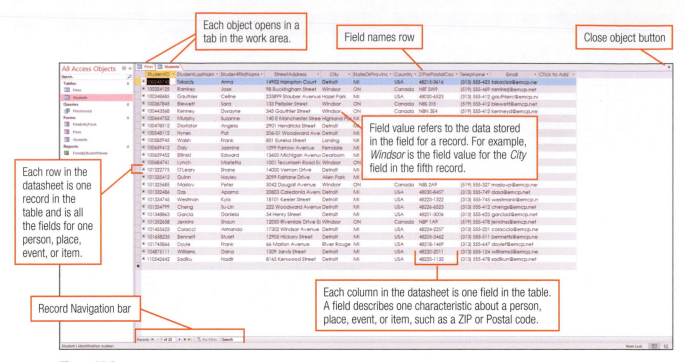

Figure 12.2

Shown is the Datasheet view for the Students table in the LibraryFines database.

Each row in the datasheet shows all the fields for one person, place, event, or item and is called a **record**. The data that is stored in one field within a record is called a **field value**.

6 Double-click *Students* in the Tables group in the Navigation pane and then compare your screen with the one shown in Figure 12.2.

7 Double-click *FinesByStudentName* in the Reports group in the Navigation pane and review the report content and layout in the work area.

A **report** is designed to view or print data from one or more tables or queries in a customized layout and with summary totals. In this report, library fines are arranged and grouped by student name.

8 Click the Close button at the top right of the work area to close the report.

9 Double-click *FineEntryForm* in the Forms group in the Navigation pane.

A **form** is used to enter, update, or view one record at a time.

10. Click the Next record button (right-pointing arrow) in the Record Navigation bar located at the bottom of the form.

Buttons in the Record Navigation bar are used to move to the first record, previous record, next record, or last record. Use the search text box to navigate to a record by typing a field value.

11. Click the Previous record button (left-pointing arrow).

12. Click the Last record button (right-pointing arrow with vertical bar) to move to the last record in the form.

13. Click the First record button (vertical bar with left-pointing arrow) to move to the first record in the form.

14. Click the Close button at the top right of the work area to close the form.

15. Double-click *FinesIssued* in the Queries group in the Navigation pane.

A **query** opens in a datasheet similar to a table. A query displays information from one or more tables and may show all the records or only a subset of records that meet a specific condition.

Overdue Book Fine Form

FineID	15-101
Student ID Look up	100548112
Date Fine Issued	10-Mar-2018
Amount	$8.75
Overdue Book ISBN	9780763843011
Book Returned?	☑
Fine Paid?	☑
Date Paid	31-Mar-2018

previous record

new (blank) record

first record   last record

Record: 1 of 17   No Filter   Search

Next record

search text box

Step 10

Quick Steps
Open a Database Object
1. Open database file.
2. Double-click object name in Navigation pane.

App Tip

Forms can be created to resemble paper-based forms used within an organization.

FineID	DateIssued	StudentFirstName	StudentLastName	Telephone	Amount	OverdueBookISBN	BookReturned	FinePaid
15-101	10-Mar-2018	Pat	Hynes	(313) 555-6569	$8.75	9780763843011	☑	☑
15-102	11-Mar-2018	Angela	Doxtator	(313) 555-4771	$2.15	9780531523541	☑	☑
15-103	12-Mar-2018	Pat	Hynes	(313) 555-6569	$1.25	9780412533145	☑	☐
15-104	12-Mar-2018	Marietta	Lynch	(519) 555-3214	$1.40	9784123524158	☑	☐
15-105	13-Mar-2018	Edward	Bilinski	(313) 555-9200	$4.25	9784125312517	☑	☑
15-106	13-Mar-2018	Stuart	Bennett	(313) 555-5112	$3.25	8745125412352	☑	☐

A query can show data from more than one table and display the fields in any order as seen in the first six rows in the FinesIssued query shown here.

16. Close the query and the two tables.

17. Click the File tab and then click *Close* at the backstage area.

Always close a database file using the backstage area before exiting Access so that all temporary files used by Access while you are viewing and updating records are properly closed.

App Tip

Close objects as soon as you are finished viewing or updating data. Some Access commands will not run if an object is open in the background.

Beyond Basics   **One Database at a Time**

Unlike Word, Excel, or PowerPoint, Access allows only one file to be open at a time in the current window. If you open a second database in the current window, Access automatically closes the existing database before opening the new one.

12.2 Adding Records to a Table Using a Datasheet

Skills

Add a new record using a datasheet

**Tutorial**
Adding and Deleting
Records in a Table

To add a new record to a table, open the table and click the **New (blank) record** button in the Record Navigation bar. Type the field values for the new record using Tab or Enter to move to the next column (field) in the datasheet. When you move past the last column in a new row in the datasheet, the record is automatically saved.

1. Open the database named **LibraryFines**.

2. Click the File tab and then click *Save As* to open the Save As backstage area.

3. With *Save Database As* already selected as the *File Types* option and *Access Database* already selected as the *Save Database As* option, click the Save As button.

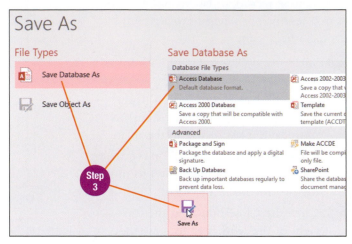

4. At the Save As dialog box, create a new folder named *Ch12* in the CompletedTopicsByChapter folder on your storage medium, change the file name to **LibraryFines-YourName**, and then click the Save button.

5. Click the Enable Content button in the SECURITY WARNING message bar. Close the SAVE CHANGES message bar if the bar appears.

6. Open the Fines table.

7. Click the New (blank) record button in the Record Navigation bar.

8. Type 15-118 in the *FineID* field and then press Tab to move to the next field.

The field *StudentID* looks up names and ID numbers in the Students table using a drop-down list.

9. Click the down-pointing arrow in the *StudentID* field, click *100478512 Angela Doxtator* in the drop-down list, and then press Tab.

The student ID number becomes the field value, which is connected to Angela Doxtator's record in the Students table.

App Tip

You can also move to the next field by pressing the Enter key or by clicking in the next column.

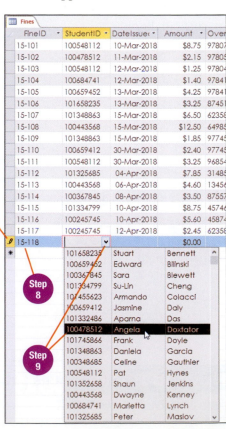

The Pencil icon indicates the record is being edited. The pencil disappears when Access saves the changes.

10 Type 12apr2018 in the *DateIssued* field and then press Tab.

The date field has been set up to display underscores and dashes as soon as you begin typing to help you enter the date in the correct format *dd-mm-yyyy*. This configuration also ensures that all dates are entered consistently in the database.

11 Type 8.75 in the *Amount* field and then press Tab.

12 Type 4348973098226 in the *OverdueBookISBN* field and then press Tab.

13 Press the spacebar to insert a check mark in the *BookReturned* field and then press Tab.

BookReturned is a field that has been set up to store only one of two possible field values: *Yes* or *No*. Inserting a check mark stores *Yes*, while an empty check box stores *No*.

14 Click the check box to insert a check mark in the *FinePaid* field and then press Tab.

15 Type 15apr2018 in the *DatePaid* field and then press Tab.

Moving to the next row in the datasheet automatically saves the record just typed and starts a new record.

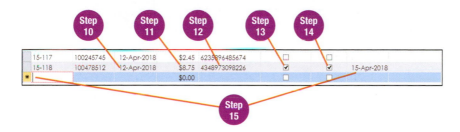

16 Add the following field values in the new row in the fields indicated pressing Tab after typing the data to move to the next field:

FineID	15–119	*OverdueBookISBN*	7349872345760
StudentID	101348863 Daniela Garcia	*BookReturned*	Yes
DateIssued	15apr2018	*FinePaid*	No (leave blank)
Amount	5.25	*DatePaid*	(leave blank)

17 With the insertion point positioned in the *FineID* field in a new row, close the Fines table. Leave the database open for the next topic.

Alternative Method **Adding New Records Using the Ribbon or Keyboard Commands**

New records can be added to a table using any of these methods:
• Click the New button in the Records group on the Home tab
• Keyboard shortcut Ctrl + + (hold down Ctrl key and press plus symbol)
• Click in the last cell in the table and then press Tab

12.3 Editing and Deleting Records Using a Datasheet

Edit a field value in a table using a datasheet by opening the table, selecting the text to be changed and then typing the new text, or by clicking in the table cell to place an insertion point and then inserting or deleting text as required. Select a record for deletion by clicking in the gray record selector bar along the left edge of the datasheet next to the record and then clicking the Delete button in the Records group on the Home tab. Access requires confirmation before deleting a record.

See Beyond Basics at the end of this topic for information about precautions to take to back up data in a database before deleting a record.

1 With the **LibraryFines–YourName** database open, open the Students table.

2 Select the text *Murphy* in the *StudentLastName* column in the sixth row in the datasheet and then type Hall as the new last name for Suzanne.

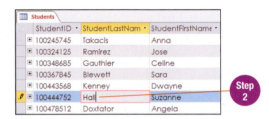

App Tip

As soon as you move to another record in the datasheet, the changes are automatically saved.

3 Press Tab eight times to move to the *Email* field.

4 Press F2 to open the field for editing, move the insertion point as needed, delete *murphy* at the beginning of the email address, and then type hall so that the email address becomes *halls@emcp.net*.

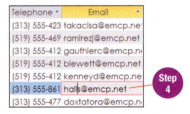

Quick Steps

Edit a Record
1. Open table.
2. Click in table cell to place insertion point.
3. Insert or delete text as required.

Delete a Record
1. Open table.
2. Select record.
3. Click Delete button.
4. Click Yes.

5 Click at the end of *N8T 3W9* in the *ZIPorPostalCode* field in the second row in the datasheet to position the insertion point, press Backspace to remove *3W9*, and then type 2E6.

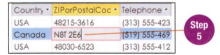

6 If necessary, scroll left until you can see the student names.

7   Click in the record selector bar next to the record for the student *Das Aparna* when you see the mouse pointer change to a right-pointing black arrow.

The gray bar at the left edge of the datasheet is used to select a record. When using a mouse, the pointer displays as a black right-pointing arrow when positioned next to a record in the gray record selector bar.

Select all button

Record selector bar

Step 7

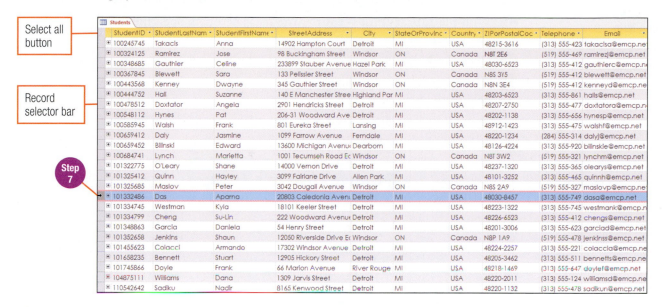

8   Click the Delete button in the Records group on the Home tab. Do *not* click the down-pointing arrow on the button.

Step 8

9   Click Yes at the message box that appears asking if you are sure you want to delete the record.

Step 9

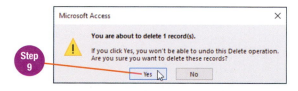

10   Close the Students table. Leave the database open for the next topic.

App Tip

Be cautious with the Delete command because Undo does not work to restore a record. Consider making a backup copy of a database before deleting records.

Beyond Basics   Best Practices for Deleting Records

Depending on the purpose of the database, deleting records is generally not done until the records to be deleted are first copied to an archive database and/or a backup copy of the database has been made. In many cases, records need to be retained for historical purposes. Always check before deleting a record to make sure you are following proper procedure.

12.4 ## Adding, Editing, and Deleting Records Using a Form

Skills

Add a record
using a form

Edit a record
using a form

Delete a record
using a form

Tutorials
Adding and Deleting
Records in a Form
Navigating in Objects

A form is an Access object that provides a different view for the data stored in a table. Generally only one record at a time is displayed in a columnar layout instead of the spreadsheet style datasheet. Forms are usually preferred over a datasheet for adding, editing, and deleting records because the user can focus on one record at a time.

1 With the **LibraryFines–YourName** database open, open the form named *FineEntryForm*.

2 Click the New (blank) record button in the Record Navigation bar.

3 Add the field values as shown in the image below, using Tab to move from one field to the next field.

A new blank form displays when you press Tab after the last field in a form.

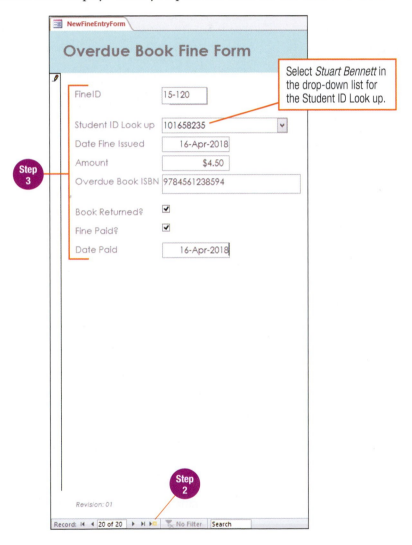

App Tip

Print the current record displayed in a form by displaying the Print backstage area, clicking *Print*, and then clicking *Selected Record(s)* in the *Print Range* section of the Print dialog box.

4 Click the First record button in the Record Navigation bar to display the first record in the form.

5 Select $8.75 in the *Amount* field and then type 7.25.

6 Click the Next record button two times to display record 3 in the form.

7 Click the Delete button arrow in the Records group on the Home tab and then click *Delete Record* at the drop-down list.

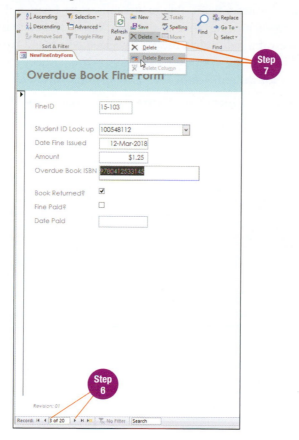

8 Click Yes at the message box that appears asking if you are sure you want to delete the record.

9 Close the form. Leave the database open for the next topic.

Alternative Method	Navigating Records in a Form Using Keyboard Shortcuts

These keyboard shortcuts can be used in a form to move to another record:

Page Down	Next record	Page Up	Previous record
Ctrl + Home	First record	Ctrl + End	Last record (last field)

Beyond Basics **Using Find to Move to a Record**

As databases expand to store hundreds or thousands of records, using the navigation buttons at the bottom of the form to locate a record that needs to be changed or deleted is not feasible. The Find feature locates a record instantly when you search by a name or ID number. You will use Find in the next topic.

12.5 Finding and Replacing Data and Adjusting Column Widths in a Datasheet

Skills

Find and replace data
using a datasheet

Adjust column widths in a
datasheet

Tutorials

Finding Data

Finding and Replacing
Data

Adjusting Field Column
Width

The Find feature locates a field value in a datasheet or form and moves the insertion point to each occurrence of the data. When a change needs to be made to all occurrences of a field value, use the Replace command to make the change automatically. The column width for a column in a datasheet can be made wider to fully display data for those fields that do not currently show all the field values.

1 With the **LibraryFines-YourName** database open, open the Fines table.

2 Click to place an insertion point within the *StudentID* field value in the first record.

3 Click the Find button in the Find group on the Home tab.

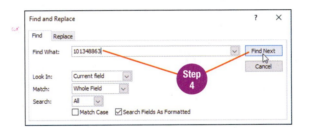

4 Type 101348863 in the *Find What* text box and then click the Find Next button.

The first record (record 6) that matches the field value is made active.

Oops! !

No records found? Tap or click OK and then check that you typed the ID number without errors.

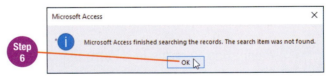

5 Continue clicking the Find Next button to review all occurrences of the matching field value.

6 Click OK at the message that Microsoft Access has finished searching records.

7 Click the Cancel button or the Close button to close the Find and Replace dialog box.

8 Click to place an insertion point within the *FineID* field in the first record.

9 Click the Replace button in the Find group.

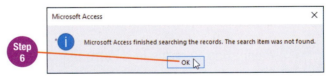

10　Type 15- in the *Find What* text box and then press Tab.

11　Type MCL- in the *Replace With* text box.

12　Click the *Match* option box arrow and then click *Any Part of Field* at the drop-down list.

13　Click the Replace All button.

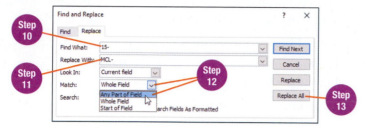

14　Click Yes at the message asking if you want to continue and informing you that the Replace operation cannot be undone.

15　Close the Find and Replace dialog box.

16　Click to place an insertion point in any record within the *DateIssued* field.

17　Click the More button in the Records group on the Home tab and then click *Field Width* at the drop-down list.

18　Click Best Fit at the Column Width dialog box.

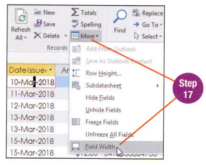

Best Fit adjusts the width of the column to accommodate the length needed to display the longest field value.

19　Close the Fines table. Click Yes when prompted to save the changes to the layout of the table. Leave the database open for the next topic.

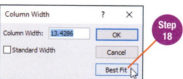

Saving changes to the layout of the table means that Access will retain the new column width for the *DateIssued* field when the table is reopened.

Quick Steps

Find a Record
1. Open table or form.
2. Click in field to be searched.
3. Click Find button.
4. Type field value to find in *Find What* text box.
5. Click Find Next until done.
6. Click OK.
7. Close dialog box.

Replace a Field Value
1. Open table or form.
2. Click in field to be searched.
3. Click Replace button.
4. Type field value to find in *Find What* text box.
5. Type new field value in *Replace With* text box.
6. Click Replace or Replace All as needed.
7. Click Yes.
8. Close dialog box.

Adjust Column Width
1. Open table.
2. Click in any record in column.
3. Click More button.
4. Click *Field Width*.
5. Type value or click Best Fit.

Alternative Method　**Adjusting the Column Width in a Datasheet**

You can also adjust column widths using the following methods:
- Drag the right column boundary in the field names row right to lengthen, or left to shorten the column width.
- Double-click the right column boundary to best fit the column width.
- Type a value in the *Column Width* text box at the Column Width dialog box.

12.6 Sorting and Filtering Records

Skills

Sort records

Filter records

**Tutorials**

Sorting Records in a
Table

Filtering Records

Records are initially arranged in the datasheet alphanumerically by the field in the table that has been defined as the primary key. A **primary key** is a field that contains the data that uniquely identifies each record in the table. Generally, the primary key is an identification number, such as *StudentID* in the Students table. To change the order of the records, click in the column by which to sort and use the Ascending or Descending buttons in the Sort & Filter group on the Home tab.

1   With the **LibraryFines–YourName** database open, open the Students table.

The primary key field in the Students table is the field named *StudentID*. Notice the records in the datasheet are arranged in order of the ID field values.

2   Click to place an insertion point within any field value in the *StudentLastName* column.

3   Click the Ascending button in the Sort & Filter group on the Home tab.

Notice the records in the table are now arranged in order by the student last name field values.

App Tip

When a datasheet is sorted by a field other than the primary key, an up-pointing arrow (ascending order indicator) or down-pointing arrow (descending order indicator) displays next to the field name used to sort.

4   Close the Students table. Click Yes when prompted to save the changes to the design of the table.

Selecting Yes to save changes to the design of the table means that the table will remain sorted by the *StudentLastName* field when you reopen the datasheet. Each object based upon the Students table can have its own sort option saved.

5   Open the Students form.

6   Click the Next record button a few times to view the first few records. Notice the records are arranged by *StudentID*.

7   Click the First record button to return the display to the first record.

8   Click to place an insertion point in the *Student Last Name* field.

9   Click the Ascending button in the Sort & Filter group.

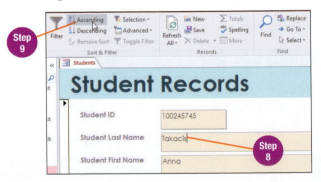

10   Scroll through the first 10 records in the form to view the sorted order and then close the form.

11 Open the Students table. Notice the up-pointing arrow next to *StudentLastName* in the field names row indicating the records are arranged alphabetically by the student last names.

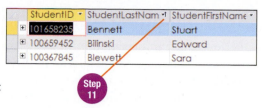

Step 11

You can filter a datasheet in Access using the same techniques you learned for filtering a table in Excel in Chapter 9. Recall that a filter temporarily hides the rows that you do not want to view.

12 Click the filter arrow (down-pointing arrow) next to *Country*.

13 Click the check box next to *USA* to clear the check mark from the box at the Sort & Filter list box and then click OK.

The datasheet is filtered to show records for students who reside in Canada only.

Step 12

Step 13

14 Click the Toggle Filter button in the Sort & Filter group to clear the filter. All records are now redisplayed.

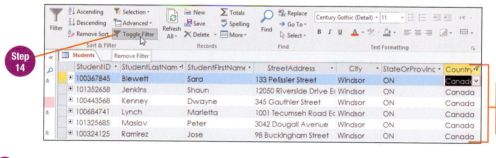

Step 14

This is the filtered list of records after Step 13.

15 Close the Students table. Click No when prompted to save changes to the design of the table. Leave the database open for the next topic.

Quick Steps

Sort a Datasheet or Form
1. Open table or form.
2. Click in field by which to sort.
3. Click Ascending or Descending button.

Filter a Datasheet
1. Open table.
2. Click filter arrow next to field.
3. Clear check boxes as needed.
4. Click OK.

App Tip

Access displays a funnel icon next to a field used to filter a datasheet and displays *Filtered* highlighted in orange in the Record Navigation bar as indicators that all records are not currently displayed.

App Tip

You can also click the funnel icon next to a field and then click the Clear Filter option to redisplay all records.

Beyond Basics   **Sorting by More Than One Field**

Sort by more than one field in a datasheet by dragging across the field names to select the columns by which to sort and then clicking the Ascending or Descending button. Access sorts fields left to right. For example, if *StudentLastName* and *StudentFirstName* columns are selected, Access sorts first by last names and then by first names when two or more records have the same last name. To sort first by a field other than the leftmost column, move the column before sorting to change the sort order by clicking the field name to select the column, and then dragging the field name left to the desired location.

12.7 Creating a Query Using the Simple Query Wizard

Skills

Create a query using wizard

Tutorial

Creating a Query Using the Simple Query Wizard

Creating a Form Using the Form Wizard

Queries extract information from one or more tables in the database and display the results in a datasheet. Some queries display fields from more than one table in the same datasheet. For example, in the LibraryFines database, the student names are in one table and the fines are in another table; a query can combine the names and fines in one datasheet. Other queries are designed to answer a question about the data; for example, *Which library fines are unpaid?* The **Simple Query Wizard** assists a user with creating a query by prompting the user to make selections in a series of dialog boxes.

1. With the **LibraryFines–YourName** database open, click the Create tab.

2. Click the Query Wizard button in the Queries group.

3. Click OK at the New Query dialog box with *Simple Query Wizard* already selected.

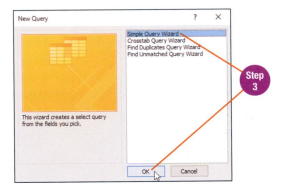

4. At the first Simple Query Wizard dialog box, click the *Tables/Queries* list box arrow and then click *Table: Students* at the drop-down list. Skip this step if *Table: Students* is already displayed in the *Tables/Queries* list box.

   The first step in creating a query is to choose the tables or queries and the fields from each table or query that you want to display in a datasheet.

5. With *StudentID* already selected in the *Available Fields* list box, click the Add Field button (displays as a right-pointing arrow) to move *StudentID* to the *Selected Fields* list box.

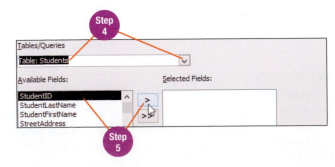

6 Double-click *StudentFirstName* in the *Available Fields* list box to move the field to the *Selected Fields* list box.

You add fields to the *Selected Fields* list box in the order that you want the fields displayed in the query results datasheet.

7 Double-click the following fields in the *Available Fields* list box to move each field to the *Selected Fields* list box.

> *StudentLastName*
> *Telephone*
> *Email*

8 Click the *Tables/Queries* list box arrow and then click *Table: Fines*.

9 Double-click the following fields in the *Available Fields* list box to move each field to the *Selected Fields* list box.

> *DateIssued*
> *Amount*
> *FinePaid*

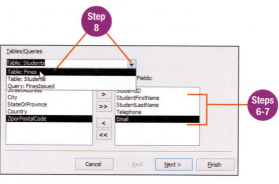

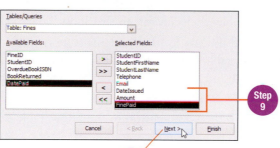

Quick Steps

Create a Query Using Simple Query Wizard
1. Click Create tab.
2. Click Query Wizard button.
3. Click OK.
4. Choose each table and/ or query and fields in required order.
5. Click Next.
6. Click Next.
7. Type title for query.
8. Click Finish.

10 Click Next.

11 Click Next at the second Simple Query Wizard dialog box to accept *Detail (shows every field of every record)* for the query results.

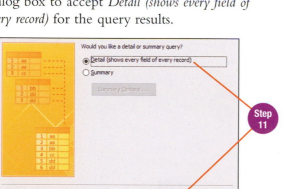

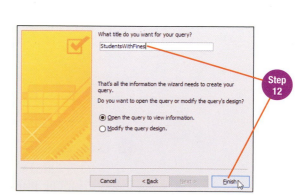

12 At the third Simple Query Wizard dialog box, select the current text in the *What title do you want for your query?* text box, type StudentsWithFines, and then click Finish.

13 Review the query results datasheet. Notice the fields are displayed in the order selected at the first Simple Query Wizard dialog box.

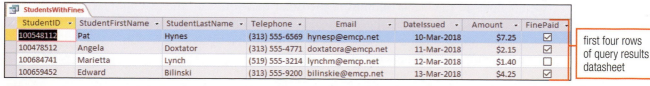

StudentID	StudentFirstName	StudentLastName	Telephone	Email	DateIssued	Amount	FinePaid
100548112	Pat	Hynes	(313) 555-6569	hynesp@emcp.net	10-Mar-2018	$7.25	☑
100478512	Angela	Doxtator	(313) 555-4771	doxtatora@emcp.net	11-Mar-2018	$2.15	☑
100684741	Marietta	Lynch	(519) 555-3214	lynchm@emcp.net	12-Mar-2018	$1.40	☐
100659452	Edward	Bilinski	(313) 555-9200	bilinskie@emcp.net	13-Mar-2018	$4.25	☑

first four rows of query results datasheet

14 Close the StudentsWithFines query. Leave the database open for the next topic.

Skills

Create a query using
Design view

Tutorials

Creating a Query in
Design View

Creating a Query in
Design View Using
Multiple Tables

12.8 Creating a Query Using Design View

Every Access object has at least two views. In one view, you browse the data in the table, query, form, or report. This is the view that is active when you open the object from the Navigation pane. Another view, called **Design view**, is used to set up or define the structure and/or layout of a table, query, form, or report. A query can be created in Design view, which displays a blank grid into which you add the fields you want to display in the query results.

1 With the **LibraryFines–YourName** database open and with the Create tab active, click the Query Design button in the Queries group.

2 At the Show Table dialog box with the *Fines* table selected, click the Add button.

A field list box for the Fines table is added to the top of the *Query1* design grid in the work area.

3 Double-click *Students* in the Show Table dialog box.

A field list box for the Students table is added to the top of the design grid beside the *Fines* table field list box. A black join line connects the two tables. The black line displays 1 and an infinity symbol (∞), which indicates the type of relationship for the two tables. You will learn about relationships in the next chapter.

4 Click the Close button in the Show Table dialog box.

Oops! !

Closed the Show Table dialog box by mistake? Reopen the dialog box using the Show Table button in the Query Setup group on the Query Tools Design tab.

Join line shows tables are connected with a relationship.

Fines table field list box is added to design grid at Step 2.

FineID is added here after you perform Step 5.

5 Double-click *FineID* in the *Fines* table field list box.

FineID is added to the *Field* text box in the first column of the design grid. The blank columns in the bottom of the window represent the query results datasheet. You build the query by adding fields to the blank columns in the order that you want them to appear in the datasheet. A field is added to the next available column by double-clicking a field name in a table field list box in the top half of the window.

6 Double-click the following fields in the *Fines* table field list box to add each field to the next available column in the query design grid.

> *DateIssued*
> *Amount*
> *BookReturned* (scroll down the table field list box to the field)

7 Double-click the following fields in the *Students* table field list box to add each field to the next available column in the query design grid.

> *StudentFirstName*
> *StudentLastName*

8 Click the Run button in the Results group on the Query Tools Design tab to view the query results datasheet.

The **Run button** instructs Access to produce the query results datasheet by assembling the data from the tables according to the query instructions. A query is a set of instructions with table names and field names to display in a datasheet. The query results datasheet is not a duplicate copy of the data—each time a query is opened or run, the data is generated by extracting the field values from the tables.

9 Click the Save button on the Quick Access toolbar.

10 Type FinesWithBookReturnsList at the Save As dialog box and then press Enter or click OK.

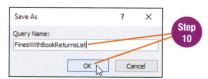

11 Close the FinesWithBookReturnsList query. Leave the database open for the next topic.

Quick Steps

Create a Query Using Design View
1. Click Create tab.
2. Click Query Design button.
3. Add tables to design grid.
4. Close Show Table dialog box.
5. Double-click field names in table field list boxes in desired order for query results datasheet.
6. Click Run button.
7. Click Save button.
8. Type query name.
9. Click OK.

Alternative Method **Adding Fields to Columns in the Query Design Grid**

Add fields to the design grid from the table field list boxes using these other methods:

- Drag a field name from the table field list box to the Field text box in the desired column; if a field already exists in the column, the field is moved to the column to the right.
- Click in a blank Field text box in the design grid, click the down-pointing arrow that appears, and then click the field name in the drop-down list.

12.9 Entering Criteria to Select Records in a Query

Both query results datasheets for the queries you created using the Simple Query Wizard and using Design view displayed all records in the tables. Queries are often created to select records from tables that meet one or more conditions. For example, in this topic you will add a criterion to show only those records in which the fines are unpaid.

1. With the **LibraryFines–YourName** database open, open the StudentsWithFines query.

2. Click the View button in the Views group on the Home tab. Do *not* click the down-pointing arrow on the button.

The View button is used to switch between the query results datasheet and Design view.

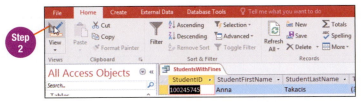

3. Click in the *Criteria* box in the *FinePaid* column in the design grid, type No, press the spacebar, and then press Enter.

Access displays functions in a drop-down list as you type text that matches the letters in a function name. As you type *No*, the function wizard displays *Now* in a drop-down list. Typing a space after *No* causes the *Now* function to disappear. *FinePaid* is a field in which the field value is either *Yes* or *No*. By typing *No* in the *Criteria* box, you are instructing Access to select the records from the *Fines* table in which *No* is the field value for *FinePaid*.

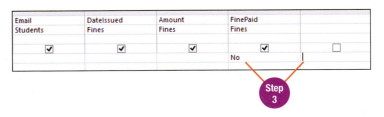

Oops!

Empty datasheet? Use the View button to return to Design view and check that you typed *No* in the *FinePaid* column and/or that *No* is entered in the *Criteria* box of the column.

4. Click the Run button.

Notice that 10 records are selected in the query results datasheet and that the check box in the *FinePaid* column for each record is empty.

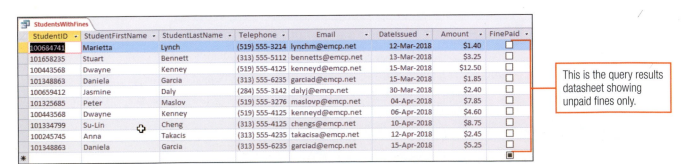

This is the query results datasheet showing unpaid fines only.

5   Click the File tab and then click *Save As* at the backstage area.

6   Click *Save Object As* at the Save As backstage area and then click the Save As button in the *Save the current database object* panel.

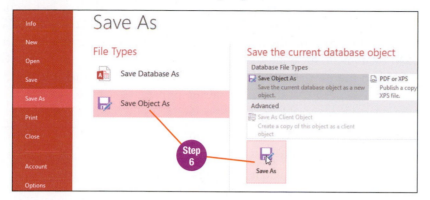

7   Type UnpaidFines in the *Save 'StudentsWithFines' to* text box at the Save As dialog box and then press Enter or click OK.

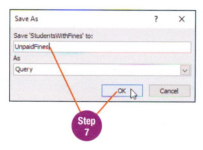

8   Close the UnpaidFines query. Leave the database open for the next topic.

See Table 12.2 for more criteria statement examples.

Table 12.2

Criteria Examples

Field	Entry Typed in *Criteria* Box	Records Selected
Amount	<=5	Fines issued that were $5.00 or less
Amount	>5	Fines issued that were more than $5.00
DateIssued	March 15, 2018 (entry converts automatically to #3/15/2018#)	Fines issued on March 15, 2018
StudentLastName	Kenney (entry converts automatically to "Kenney")	Fines issued to student with the last name *Kenney*.

Beyond Basics   **Selecting Records Using a Range of Dates**

Table 12.2 provides the example that typing March 15, 2018 in the *DateIssued* field selects records of fines issued on March 15, 2018. What if one wanted to view a list of all fines issued in the month of March? To do this, type *Between March 1, 2018 and March 31, 2018* in the *Criteria* box of the *DateIssued* column. Access converts the entry to Between #3/1/2018# And #3/31/2018#.

12.10 Entering Multiple Criteria to Select Records and Sorting a Query

Skills

Select records using AND criteria

Select records using OR criteria

Sort query results

Tutorials

Designing a Query with an AND Criteria Statement

Designing a Query with an OR Criteria Statement

Sorting Data and Showing/Hiding Fields in Query Results

More than one criterion can be entered in the query design grid to select records. For example, you may want a list of all unpaid fines that are more than $5.00. When more than one criterion is on the same row in the query design grid, it is referred to as an *AND* statement, meaning that each criterion must be met for a record to be selected. When more than one criterion is on different rows in the query design grid, it is referred to as an *OR* statement, meaning that any criterion can be met for a record to be selected.

1. With the **LibraryFines–YourName** database open, open the UnpaidFines query.

2. Click the View button to switch to Design view.

3. Click in the *Criteria* box in the *Amount* column, type >5, and then press Enter.

4. Click the Run button.

Step 3

Multiple criteria typed in the same *Criteria* row means each condition must be met for a record to be selected.

StudentID	StudentFirstName	StudentLastName	Telephone	Email	DateIssued	Amount	FinePaid
100443568	Dwayne	Kenney	(519) 555-4125	kenneyd@emcp.net	15-Mar-2018	$12.50	☐
101325685	Peter	Maslov	(519) 555-3276	maslovp@emcp.net	04-Apr-2018	$7.85	☐
101334799	Su-Lin	Cheng	(313) 555-4125	chengs@emcp.net	10-Apr-2018	$8.75	☐
101348863	Daniela	Garcia	(313) 555-6235	garciad@emcp.net	15-Apr-2018	$5.25	☐

This is the query results datasheet showing unpaid fines over $5.00.

5. Use *Save Object As* at the Save As backstage area to save the revised query as *UnpaidFinesOver$5*.

6. Close the UnpaidFinesOver$5 query.

7. Click the Create tab and then click the Query Design button.

8. Double-click *Students* in the Show Table dialog box and then click the Close button.

9. Double-click the following fields in the *Students* table field list box to add each field to the next available column in the query design grid.

> *City*
> *StudentID*
> *StudentFirstName*
> *StudentLastName*
> *Telephone* (scroll down the table field list box to the field)

10. Click in the *Criteria* box in the *City* column, type Detroit, click in the row below *Detroit* next to *or*, type Windsor, and then press Enter.

Access inserts double quotation marks at the beginning and end of a criterion for a field that contains text such as a city, name, or other field not used for calculating values.

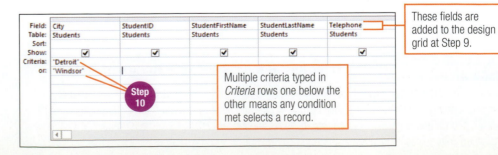

These fields are added to the design grid at Step 9.

Step 10

Multiple criteria typed in *Criteria* rows one below the other means any condition met selects a record.

11 Click the Run button.

Students who reside in Detroit *or* Windsor are shown in the query results datasheet.

12 Click the View button to return to Design view.

A query is sorted by choosing *Ascending* or *Descending* in the *Sort* list box of the column by which you want to sort. At Step 9, City was placed first in the design grid because Access sorts query results by column left to right. To arrange the records alphabetically by student last name grouped by cities, the *City* field needed to be positioned left of the *StudentLastName* field.

13 Click in the *Sort* list box in the *City* column to place an insertion point and display the list box arrow, click the *Sort* list box arrow, and then click *Ascending*.

14 Click in the *Sort* list box in the *StudentLastName* column, click the list box arrow that appears, and then click *Ascending*.

15 Click the Run button.

The query results datasheet is sorted alphabetically by city and then by the student last name within each city.

16 Save the query and name it **DetroitAndWindsorStudents**.

17 Close the DetroitAndWindsorStudents query. Leave the database open for the next topic.

<table>
<tr><td colspan="6">**Quick Steps**</td></tr>
</table>

Quick Steps

Select Records Using AND
1. Open query in Design view.
2. Type criterion in *Criteria* box of first field by which to select.
3. Type criterion in *Criteria* box of second field by which to select.
4. Run query.

Select Records Using OR
1. Open query in Design view.
2. Type criterion in *Criteria* box of first field by which to select.
3. Type criterion in *or* box of second field by which to select.
4. Run query.

Sort a Query
1. Open query in Design view.
2. Click in *Sort* list box of column by which to sort.
3. Click *Sort* list box arrow.
4. Click *Ascending* or *Descending*.

Sorted query results datasheet showing students who reside in either Detroit or Windsor.

Beyond Basics   **Selecting Using a Wildcard Character**

A criterion can be entered that provides Access with a partial entry to match for selecting records. The asterisk is a wildcard character that can be inserted in a criterion in place of characters that you do not want to specify. For example, to select all students with the last name beginning with *C*, type *C** in the *Criteria* box in the *StudentLastName* column.

12.11 Inserting and Deleting Columns, Creating a Calculated Field in a Query, and Previewing a Datasheet

Skills

Delete and insert columns in a query

Create a calculated field

Format a field

Print Preview a datasheet

Tutorials

Modifying Field Properties in Datasheet View

Performing Calculations in a Query

Previewing and Printing a Table

Oops! !

Error message appears? Check your typing to make sure you typed a colon after *Fine with Admin Fee*, used square brackets, and that the entry has no other spelling errors.

App Tip ▶

Use the same mathematical operators in Access that you would use in a formula in Excel: + to add, - to subtract, * to multiply, and / to divide.

A calculated field can be created in a query that performs a mathematical operation on a numeric field. A database design best practice is to avoid including a field in a table for storing data that can be generated by performing a calculation on another field. For example, assume that in the LibraryFines database, each fine is assessed a $2.50 administrative fee. Because the fee is a constant value, adding a field in the table to store the fee is not necessary. In this topic, you will use a query to calculate the total fine, including the administrative fee.

1. With the **LibraryFines–YourName** database open, open the FinesIssued query.

2. Switch to Design view.

3. Click in any cell in the *Telephone* column in the query design grid.

4. Click the Delete Columns button in the Query Setup group on the Query Tools Design tab.

   Buttons in the Query Setup group are used to modify a query by deleting columns or inserting new columns between existing fields.

5. Delete the *OverdueBookISBN* and *BookReturned* fields by completing steps similar to Steps 3 and 4.

6. With *FinePaid* the active field, click the Insert Columns button in the Query Setup group.

7. With an insertion point positioned in the *Field* box in the new column between *Amount* and *FinePaid*, type Fine with Admin Fee: [Amount]+2.50 and then press Enter.

8. Drag the right column boundary line in the gray field selector bar at the top of the design grid to widen the column as shown in the image below. Note that Access drops the zero at the end of the formula.

9. Click the Run button.

   Notice that the calculated field is not formatted the same as the *Amount* field, and that the column needs to be widened to show the entire column heading.

The text before the colon is the column heading for the new field *Fine with Admin Fee*. After the colon the mathematical expression *[Amount]+2.5* is stored. A field name used in a formula is typed within square brackets.

10. Click in any cell in the *Fine with Admin Fee* column, click the More button in the Records group, click *Field Width*, and then click the Best Fit button in the Column Width dialog box.

11. Switch to Design view.

12 Click in any cell in the calculated column in the query design grid and then click the Property Sheet button in the Show/Hide group on the Query Tools Design tab.

Step 12

13 Click in the *Format* box in the Property Sheet task pane, click the list box arrow that appears, and then click *Currency* at the drop-down list.

Step 13

14 Close the Property Sheet task pane.

15 Click the Run button.

16 Use *Save Object As* at the Save As backstage area to save the revised query as *FinesWithAdminFee*.

Quick Steps

Create a Calculated Field in a Query
1. Open query in Design view.
2. Insert column if necessary.
3. Click in *Field* box for calculated column.
4. Type formula.
5. Press Enter.
6. Run query.
7. Save query.

A table datasheet or query results datasheet should be viewed in Print Preview before printing to make adjustments as necessary to the orientation and margins.

These are the first five records in the FinesWithAdminFee query showing the *Fine with Admin Fee* column.

17 Click the File tab, click *Print*, and then click *Print Preview* at the Print backstage area.

The entire datasheet does not fit on one page in the default Portrait orientation.

18 Click the Landscape button in the Page Layout group on the Print Preview tab.

Notice that Access prints the query name and the current date at the top of the page and the page number at the bottom of the page.

19 Click the Close Print Preview button in the Close Preview group.

App Tip

Access does not save page layout options. If you change the datasheet orientation to landscape for printing purposes and then close the table, the datasheet reverts to the default portrait orientation when the table is reopened.

Step 18

Step 19

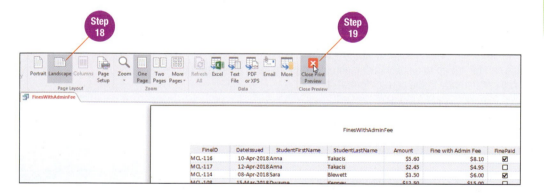

20 Close the FinesWithAdminFee query and then close the LibraryFines database.

View

Model Answer
Compare your completed file with the model answer.

Beyond Basics   Exporting Data from Access

Buttons in the Data group on the Print Preview tab are used to export the active table, query, form, or report shown in the work area. For example, click the PDF or XPS button to save a copy of a query results datasheet in a PDF file.

Topics Review

Topic	Key Concepts	Key Terms
12.1 Understanding Database Objects and Terminology	A database includes data stored in an organized manner to provide information for a variety of purposes.	database
	A database management system (DBMS) is a software program designed to organize, store, and maintain a database.	database management system (DBMS)
	An Access database is a collection of objects used to enter, maintain, and view data.	table
	Access opens with a Navigation pane along the left side of the window used to select an object to open and view data.	field
	A table stores data about a single topic or subject such as people, places, events, items, or other category.	record
	Each characteristic about the subject or topic of a table is called a *field*.	field value
	The data stored in a field is called a *field value*.	report
	A set of fields for one person, place, event, item, or other subject of a table is called a *record*.	form
	A table opens in a datasheet where the columns are fields and the rows are records.	query
	A report is an object used to view or print data from tables in a customized layout and with summary totals.	
	A form is another interface to view, enter, or edit data that shows only one record at a time in a customized layout.	
	Use buttons in the Record Navigation bar to scroll records in a form.	
	A query is used to combine fields from one or more tables in a single datasheet and may show all records or only some records that meet a condition.	
12.2 Adding Records to a Table Using a Datasheet	New records are added to a table by opening the table datasheet and then clicking the New (blank) record button in the Record Navigation bar.	New (blank) record
	Type field values in a new row at the bottom of the table datasheet, pressing Tab to move from one field to the next field.	
	A field that displays with a down-pointing arrow means that you can enter the field value by selecting an entry from a drop-down list.	
	Date fields can be set up to display underscores and hyphens to make sure dates are entered consistently in the correct format.	
	A field that displays with a check box stores *Yes* if the box is checked and *No* if the box is left empty.	
	Access automatically saves a new record as soon as you press Tab to move past the last field.	
12.3 Editing and Deleting Records Using a Datasheet	Edit a field value in a datasheet by selecting text to be changed and type new text, or by clicking to place an insertion point within a field (cell) and inserting or deleting text as required.	
	Function key F2 opens a field for editing.	
	The gray record selector bar along the left edge of a datasheet is used to select a record.	
	Click the Delete button in the Records group on the Home tab to delete the selected record from the table.	
	Access requires that you confirm a deletion before the record is removed.	
	Generally, records are not deleted until data has been copied to an archive database and/or a backup copy of the database has been made.	

continued…

Topic	Key Concepts	Key Terms
12.4 Adding, Editing, and Deleting Records Using a Form	A form is the preferred object for adding, editing, and deleting records because one record at a time is displayed in a columnar layout in the work area. Add new records and edit field values in records using the same techniques that you used for adding and editing records using a datasheet. To delete a record using a form, display the record, click the Delete button arrow in the Records group on the Home tab, and then click *Delete Record* at the drop-down list. In a database with many records, navigating to a record using the Find feature is more efficient.	
12.5 Finding and Replacing Data and Adjusting Column Widths in a Datasheet	Click in the column in a datasheet that contains the field value you want to locate and use the Find command to move to all occurrences of the *Find What* text. Use the Replace command to find all occurrences of an entry and replace the field value with new text. Activate any cell in a column for which the width needs to be adjusted, click the More button in the Records group on the Home tab, click *Field Width*, and then enter the desired width or click the Best Fit button. The Best Fit button in the Column Width dialog box adjusts the width of the column to accommodate the longest entry in the field.	Best Fit
12.6 Sorting and Filtering Records	Initially, a table is arranged alphanumerically by the primary key field values. A primary key is a field in the table that uniquely identifies each record such as *StudentID*. To sort by a field other than the primary key, click in the field by which to sort and then click the Ascending or Descending button in the Sort & Filter group on the Home tab. Filter a datasheet by clearing check boxes for items you do not want to view in the Sort & Filter list box accessed from the filter arrow next to the field name. Use the Toggle Filter button in the Sort & Filter group to redisplay all records. Sort by more than one field by selecting the columns before clicking the Ascending or Descending button. Change the sort order by moving a column left if you want to sort first by a field other than the leftmost column.	primary key
12.7 Creating a Query Using the Simple Query Wizard	Queries extract information from one or more tables in a single datasheet. The Simple Query Wizard helps you build a query by making selections in three dialog boxes. At the first Simple Query Wizard dialog box, choose each table or query and the fields in the order that you want them in the query results datasheet. At the second Simple Query Wizard dialog box, choose a detail or summary query. Assign a name to the query at the third Simple Query Wizard dialog box.	Simple Query Wizard

continued…

Topic	Key Concepts	Key Terms
12.8 Creating a Query Using Design View	Every object in Access has at least two views. Opening an object from the Navigation pane opens the table, query, form, or report in Datasheet view. Design view is used to set up or define the structure or layout of an object. Design view for a query presents a blank grid of columns in which you add the fields in the order you want them in the query results datasheet. Add table field list boxes to the query design grid at the Show Table dialog box. Double-click field names in the table field list boxes in the order you want the columns in the query results datasheet. The Run button is used after building a query in Design view to instruct Access to generate the query and show the query results datasheet.	Design view Run button
12.9 Entering Criteria to Select Records in a Query	Queries can be created that show only those records that meet one or more conditions. Use the View button in the Views group on the Home tab to switch between Datasheet view and Design view in a query. Type the criterion by which you want records selected in the *Criteria* box of the column by which records are to be selected. In a field that displays check boxes, the criterion is either *Yes* or *No*.	
12.10 Entering Multiple Criteria to Select Records and Sorting a Query	More than one criterion entered in the same *Criteria* row in the query design grid is an AND statement, which means each criterion must be met for the record to be selected. More than one criterion entered in *Criteria* rows one below the other is an OR statement, which means that any condition can be met for the record to be selected. Choose *Ascending* or *Descending* in the *Sort* list box for the column by which to sort query results in Design view. Access sorts a query by column left to right. If necessary, position the field to be sorted first to the left of another field that is to be sorted.	
12.11 Inserting and Deleting Columns, Creating a Calculated Field in a Query, and Previewing a Datasheet	A calculated field can be created in a query that generates values using a mathematical expression. Use the Delete Columns and Insert Columns buttons in the Query Setup group on the Query Tools Design tab to remove or add new columns in a query. A calculated column is created by typing in the *Field* box a column heading, a colon (:), and then the mathematical expression. Type a field name in a mathematical expression within square brackets. Open the Property Sheet task pane to change the format of a calculated field. A table or query results datasheet should be previewed before printing to make adjustments to page orientation and/or margins.	

 Recheck
Recheck your understanding of the topics covered in this chapter.

 **Workbook**
Chapter review and assessment resources are available in the *Workbook* ebook.

Chapter 13

Creating a Table, Form, and Report in Access

Precheck
Check your understanding of the topics covered in this chapter.

Creating a new database involves understanding the purpose of the database and the information the database user will need. The database designer needs to carefully analyze the collected data and decide how best to define and group the elements into logical units. Tables are created first because they are the basis for all other objects. Tables that need to be connected for queries, forms, or reports are joined in a relationship. Objects such as queries, forms, and reports are created after the tables and relationships are defined.

In Chapter 12, you examined an existing database and added and edited data in a table and form. You also created queries to select records for a variety of purposes. Now that you have seen how Access data interacts with objects, you are ready to build a new database on your own. In this chapter you will learn to create a new database, create a new table, assign a primary key, modify field properties, edit relationships, create a form, create a report, compact and repair a database, and create a backup copy of a database.

Learning Objectives

13.1 Create a new database and describe guidelines for designing tables

13.2 Create a new table using Datasheet view and assign a caption to a field

13.3 Create a new table using Design view and assign a primary key

13.4 Add a field to a table

13.5 Change the field size and add a default value for a field using Design view

13.6 Create a lookup list for a field

13.7 Identify a one-to-one relationship and a one-to-many relationship, and edit a relationship

13.8 Create and edit a form

13.9 Create, edit, and view a report

13.10 Compact and repair, and back up a database

 SNAP If you are a SNAP user, go to your SNAP Assignments page to complete the Precheck, Tutorials, and Recheck.

 Data Files
Before beginning this chapter, be sure you have copied the student data files for this course to your storage medium. Steps on downloading and extracting the data files are provided in Chapter 1, Topic 1.8, on pages 22–23.

13.1 Creating a New Database File and Understanding Table Design Guidelines

The first step in creating a database is to assign a name and storage location for the new database file. Because Access saves records automatically as data is added to a table, the file name and storage location are required in advance. Once the file is created, Access displays a blank table for you to fill in. Before you create a new table, you must carefully plan the fields and field names and identify a primary key. Although the tables you will create in this chapter have already been planned, the guidelines in Table 13.1 provide you with an overview of the table design process.

Table 13.1

Guidelines for Planning a New Table

Guideline	Description
Divide data into the smallest possible units	A field should be segmented into the smallest units of information to facilitate sorting and filtering. For example, a person's name could be split into three fields: first name, middle name, and last name.
Assign each field a name	Up to 64 characters can be used in a field name with a combination of letters, numbers, spaces, and some symbols. Database programmers prefer short field names with no spaces. A field to store a person's last name could be assigned the name *LName*, *Last*, *LastName*, or *Last_Name*. Short names are preferred because Access provides the ability to enter a longer descriptive title for column headings in datasheets, forms, and reports that is separate from the field name.
Assign each field a data type	Data type refers to the type of information that will be entered as field values. Look at examples of data to help you determine the data type. By assigning the most appropriate data type, Access can verify data as it is being entered for the correct format or type of characters. For example, a field defined as a Number field will cause Access to reject alphabetic letters typed into the field. The most common data types are Short Text, Number, Currency, Date/Time, and Yes/No. Data types are described in Table 13.2 in the next topic.
Decide the field to be used as a primary key	Each table should have one field that uniquely identifies a record, such as a student number, receipt number, or email address. Access creates an ID field automatically in a blank datasheet that can be used if the table data does not have a unique identifier. In some cases, a combination of two or more fields is used as a primary key.
Include a common identifier field in a table that will be joined to another table	Data should not be duplicated in a database. For example, a book title would not be stored in both the Books table and the Sales table. Instead, the book title is stored in the Books table only and a book ID field in the Sales table is used to join the two tables in a relationship. You will learn more about relationships in a later topic.

① Start Access 2016.

② Click *Blank desktop database* at the Access Start screen.

Step 2

At the Access Start screen, you can also choose to create a new database based on a template. Database templates have a set of predefined tables, queries, forms, and reports.

③ Select the current text in the *File Name* text box, type UsedBooks-YourName, and then click the Browse button (file folder icon).

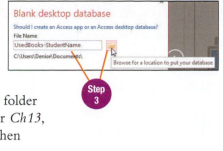

Step 3

④ At the File New Database dialog box, navigate to the CompletedTopicsByChapter folder on your storage medium, create a new folder *Ch13*, double-click to open the Ch13 folder, and then click OK.

⑤ Click the Create button.

Access creates the database file and opens a new table datasheet named *Table1* in the work area, as shown in Figure 13.1. You can create a new table using the blank datasheet.

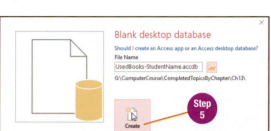

Step 5

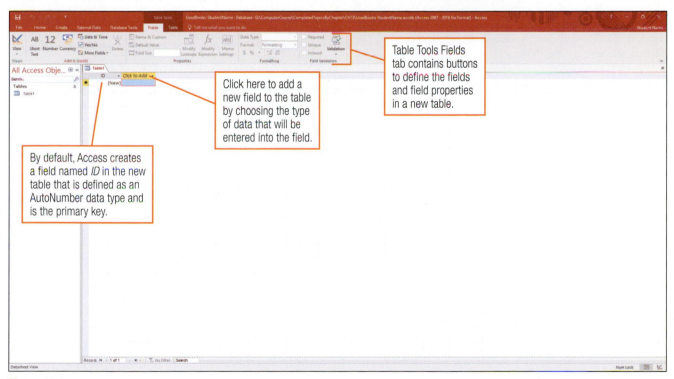

Table Tools Fields tab contains buttons to define the fields and field properties in a new table.

Click here to add a new field to the table by choosing the type of data that will be entered into the field.

By default, Access creates a field named *ID* in the new table that is defined as an AutoNumber data type and is the primary key.

Figure 13.1
Access creates a blank table datasheet in a new database file.

⑥ Leave the blank table datasheet open for the next topic.

Each table in a database should contain information about one subject only. In this chapter, you will create tables for a used-textbook database that a student organization may use to keep track of students, textbooks, and sales. The three tables you will create in this chapter are described as follows:

Books: shows the title, author, condition, and asking price for each book
Sales: tracks each sale with the date, amount, and payment method
Students: shows information about each student with textbooks for sale

App Tip

A database designer may use data models by creating sample forms and reports before creating tables to make sure all data elements are included in the table design.

13.2 Creating a New Table Using a Datasheet

Skills

Create a new table using a Datasheet

Add a caption for a field

Tutorial

Creating a Table in Datasheet View

Create a new table in a blank datasheet in Datasheet view by adding a column for each field. Begin by specifying the data type for a field and then typing the field name. Data types are described in Table 13.2. Once the fields have been defined, use the Save button on the Quick Access Toolbar to assign the table a name.

Table 13.2

Field Data Types

Data Type	Use for This Type of Field Value
Short Text	Alphanumeric text up to 255 characters for names, identification numbers, telephone numbers, or other similar data.
Number	Numeric data other than monetary values.
Currency	Monetary values such as sales, costs, or wages.
Date & Time or Date/Time	Dates or times that you want to verify, sort, select, or calculate.
Yes/No	Data that can only be Yes or No, or True or False.
Lookup & Relationship or Lookup Wizard	A drop-down list with field values from another table, or from a predefined list of items.
Long Text or Rich Text	Alphanumeric text of more than 255 characters. Select Rich Text to enable formatting options such as font, font color, bold, and italic in the field values.
AutoNumber	A unique number generated by Access to be used as an identifier field. Access generates sequential numbers starting from 1.
Hyperlink	Stores web addresses.
Attachment	Attach a file such as a picture to a field in a record.
Calculated Field	A formula calculates the field value using data in other fields.

1 With the **UsedBooks-YourName** database open and with the blank datasheet for Table1 open, click the Date & Time button in the Add & Delete group on the Table Tools Fields tab.

Step 1

Choosing the most appropriate data type for a field is important for sorting, calculating, and verifying data. Access expects dates to be entered in the format m/d/y unless the region setting in the Control Panel is changed to another format.

2 With *Field1* selected in the field name box for the new column, type SaleDate and then press Enter.

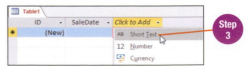

Step 2

The *Click to Add* column opens the data type drop-down list for the next new field. Add a new field using either the *Click to Add* drop-down list or the buttons in the Add & Delete group on the Table Tools Fields tab.

3 Click *Short Text* in the *Click to Add* drop-down list.

Step 3

④ Type BookID as the field name and then press Enter.

⑤ Click *Currency* in the *Click to Add* drop-down list, type Amount, and then press Enter.

⑥ Click *Short Text*, type PayMethod, and then press Enter.

⑦ Click the *SaleDate* field name to select the field.

⑧ Click the Name & Caption button in the Properties group on the Table Tools Fields tab.

The **Caption property** is used to type a descriptive title for a field that includes spaces between words or the full text of an abbreviated field name.

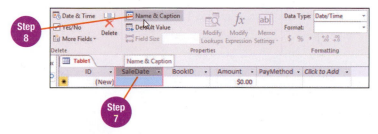

⑨ Click in the *Caption* text box, type Sale Date, and then press Enter or click OK at the Enter Field Properties dialog box.

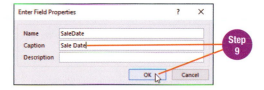

⑩ Click to select the *Amount* field, click the Name & Caption button, click in the *Caption* text box, type Sale Amount, and then press Enter or click OK.

⑪ Drag the right column boundary of the *Sale Amount* column until the entire column heading is visible.

⑫ Click to select the *PayMethod* field, click the Name & Caption button, click in the *Caption* text box, type Payment Method, and then press Enter or click OK.

⑬ Drag the right column boundary of the *Payment Method* column until the entire column heading is visible.

⑭ Click the Save button on the Quick Access Toolbar.

⑮ Type Sales in the *Table Name* text box in the Save As dialog box and then press Enter or click OK.

completed Sales table datasheet

⑯ Close the Sales table. Leave the database open for the next topic.

Quick Steps

Create a Field in a New Table
1. With blank table datasheet open, click required data type button in Add & Delete group on Table Tools Fields tab.
2. Type field name.
3. Press Enter.
OR
1. With blank datasheet open, click *Click to Add*.
2. Click required data type.
3. Type field name.
4. Press Enter.

Add a Caption to a Field
1. Select field.
2. Click Name & Caption button.
3. Click in *Caption* text box.
4. Type caption text.
5. Click OK.

Save a Table
1. Click Save button.
2. Type table name.
3. Click OK.

App Tip

Double-click the right column boundary to best fit the column width.

13.3 Creating a New Table Using Design View and Assigning a Primary Key

Skills

Create a new table using Design view

Assign a primary key

 Tutorials

Setting the Primary Key Field

Creating a Table in Design View

Managing Fields in Design View

A new table can be created in Design view in which fields are defined in rows in the top half of the work area. In the previous topic, the Sales table, created in a blank datasheet, had an ID field automatically created and designated as the primary key. In Design view, an ID field is not created for you. After creating the fields in the Design view window, assign the primary key to the field that will uniquely identify each record and then save the table.

1. With the **UsedBooks-YourName** database open, click the Create tab.

2. Click the Table Design button in the Tables group.

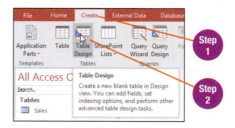

3. With the insertion point positioned in the first row of the *Field Name* column, type StudentID and then press Enter.

4. With *Short Text* in the *Data Type* column, press Enter to accept the default data type.

Press Enter at Step 5 to move past the optional *Description* entry.

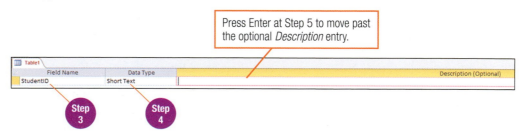

5. Press Enter to move past the *Description* column and move down to the next row to start a new field.

    Descriptions are optional entries. A description can be used to type additional information about a field or to enter instructions to end users who will see the description in the Status bar of a datasheet when the field is active.

6. Type LName in the *Field Name* column and then press Enter three times to move to the next row.

7. Enter the remaining fields as shown in the image below by completing a step similar to Step 6.

App Tip

You can also use the Tab key to move to the next column in Design view.

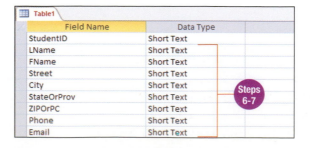

⑧ Click to place an insertion point within the *StudentID* field name.

⑨ Click the Primary Key button in the Tools group on the Table Tools Design tab.

A key icon in the field selector bar (gray bar along left edge of *Field Name* column) indicates the field is designated as the primary key for the table.

⑩ Click the Save button on the Quick Access Toolbar, type Students, and then press Enter or click OK.

⑪ Close the Students table.

⑫ Click the Create tab and then click the Table Design button.

App Tip

Access will not allow field values to be duplicated in a primary key—always choose a primary key field that you are sure will never have field values that repeat.

Quick Steps

Create a Table in Design View
1. Click Create tab.
2. Click Table Design button.
3. Type field name.
4. Press Enter.
5. If necessary, change data type.
6. Press Enter until new row is active.
7. Repeat Steps 3–6 until finished.
8. Assign primary key.
9. Save table.

Assign a Primary Key
1. If necessary, open table in Design view.
2. Click to place an insertion point in primary key field name.
3. Click Primary Key button.
4. Save table.

⑬ Create the first five fields in the new table using the default Short Text data type and without descriptions as follows:

> *BookID*
> *StudentID*
> *Title*
> *Author*
> *Condition*

⑭ Type AskPrice as the field name in the sixth row and then press Enter.

⑮ Click the *Data Type* list box arrow and then click *Currency* at the drop-down list.

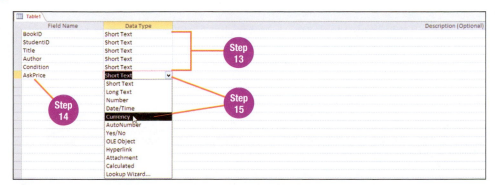

⑯ Click to place an insertion point within the *BookID* field name and then click the Primary Key button.

BookID is shown as the primary key field after Step 16 is completed.

⑰ Click the Save button on the Quick Access Toolbar, type Books, and then press Enter or click OK.

⑱ Close the Books table. Leave the database open for the next topic.

13.4 Adding Fields to an Existing Table

Skills

Add a field to a table

Tutorial

Editing and Deleting a Relationship

Open a table in Datasheet view and use the *Click to Add* column to add a new field, or make active a field in the datasheet and use the buttons in the Table Tools Fields tab to add a new field after the active field.

1. With the **UsedBooks-YourName** database open, open the Books table.

2. Click the *Click to Add* column heading and then click *Currency* at the drop-down list.

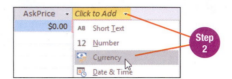

3. Type StopPrice as the field name and then press Enter.

4. Click the top part of the View button in the Views group on the Table Tools Fields tab to switch to Design view.

5. Click in the *Description* column in the *StopPrice* field row and then type Do not sell for lower than the student's stop price value.

6. Save and then close the Books table.

7. Open the Students table.

8. Click the *Phone* field name to select the column.

9. Click the Table Tools Fields tab and then click the Yes/No button in the Add & Delete group to add the new field in the column to the right of the *Phone* field.

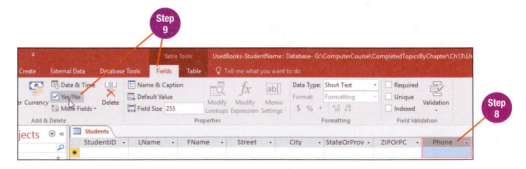

10. Type DirectDeposit as the field name and then press Enter.

11. Drag the right column boundary of the *DirectDeposit* field to the right until the entire column heading is visible.

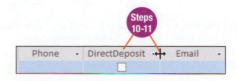

12 Save and then close the Students table.

13 Open the Books table.

14 Click the *StopPrice* field name to select the column.

15 Look at the message that displays in the Status bar.

Notice that the text typed in the *Description* column for the field in Design view at Step 5 displays here. Description entries also display in the Status bar when a form is open that is based upon the same table.

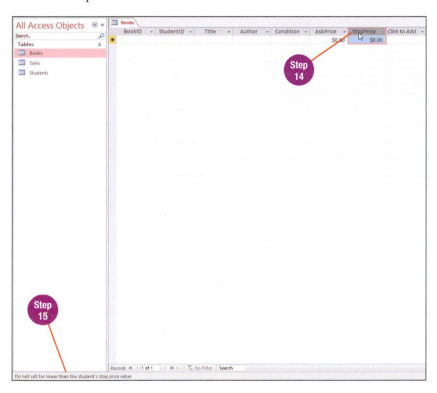

16 Close the Books table. Leave the database open for the next topic.

Alternative Method **Adding a New Field to a Table Using Design View**

Open a table in Design view and type a new field name in the next available row or make a field active and use the Insert Rows button in the Tools group on the Table Tools Design tab to add a new field above the active field.

Beyond Basics   **Deleting Fields**

Generally, a field that contains data should not be deleted, because deleting a field causes all field values to be removed from the database. However, if a field added to a table is considered unnecessary, remove the field by opening the table in either Datasheet view or Design view. In Datasheet view, make the field active and use the Delete button in the Add & Delete group on the Table Tools Fields tab. In Design view, make the field active and use the Delete Rows button on the Table Tools Design tab.

13.5 Modifying Field Properties Using Design View

Skills

Change the field size

Add a default value

Add a caption in Design view

Tutorials

Modifying Field Properties in Design View

Applying a Validation Rule in Design View

Applying a Validation Rule in Datasheet View

App Tip

Options in the Field Properties pane vary by the field data type. For example, a Date & Time field does not have the Field Size property.

App Tip

Adding an entry to the Default Value property not only saves time when a new record is added to the table, it also makes sure a field value is entered consistently and with correct spelling.

Each field in a table has a set of field properties associated with the field. A **field property** is a single characteristic or attribute of a field. For example, the field name is a field property, and the data type is another field property. The properties in each field can be modified to customize, format, or otherwise change the behavior of a field. The lower half of the work area of a table in Design view contains the **Field Properties pane** that is used to modify properties other than the field name, data type, and description.

1. With the **UsedBooks-YourName** database open, right-click the Students table name in the Navigation pane and then click *Design View* at the shortcut menu.

2. Click in the *StateOrProv* field name to select the field.

3. Double-click or drag to select *255* in the *Field Size* property box in the Field Properties pane, type *2*, and then press Enter.

Setting a field size for a state or province field ensures that all new field values use the two-character abbreviation for addressing letters or creating labels from the database.

4. Click in the *Default Value* property box, type MI, and then press Enter.

Access automatically adds quotation marks to the text in a *Default Value* property box for a Short Text data type. The text entered into a *Default Value* property box is automatically entered as the field value in new records; the end user presses Enter to accept the value, or types an alternative entry.

5. Click in the *Caption* property box and then type State or Province.

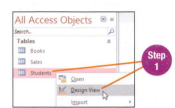

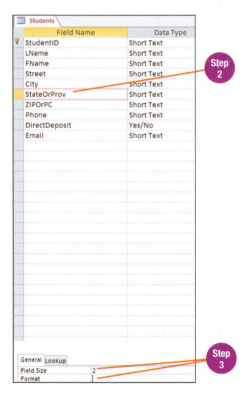

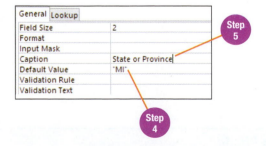

6. Click in the *StudentID* field name to select the field, click in the *Caption* property box, type Student ID, and then press Enter.

7. Add the following caption properties by completing a step similar to Step 6.

Field Name	Caption
LName	Last Name
FName	First Name
Street	Street Address
ZIPOrPC	ZIP or Postal Code
Phone	Telephone
DirectDeposit	Direct Deposit
Email	Email Address

8. Save the table.

9. Click the top part of the View button in the Views group to switch to Datasheet view.

10. Double-click the right column boundary of column headings that are not entirely visible to best fit the column widths.

   Notice that *MI* appears in the *State or Province* column by default.

11. Type your name and a fictitious student ID and address into a new record in the datasheet. If necessary, adjust column widths to show all data in all columns.

<table>
<thead>
<tr><th colspan="2">Quick Steps</th></tr>
</thead>
</table>

Quick Steps

Modify a Field Property
1. Open table in Design view.
2. Click in field name in top half of work area.
3. Click or select current value in field property box.
4. Type or select option from drop-down list.
5. Save table.

Steps 10-11

12. Close the table, saving the changes to the table layout. Leave the database open for the next topic.

Alternative Method   Modifying Field Properties Using a Datasheet

Field properties can also be changed for a table open in Datasheet view. Buttons in the Properties group on the Table Tools Fields tab can be used to enter a caption, default value, or field size. Modify or apply format, data validation, or required properties (see Beyond Basics) with buttons in the Formatting and Field Validation groups on the Table Tools Fields tab.

Beyond Basics   Formatting and Data Validation Field Properties

There are other properties that are often changed for a field:

- *Format.* Modifies the display of the field value. For example, a date can be formatted to display as a long date or a medium date.

- *Validation.* Use a validation rule to enter an expression that is tested as each new field value is typed into a record. For example, the expression *>=5* in the *SaleAmount* field would ensure no amounts less than $5.00 are entered.

- *Required.* Select *Yes* to ensure that the field is not left blank in a new record. For example, a ZIP or Postal Code field should not be left blank.

13.6 Creating a Lookup List

A **lookup list** is a drop-down list of field values that appears when a field is made active while new records are added in a datasheet or form. The list entries can be a fixed list or field values from another table can be shown in the list. A lookup list has many advantages, including consistency, accuracy, and efficiency when adding data in new records. Access provides the **Lookup Wizard** to assist with creating a lookup list's field properties by choosing options and typing values using dialog boxes.

1. With the **UsedBooks–YourName** database open, right-click the Books table name in the Navigation pane and then click *Design View* at the shortcut menu.

2. Click in the *Condition* field name to select the field.

3. Click the *Data Type* list box arrow and then click *Lookup Wizard* at the drop-down list.

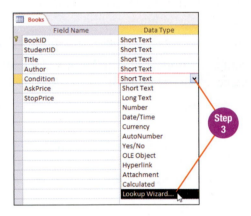

4. Click *I will type in the values that I want.* at the first Lookup Wizard dialog box and then click Next.

5. Click in the first blank row below *Col1* at the second Lookup Wizard dialog box and then type Excellent – No wear or markings.

6. Drag the right column boundary approximately two inches to the right to increase the column width.

7. Click in the second row and then type Very Good – Minor wear to cover.

8. Type the remaining entries in the list as shown in the image at right.

9. Click Next.

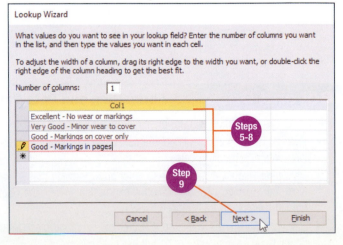

10 Click Finish at the last Lookup Wizard dialog box.

11 Click the Lookup tab in the Field Properties pane.

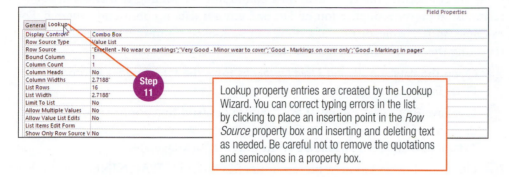

Field Properties

General	Lookup
Display Control	Combo Box
Row Source Type	Value List
Row Source	"Excellent - No wear or markings";"Very Good - Minor wear to cover";"Good - Markings on cover only";"Good - Markings in pages"
Bound Column	1
Column Count	1
Column Heads	No
Column Widths	2.7188"
List Rows	16
List Width	2.7188"
Limit To List	No
Allow Multiple Values	No
Allow Value List Edits	No
List Items Edit Form	
Show Only Row Source V	No

Step 11

Lookup property entries are created by the Lookup Wizard. You can correct typing errors in the list by clicking to place an insertion point in the *Row Source* property box and inserting and deleting text as needed. Be careful not to remove the quotations and semicolons in a property box.

12 Save the table and then switch to Datasheet view.

13 Enter the following record as shown in the image below with the fictitious ID you created for yourself in the *StudentID* column. At the *Condition* field, click the down-pointing arrow that appears and then click *Good - Markings on cover only* at the drop-down list.

14 Adjust column widths as needed to show all data in each column.

Books						
BookID	StudentID	Title	Author	Condition	AskPrice	StopPrice
DJ-1	999	Pride and Prejudice	Austen	Good - Markings on cover only	$15.00	$10.00
*					$0.00	$0.00

Steps 13-14

15 Close the table, saving the changes to the table layout.

16 Close the database.

Quick Steps

Create a Lookup List
1. Open table in Design view.
2. Click *Data Type* list arrow for field.
3. Click *Lookup Wizard*.
4. Click *I will type in the values that I want.*
5. Click Next.
6. Type list entries in *Col1* column.
7. Adjust column width.
8. Click Next.
9. Click Finish.
10. Save table.

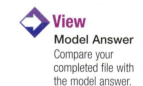

View

Model Answer
Compare your completed file with the model answer.

Beyond Basics   **Creating a Lookup List with Field Values in Another Table**

To create a lookup list in which the entries are field values from a field in another table, proceed through the dialog boxes in the Lookup Wizard as follows:

1. Select *I want the lookup field to get the values from another table or query.*

2. Select the table or query name that contains the field values you want to use in the list.

3. Move the fields you want displayed in the drop-down list from the Available Fields list box to the Selected Fields list box.

4. Select a field to sort the list entries or leave empty for an unsorted list.

5. Adjust column widths as needed and/or uncheck *Hide key column*.

6. Select the field that contains the field value you want to store if more than one field was chosen at Step 3.

13.7 Displaying and Editing a Relationship

Skills

Identify a one-to-one relationship

Enforce referential integrity

Identify a one-to-many relationship

Tutorials

Creating a One-to-One Relationship

Creating a One-to-Many Relationship

Creating a Relationship Report

A relationship allows you to create queries, forms, or reports with fields from two tables by connecting the two tables on a common field. Joining two tables in a relationship prevents duplication of data and ensures data is consistently entered. For example, an ID, name, or title of a book can be looked up in one table rather than repeating the information in another table.

1. Open the database named *UsedTextbooks* from the Ch13 folder in Student_Data_Files.

2. Use Save As to save a copy of the database as **UsedTextbooks-YourName** in the Ch13 folder in CompletedTopicsbyChapter. Accept the default options *Save Database As* and *Access Database* at the Save As backstage area.

3. Click the Enable Content button in the SECURITY WARNING message bar and close the SAVE CHANGES message bar if the message bar appears.

This file is similar to the database you have been working on in this chapter but with the Books table modified, additional lookup lists, and with 10 records added to each table.

4. Open the Books table, review the datasheet, and then close the table.

5. Open the Sales table, review the datasheet, and then close the table.

6. Open the Students table and then change *Doe* in the last record of the *Last Name* field to your last name.

7. Change *Jane* in the last record of the *First Name* field to your first name and then close the table.

8. Click the Database Tools tab and then click the Relationships button in the Relationships group.

A field list box for each table is located in the Relationships window. A black join line connecting two table field list boxes indicates a relationship. Observe that each join line connects the tables on a common field name.

9. Click to select the black join line that connects the Books table field list box to the Sales table field list box, and then click the Edit Relationships button in the Tools group on the Relationship Tools Design tab.

In the Edit Relationships dialog box that appears, *One-To-One* is shown in the *Relationship Type* section. A **one-to-one relationship** means that the two tables are joined on the primary key in each table. (*BookID* displays a key next to the field name in each table field list box.) In this type of relationship, only one record can exist for the same *BookID* in each table.

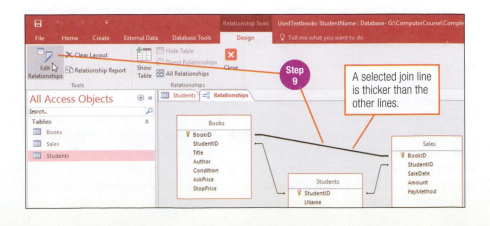

A selected join line is thicker than the other lines.

10 Click to insert a check mark in the *Enforce Referential Integrity* check box and then click OK.

Turning on **Enforce Referential Integrity** means that a record in Books is entered first before a record with a matching *BookID* is entered in Sales. Books is the left table name below *Table /Query*, and Sales is the right table name below *Related Table/Query*. The table below *Table /Query* is the one for which referential integrity is applied—the table in which new records are entered first. The table shown at the left is also referred to as the **primary table** (the table in which the joined field is the primary key and in which new records should be entered first).

11 Click to select the black join line that connects the Books table field list box to the Students table field list box and then click the Edit Relationships button.

12 Click to insert a check mark in the *Enforce Referential Integrity* check box and then click OK.

A **one-to-many relationship** occurs when the common field used to join the two tables is the primary key in only one table (the primary table). *One* student can have *many* textbooks for sale. In this instance, a record must first be entered into Students (primary table) before a record with a matching student ID can be entered into Books (related table). A field added to a related table that is not a primary key and is included for the purpose of creating a relationship is called a **foreign key**.

13 Click the Close button in the Relationships group. Click Yes if prompted to save changes to the layout of the relationships and leave the database open for the next topic.

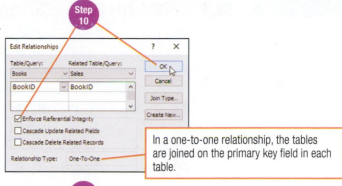

Step 10

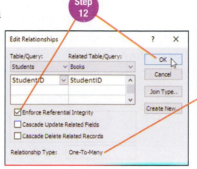

Step 12

In a one-to-one relationship, the tables are joined on the primary key field in each table.

In a one-to-many relationship, the field that joins the two tables is the primary key in one table and the foreign key in the other table.

Oops! !

Message appears that the database engine could not lock table? This happens when the Students table is currently open. Click Cancel, close the Students table and then try Step 12 again.

Quick Steps

Display Relationships
1. Click Database Tools tab.
2. Click Relationships button.

Enforce Referential Integrity
1. Display Relationships.
2. Click to select black join line.
3. Click Edit Relationships button.
4. Click *Enforce Referential Integrity* check box.
5. Click OK.

App Tip
A one-to-many relationship is the most common type of relationship in databases.

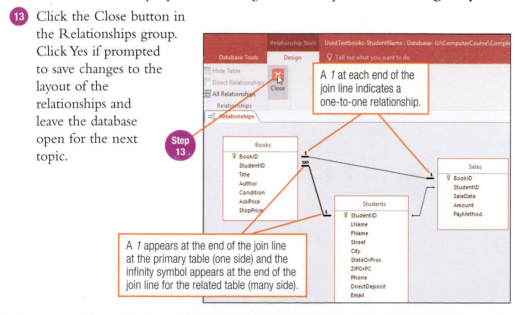

A *1* at each end of the join line indicates a one-to-one relationship.

Step 13

A *1* appears at the end of the join line at the primary table (one side) and the infinity symbol appears at the end of the join line for the related table (many side).

Beyond Basics   Creating Relationships

To create a new relationship, open the Relationships window and then drag the common field name from the primary table field list box to the related table field list box. Always drag the field name starting from the primary table. Use the Show Table button in the Relationships group if you need to add a table to the Relationships window in order to create a relationship.

13.8 Creating and Editing a Form

Skills

Create a form

Apply a theme

Add and format
a picture on a form

Format the form title

Tutorials

Formatting Table Data

Managing Control
Objects in a Form

Applying Conditional
Formatting to a Form

Creating a Form Using
the Form Button

The Forms group on the Create tab includes buttons to create forms ranging from a tool to create a simple form that adds all the fields in the selected table, to tools for more complex forms that work with multiple tables. Once created, a form can be modified using buttons on the Form Layout Tools Design, Arrange, and Format tabs.

1 With the **UsedTextbooks–YourName** database open, click to select the Books table name in the Navigation pane if Books is not already selected.

2 Click the Create tab.

3 Click the Form button in the Forms group.

A form is created with all the fields in the selected table arranged in a vertical layout and displayed in Layout view. **Layout view** is the view in which you edit a report or form structure and appearance using buttons on the Form Layout Tools tabs. **Form view** is the view in which data is viewed, entered, and updated in the form and is the view in which a form is opened from the Navigation pane.

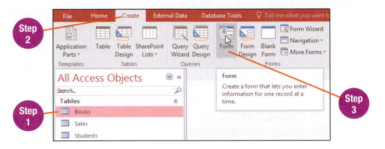

App Tip

Buttons to switch between views are at the right end of the Status bar.

4 Click the Form Layout Tools Design tab if the tab is not already active and then click the Themes button in the Themes group.

5 Click *Slice* in the *Office* section.

6 Click the Logo button in the Header/Footer group.

7   At the Insert Picture dialog box, navigate to the Ch13 folder in Student_Data_ Files and then double-click the file named *Textbooks*.

The picture is inserted into the selected logo **control object** near the top left of the form. A control object is a rectangular content placeholder in a form or report. Each control object can be selected and edited to modify the appearance of the content.

8   With the logo control object still selected, click the Property Sheet button in the Tools group.

9   Click the Format tab in the Property Sheet task pane, click in the *Size Mode* property box, click the down-pointing arrow that appears, and then click *Zoom*.

10   Select the current value in the *Width* property box and then type 1.75.

11   Select the current value in the *Height* property box and then type 1.25.

12   Close the Property Sheet task pane.

13   Click the *Books* title to select the control object.

An orange border around the control object indicates the object is selected.

14   Click the Form Layout Tools Format tab.

15   Click the Font Size option box arrow and then click *48* at the drop-down list.

16   Click the Save button on the Quick Access Toolbar and then click OK at the Save As dialog box to accept the default form name *Books*.

17   Close the Books form.

18   Double-click the Books form name in the Navigation pane to reopen the form, scroll through a few records, and then close the form. Leave the database open for the next topic.

App Tip

Control objects are placeholders for pictures, text, field names, and field values.

App Tip

The *Zoom* option for the *Size Mode* property fits the picture to the control object size maintaining proportions.

App Tip

You can also drag the orange border on a selected control object to resize the object.

Quick Steps

Create a Form
1.   Click to select table or query name in Navigation pane.
2.   Click Create tab.
3.   Click Form button.
4.   Modify form as required.
5.   Click Save button.
6.   Type form name.
7.   Click OK.

Beyond Basics   Creating a Form Using the Form Wizard

Use the Form Wizard button in the Forms group on the Create tab to create a form that uses fields from one or more tables. With the wizard, you also have control over the fields from each table that are included on the form and the form layout. Fields from related tables can be arranged in a columnar, tabular, datasheet, or justified layout.

13.9 Creating, Editing, and Viewing a Report

Skills

Create a report

Resize control objects

**Tutorials**

Creating a Report

Creating a Report
Using the Report
Wizard

Formatting a Report

Managing Control
Objects in a Report

Customizing a Report
in Print Preview

Create a report using techniques similar to those used to create a form. The Reports group on the Create tab has a Report tool similar to the Form tool. Other buttons in the Reports group include options to design a report from a blank page, create a report using the Report Wizard, or generate mailing labels using the Label Wizard. Modify a report with buttons in the Report Layout Tools tabs. Change the page layout options for printing purposes with buttons in the Report Layout Tools Page Setup tab.

1 With the **UsedTextbooks-YourName** database open, click to select the Sales table name in the Navigation pane.

2 Click the Create tab and then click the Report button in the Reports group.

A report is created with all the fields in the Sales table arranged in a tabular layout. By default, Access includes the current date and time, page numbering, and totals for numeric fields in all reports.

App Tip

Reports use the same theme as forms so that all objects have a consistent look.

Oops! !

Textbooks image not shown? Access defaults to the last folder used at the Insert Picture dialog box. If necessary, navigate to the Ch13 folder in Student_ Data_Files.

App Tip

A control object is filled with pound symbols (#) if the object is made too narrow for the content to be fully displayed. In that case, increase the width of the control object until the content redisplays.

3 Click the Logo button in the Header/Footer group and then double-click the file named *Textbooks*.

4 Open the Property Sheet, change the *Size Mode*, *Width*, and *Height* properties to the same settings that you applied to the picture in the previous topic, and then close the Property Sheet task pane.

5 Click to select the *Sales* report title control object, click the Report Layout Tools Format tab, click the Font Size option box arrow, and then click *48*.

6 Click to select the current date control object near the top right of the report.

7 Drag the right border of the control object to the left until the control ends just left of the vertical dashed line that extends the height of the report. If pound symbols display after resizing the control, drag the left border to the left until the date displays again.

The vertical dashed line indicates a page break. Resize control objects so that all objects are to the left of the vertical dashed line to fit on one page.

Step 7

Steps 3-4   **Step 5**

8 Click to select the *Book ID* column heading control object.

9 Drag the right border of the control object left approximately one-half inch to resize the object to the approximate width shown in the image at right.

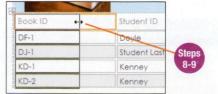

Steps 8-9

Notice that resizing a column heading control object resizes the entire column.

10 Click to select the *Page 1 of 1* control object and resize the control until the right border is just left of the vertical dashed line.

11 Click to select the control object with the total at the bottom of the *Sale Amount* column and then drag the bottom border of the control down until the value is entirely visible within the object.

12 Click the Print Preview button near the right end of the Status bar. Compare your report with the one shown in Figure 13.2. If necessary, switch to Layout view, resize control objects, and then switch back to Print Preview.

13 Click the Close Print Preview button in the Close Preview group.

Figure 13.2
Shown is the Sales report for Topic 13.9.

14 Click the Report View button near the right end of the Status bar.

  Report view is the view in which a report opens from the Navigation pane. Report view is used to view data on the screen instead of printing a hard copy. A report cannot be edited in Report view.

15 Close the Sales report, saving changes to the report design and accepting the default report name of *Sales*. Leave the database open for the next topic.

Oops!

Having difficulty resizing controls using touch? Open the Property Sheet for a selected control and change the *Width* and *Height* values. Use *.25* for the *Height* of the column total control.

Quick Steps

Create a Report
1. Click to select table or query name in Navigation pane.
2. Click Create tab.
3. Click Report button.
4. Modify report as required.
5. Click Save button.
6. Type report name.
7. Click OK.

App Tip

Select the control object for a column heading to change color or alignment options on the Report Layout Tools Format tab to make column headings stand out from the report. You can also select the control objects for the data below the column heading and change format options so that the data is formatted differently than the headings. For example, column data can be centered below a long column heading.

Beyond Basics   Grouping and Sorting a Report

The Group & Sort button in the Grouping & Totals group on the Report Layout Tools Design tab toggles on and off the Group, Sort, and Total pane at the bottom of the work area. Turn on the pane and use the Add a group and Add a sort buttons to change the arrangement of records in the report.

13.10 Compacting, Repairing, and Backing Up a Database

Skills

Compact and repair a database

Create a backup copy of a database

A database file becomes larger and fragmented over time as new records are added, edited, and deleted. The file size for the database may become larger than is necessary if the space previously used by records that have since been deleted is not compacted. Access provides a **Compact & Repair Database button** that compresses unused space reducing the file size. The compacting process eliminates unused space in the file. Backing up a database file should be done regularly for historical record keeping and data loss prevention purposes.

1 With the **UsedTextbooks–YourName** database open, click the File tab.

2 At the Info backstage area, click the Compact & Repair Database button.

Access closes all objects and the Navigation pane during a compact and repair routine. The Navigation pane redisplays when the compacting and repairing is complete. For larger database files, compacting and repairing may take a few moments to process.

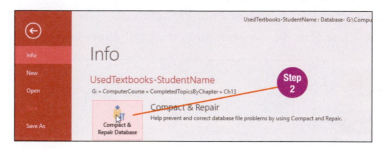

App Tip

If a database is shared, make sure no one else is using the database before starting a compact and repair operation.

3 Click the File tab and then click *Options*.

At the Access Options dialog box, you can set the database file to compact and repair each time the file is closed.

4 Click *Current Database* in the left pane of the Access Options dialog box.

5 Click to insert a check mark in the *Compact on Close* check box in the *Application Options* section.

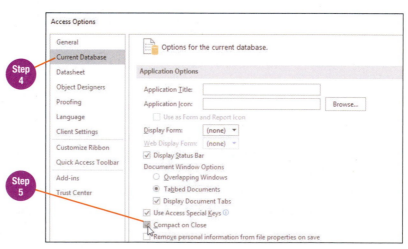

6 Click OK to close the Access Options dialog box.

7 Click OK at the message box that says the database must be closed and reopened for the option to take effect.

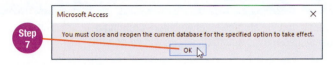

8 Click the File tab and then click *Save As*.

9 At the Save As backstage area, click *Back Up Database* in the *Advanced* section of the Save Database As panel.

10 Click the Save As button.

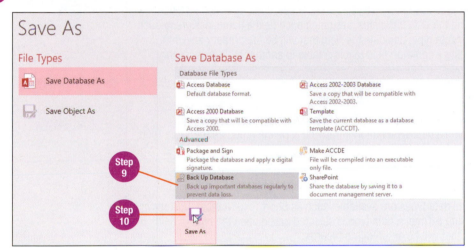

11 Click the Save button at the Save As dialog box.

    By default, the backup copy of the database is saved in the same folder as the current database and the file name is the same database file name with the current date added after an underscore at the end of the name, for example, UsedTextbooks-YourName_*currentdate*.

12 Close the database.

Beyond Basics **Encrypting a Database with a Password**

Assign a password to a database to prevent unauthorized access to confidential data stored in a database. A database has to be opened in exclusive mode to assign a password. To do this, close the current database and then display the Open dialog box. Navigate to the location of the database, select the database file, and then use the down-pointing arrow on the Open button to choose *Open Exclusive*. Enable content and then display the Info backstage area. Click the Encrypt with Password button and then type the password twice in the Set Database Password dialog box.

Topics Review

Topic	Key Concepts	Key Terms
13.1 Creating a New Database File and Understanding Table Design Guidelines	Access requires the name and file storage location before creating a new database because Access saves automatically as you work with data.	
	To create a new database file, choose *Blank desktop database* at the Access Start screen, type the file name, and browse to the desired drive and/or folder.	
	When a new database is created, Access displays a blank table datasheet.	
	Planning a new table involves several steps, some of which include dividing the data into fields, assigning each field a name, assigning each field a data type, deciding the field that will be the primary key, and including a common field to join a table to another table if necessary.	
13.2 Creating a New Table Using a Datasheet	In a blank table datasheet, begin a new field by first selecting the data type and then typing the field name.	Caption property
	Choose the data type from the *Click to Add* drop-down list or by choosing a data type button in the Add & Delete group on the Table Tools Fields tab.	
	An entry in the Caption property is used as a descriptive title that becomes the column heading for a field in a datasheet.	
	Several data types are available, such as Short Text, Number, Currency, and Date & Time. A data type is selected based upon the type of field value that will be entered into records.	
	Save a table by clicking the Save button on the Quick Access Toolbar and then entering a name for the table.	
13.3 Creating a New Table Using Design View and Assigning a Primary Key	In Design view, a new table is created by defining fields in rows in the top half of the work area.	
	Type a field name in the first row in the *Field Name* column and then specify the data type using the data type drop-down list.	
	An optional description can be added for a field with additional information about the purpose of the field or with instructions on what to type into the field.	
	Once the fields are defined, assign the primary key by placing an insertion point anywhere within the field name and then clicking the Primary Key button in the Tools group on the Table Tools Design tab.	
13.4 Adding Fields to an Existing Table	Open a table and use the *Click to Add* column to add a new field to the end of an existing table datasheet.	
	Select a column in a datasheet and use the buttons in the Add & Delete group on the Table Tools Fields tab to add a new field to the right of the selected field.	

continued…

Topic	Key Concepts	Key Terms
13.5 Modifying Field Properties Using Design View	Each field in a table has a set of associated field properties.	field property
	A field property is a single characteristic or attribute of a field that customizes, formats, or changes the behavior of the field.	Field Properties pane
	The lower half of the Design view window is the Field Properties pane in which properties for a selected field are modified.	
	The Field Size property is used to limit the number of characters that can be entered into a field.	
	The Default Value property is used to specify a field value that is automatically entered in the field in new records.	
	Other field properties that are often modified are the *Format*, *Validation*, and *Required* field properties.	
13.6 Creating a Lookup List	A lookup list is a drop-down list of items that displays when a field is made active in a datasheet or form.	lookup list
	Items in the lookup list can be predefined or extracted from one or more fields in another table.	Lookup Wizard
	The Lookup Wizard presents a series of dialog boxes to help create the field properties for a lookup list field.	
	Create a custom list of predefined entries by choosing *I will type in the values that I want.* at the first Lookup Wizard dialog box.	
	Type the list entries in the *Col1* column and adjust the column width at the second Lookup Wizard dialog box. Click Finish at the third dialog box.	
13.7 Displaying and Editing a Relationship	A relationship is when two tables are joined together on a common field.	one-to-one relationship
	Black join lines connecting a common field between two table field list boxes indicates a relationship has been created.	Enforce Referential Integrity
	A one-to-one relationship means that the two tables are joined on the primary key in each table.	primary table
	Enforce Referential Integrity causes Access to check that new records are entered into the primary table first before records with a matching field value can be entered into the related table.	one-to-many relationship
	A primary table is the table in which the common field is the primary key and into which new records are entered first. The primary table name is the table name at the left below *Table/Query* in the Edit Relationships dialog box.	foreign key
	In a one-to-many relationship, the common field used to join the tables is the primary key in only one table (the primary table).	
	A field added to a table that is not a primary key and is added for the purpose of creating a relationship is called a *foreign key*.	

continued…

Topic	Key Concepts	Key Terms
13.8 Creating and Editing a Form	The Form button in the Forms group on the Create tab creates a new form with all fields in the selected table or query arranged in a columnar layout.	Layout view Form view control object
	Layout view is the view in which you modify a report or form structure and appearance using buttons in the three Form Layout Tools tabs.	
	Form view is the view in which a form is displayed when opened from the Navigation pane and is used to add, edit, and delete data.	
	The Themes button is used to change the color scheme and fonts for a form.	
	Use the Logo button to choose a picture to display in the Logo control object near the top left of the form.	
	A control object is a rectangular placeholder for content.	
	Each control object can be selected and modified to change the appearance of the control content.	
	Open the Property Sheet task pane to make changes to the appearance of a selected picture.	
	Change the *Size Mode* property of a picture to *Zoom* to fit the content to the object size with the height and width proportions maintained.	
	A control object can be resized by changing the values for the *Width* and *Height* in the Property Sheet task pane.	
13.9 Creating, Editing, and Viewing a Report	The Report tool in the Reports group on the Create tab creates a report with all the fields in the selected table or query in a tabular arrangement.	Report view
	Buttons in the Report Layout Tools Page Setup tab are used to change page layout options for printing purposes.	
	Access creates a current date and time control, a page number control, and a total control for each numeric column in a new report.	
	The vertical dashed line in a report indicates a page break.	
	Resizing a column heading control object resizes the entire column.	
	Report view is the view in which a report is displayed when opened from the Navigation pane and is the view that displays the data in the report.	
13.10 Compacting, Repairing, and Backing Up a Database	The compact and repair process eliminates unused space in the database file.	Compact & Repair Database button
	Use the Compact & Repair Database button at the Info backstage area to perform a compact and repair operation.	
	During the compact and repair routine, Access closes all objects and the Navigation pane.	
	Turn on the *Compact on Close* option at the Access Options dialog box with *Current Database* selected to instruct Access to perform a compact and repair operation each time the database is closed.	
	Display the Save As backstage area and choose *Back Up Database* in the *Advanced* section to create a backup copy of the current database.	
	Access adds the current date after an underscore character to the end of the current database file name when a backup is created.	

 Recheck
Recheck your understanding of the topics covered in this chapter.

 **Workbook**
Chapter review and assessment resources are available in the *Workbook* ebook.

Chapter 14

Integrating Word, Excel, PowerPoint, and Access Content

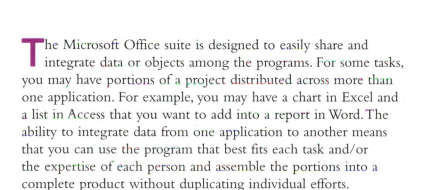

 Precheck

Check your understanding of the topics covered in this chapter.

The Microsoft Office suite is designed to easily share and integrate data or objects among the programs. For some tasks, you may have portions of a project distributed across more than one application. For example, you may have a chart in Excel and a list in Access that you want to add into a report in Word. The ability to integrate data from one application to another means that you can use the program that best fits each task and/or the expertise of each person and assemble the portions into a complete product without duplicating individual efforts.

In Chapter 3, you used the Copy and Paste buttons in the Clipboard group to copy text, a picture, and a chart between Word, Excel, and PowerPoint. Copy and paste is the best method for situations when the data to be shared is not large and is not likely to need updating. In this chapter, you will learn other methods for integrating data and objects that include importing, exporting, embedding, and linking.

Learning Objectives

14.1 Import Excel worksheet data into a table in Access

14.2 Export an Access query to Excel

14.3 Embed an Excel chart into a Word document

14.4 Embed Excel data into a PowerPoint presentation and edit the embedded data

14.5 Link an Excel chart with a PowerPoint presentation, edit the chart, and update the link

 SNAP If you are a SNAP user, go to your SNAP Assignments page to complete the Precheck, Tutorials, and Recheck.

 Data Files

Before beginning this chapter, be sure you have copied the student data files for this course to your storage medium. Steps on downloading and extracting the data files are provided in Chapter 1, Topic 1.8, on pages 22–23.

14.1 Importing Excel Worksheet Data into Access

Tutorials

Importing Data to a New Table

Linking Data to a New Table

A new Access table can be created from data in an Excel worksheet, or Excel data can be appended to the bottom of an existing Access table. Because an Excel worksheet and an Access datasheet use the same column and row structure, the two programs are often used to interchange data. To facilitate the import, the Excel worksheet should be set up like an Access datasheet, with the field names in the first row and with no blank rows or columns within the data. The **Import Spreadsheet Wizard** in Access is used to facilitate the import operation by providing a series of dialog boxes that prompt the user through the steps to create the table or add data to the bottom of an existing datasheet.

1. Start Access 2016 and open the **Parking** database from the Ch14 folder in Student_Data_Files.

2. Use Save As to save a copy of the database as **Parking-YourName** in a new folder *Ch14* within CompletedTopicsByChapter. Accept the default options *Save Database As* and *Access Database* at the Save As backstage area.

3. Click the Enable Content button in the SECURITY WARNING message bar.

4. Click the External Data tab.

5. Click the Excel button in the Import & Link group.

6. Click the Browse button in the Get External Data – Excel Spreadsheet dialog box.

7. Navigate to the Ch14 folder in Student_Data_Files at the File Open dialog box and then double-click *ParkingRecords*.

8. Click OK to accept the default option *Import the source data into a new table in the current database*.

Use the Append option if the table already exists in the database and you want to add new records from an Excel worksheet to the end of the existing datasheet table.

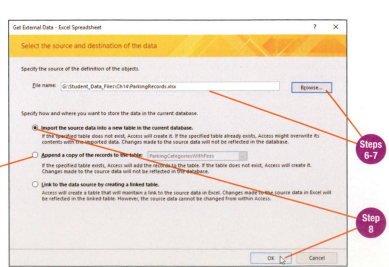

9   At the first Import Spreadsheet Wizard dialog box, Click Next to accept the worksheet labeled *Student Parking Records*.

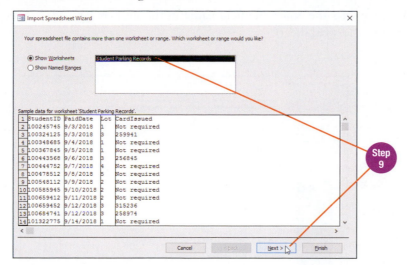

10   At the second Import Spreadsheet Wizard dialog box, click Next with a check mark already inserted in the *First Row Contains Column Headings* check box.

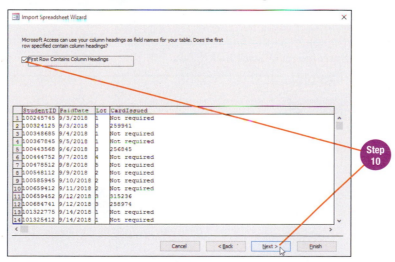

11   At the third Import Spreadsheet Wizard dialog box, click the *PaidDate* column heading and look at the option selected in the *Data Type* list box.

Notice that Access has correctly identified the data as a Date field. At this dialog box, you can review each column and modify the options in the *Field Options* section as needed, or you can elect to make changes in Design view after the import is completed. If a column exists in the Excel worksheet that you do not wish to import into the table, select the column and insert a check mark in the *Do not import field (Skip)* check box.

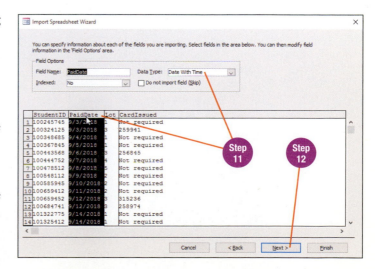

12   Click Next.

13 At the fourth Import Spreadsheet Wizard dialog box, click *Choose my own primary key*.

Access inserts the *StudentID* field name in the list box next to the option (the first column in the worksheet).

14 Click Next.

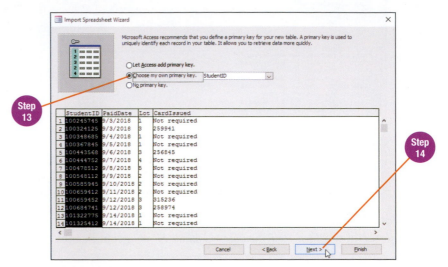

15 Type ParkingSales in the *Import to Table* text box and then click Finish at the last Import Spreadsheet Wizard dialog box.

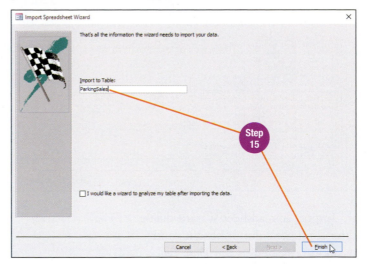

16 Click Close to finish the import without saving the import steps at the Get External Data – Excel Spreadsheet dialog box.

For situations in which you frequently import from Excel to Access, you can save the import specifications so that you can repeat the import later using the same settings.

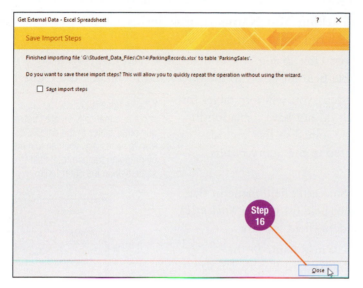

17 Open the ParkingSales table from the Navigation pane and review the datasheet.

18 Switch to Design view and review the field names and data types for the new table.

19 Click the *Data Type* list box arrow for the *Lot* field name and then click *Short Text*.

In the Parking database, the Lot field should be defined as Short Text, because lot numbers are not field values that you would add or subtract.

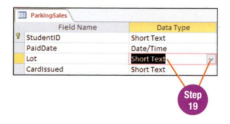

20 Save and then close the table. Leave the database open for the next topic.

App Tip

You can import text into a Word document from another source at the Open dialog box. Word tries to automatically convert text from another file type into a new Word document. Word successfully imports text from text files, rich text format files, OpenDocument text files, and WordPerfect files. You can even open a PDF document in Word, and Word converts the PDF file into an editable Word document.

View
Model Answer
Compare your completed file with the model answer.

Beyond Basics   **Linking an Excel Worksheet to an Access Table**

The option *Link to the data source by creating a linked table* at the Get External Data – Excel Spreadsheet dialog box is used when the data that is being imported is likely to be updated within Excel after the import is performed. Access will create a link between the source Excel worksheet and the Access table. Changes made to the Excel data will be automatically reflected in Access. Note that with this option the data cannot be changed from within Access.

14.2 Exporting an Access Query to Excel

Skills

Export an Access query to Excel

Tutorials

Exporting Access Data to Excel

Exporting Access Data to Word

App Tip

Create relationships in the Relationships window if the tables to be joined will be used together in other objects such as a form or report.

Access table data can be exported to use the mathematical analysis tools in Excel. Access creates a copy of the selected table or query data in an Excel worksheet file in the drive and/or folder that you specify. Buttons in the Export group on the External Data tab provide options to send a copy of Access data in a variety of file formats.

1. With the **Parking-YourName** database open, click the Create tab and then click the Query Design button.

2. At the Show Table dialog box, double-click each of the four table names to add all four table field list boxes to the query and then click the Close button.

In the next steps, you will join tables for those tables that do not have a relationship. Tables should be joined so that records are not duplicated in the query results datasheet.

3. Drag the *StudentID* field name in the ParkingSales field list box to *StudentID* in the Students field list box.

4. Drag the *Lot* field name in the ParkingSales field list box to *LotNo* in the ParkingLots field list box.

When a table is created by importing, a relationship is not automatically created between the new table and other tables in the database. When a relationship is necessary, you can join the tables in a query by dragging the common field from one table to the same field in the other table.

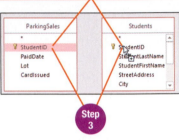

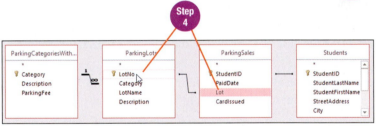

5. Double-click the following fields to add the fields to the query design grid. (Note that you are selecting fields in all four table field list boxes.)

Field Name	Table Name
StudentID	ParkingSales
StudentFirstName	Students
StudentLastName	Students
PaidDate	ParkingSales
Lot	Parking Sales
Description	ParkingLots
ParkingFee	ParkingCategoriesWithFees

StudentID	StudentFirstName	StudentLastName	PaidDate	Lot	Description	ParkingFee
ParkingSales	Students	Students	ParkingSales	ParkingSales	ParkingLots	ParkingCategoriesWif
☑	☑	☑	☑	☑	☑	☑

Step 5

6. Click the Run button in the Results group on the Query Tools Design tab.

7. Save the query as **ParkingSales2018** and then close the query.

8 Click to select the *ParkingSales2018* query name in the Navigation pane.

9 Click the External Data tab and then click the Excel button in the Export group.

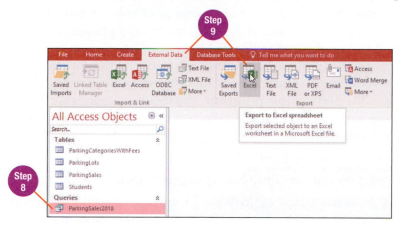

Quick Steps

Export a Query to Excel
1. Open database.
2. Select query name.
3. Click External Data tab.
4. Click Excel button in Export group.
5. Click Browse button.
6. Type file name and navigate to destination folder.
7. Click Save.
8. Specify export options.
9. Click OK.
10. If Excel opened, review worksheet and close Excel.
11. Click Close.

10 At the Export – Excel Spreadsheet dialog box, click the Browse button, type *ParkingSales-YourName* in the *File name* text box, navigate to the Ch14 folder in CompletedTopicsByChapter, and then click Save.

11 Click to insert a check mark in the *Export data with formatting and layout* check box.

12 Click to insert a check mark in the *Open the destination file after the export operation is complete* check box and then click OK.

Excel is started automatically with the data from the query results datasheet shown in a worksheet. Notice the first row contains the field names from the query and the worksheet tab is renamed to the query name.

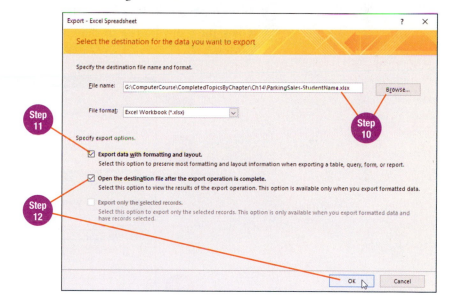

13 Close Excel to return to Access.

	A	B	C	D	E	F	G
1	StudentID	StudentFirstName	StudentLastName	PaidDate	Lot	Description	Parking Fee
2	100245745	Anna	Takacis	9/3/2018	1	Main campus north of Administration building	$250.00
3	100324125	Jose	Ramirez	9/3/2018	3	Main campus south of Technology wing	$200.00
4	100348685	Celine	Gauthier	9/4/2018	1	Main campus north of Administration building	$250.00
5	100367845	Sara	Blewett	9/5/2018	1	Main campus north of Administration building	$250.00
6	100443568	Dwayne	Kenney	9/6/2018	3	Main campus south of Technology wing	$200.00

First six rows in Excel after query is exported.

Green triangles are shown in the *StudentID* and *Lot* columns because the data is numeric but was exported from Access as text. Green triangles flag data that is a potential error. You can ignore the error flags here.

14 Click Close at the Export - Excel Spreadsheet dialog box to finish the export without saving the export steps.

15 Close the database and then close Access.

App Tip

Go to the Export backstage area in Word, Excel, and PowerPoint to find options for sending data outside the source program.

View

Model Answer

Compare your completed file with the model answer.

14.3 Embedding an Excel Chart into a Word Document

Skills

Embed an Excel chart into a document

Tutorials

Copying and Pasting Data into Word

Editing Chart Data

Formatting with Chart Buttons

Pasting Data Using Paste Special Options

In Chapter 3, you used Copy and Paste features to duplicate text and a chart between programs. You can also embed content as an object within a document, worksheet, or presentation. Embedding, like copying and pasting, inserts a duplicate of the selected text or object at the desired location. The program in which the data originally resides is called the **source program**, and the data that is copied is referred to as the **source data**. The program in which the data is embedded is referred to as the **destination program**, and the document, worksheet, or presentation into which the embedded object is placed is referred to as the **destination document**.

1. Start Excel 2016 and then open **SocialMediaStats**.

2. Start Word 2016 and then open **SocialMediaProject**.

3. Use Save As to save a copy of the Word document as **SocialMediaProject–YourName** in the Ch14 folder within CompletedTopicsByChapter.

4. Click the Excel button on the taskbar to switch to Excel and then click to select the pie chart with the title *Global Market Share*.

5. If necessary, click the Home tab.

6. Click the Copy button in the Clipboard group. (Do *not* click the arrow on the button.)

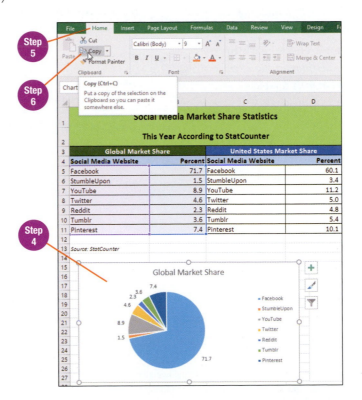

7. Click the Word button on the taskbar to switch to Word and then click to position the insertion point at the left margin on the blank line a double-space below the first table.

8 Click the Paste button arrow and then click *Use Destination Theme & Embed Workbook* (first button in *Paste Options* section) at the drop-down list.

9 Click to select the chart object and then click the Center button in the Paragraph group on the Home tab.

The Chart feature is standardized in Word, Excel, and PowerPoint. A chart embedded within any of the three programs offers the Chart Tools tabs and three chart editing buttons with which the chart can be modified after being embedded.

10 Click the Chart Elements button (plus symbol), point at *Legend* (right-pointing arrow appears), click the right-pointing arrow, click *Bottom*, and then click in the document outside the chart to deselect the chart object.

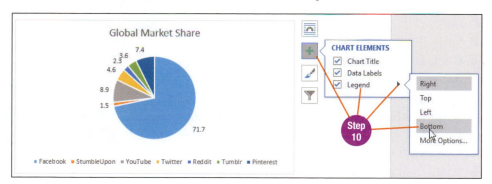

11 Switch to Excel and select and copy the pie chart with the title *United States Market Share*.

12 Switch to Word, position the insertion point at the bottom of the document, and then embed, center, and format the pie chart by completing steps similar to Steps 8 through 10.

13 Save the revised document using the same name (**SocialMediaProject-YourName**) and then close Word. Leave Excel and the **SocialMediaStats** workbook open for the next topic.

Quick Steps

Embed an Excel Chart into Word
1. Open worksheet in Excel.
2. Open document in Word.
3. Make Excel active, select and copy chart.
4. Switch to Word.
5. Position insertion point.
6. Click Paste button arrow.
7. Click *Use Destination Theme & Embed Workbook*.

View

Model Answer
Compare your completed file with the model answer.

Alternative Method   **Embedding Copied Data Using Paste Special Dialog Box**

Another way to embed copied data is to select *Paste Special* at the Paste button arrow drop-down list in the destination document. This opens the Paste Special dialog box in which you select the source object in the *As* list box and then click OK.

14.4 Embedding Excel Data into PowerPoint and Editing the Embedded Data

Tutorials

Embedding Objects

Using Paste Options

Embedding text or worksheet data uses the same process as embedding a chart. Double-click an embedded object to edit text or worksheet data in the destination location. Embedded text or cell data is edited using the tools on the ribbon from the source program. Click outside the embedded object to end editing and to restore the ribbon of the destination program.

1. With Excel active and the **SocialMediaStats** workbook open, select and copy A3:B11.

2. Start PowerPoint 2016 and open **SocialMediaPres**.

3. Use Save As to save a copy of the presentation as **SocialMediaPres-YourName** in the Ch14 folder within CompletedTopicsByChapter.

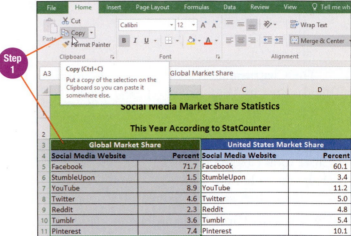

4. Make Slide 3 the active slide.

5. Click the Paste button arrow.

6. Click *Embed* (third button in *Paste Options* section).

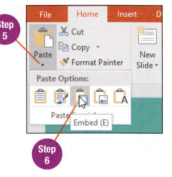

7. Click the Drawing Tools Format tab, click the Shape Fill button in the Shape Styles group, and then click *White, Text 1* (second option in *Theme Colors* section).

8. Resize and position the embedded object to the approximate size and position shown in the image below.

Global Social Media Market Share
This Year

Global Market Share	
Social Media Website	**Percent**
Facebook	71.7
StumbleUpon	1.5
YouTube	8.9
Twitter	4.6
Reddit	2.3
Tumblr	3.6
Pinterest	7.4

9 Double-click the inserted cells to open the embedded object for editing.

Notice the embedded cells open in an Excel worksheet and the ribbon changes to Excel's ribbon.

10 Select B5:B11 and then click the Decrease Decimal button in the Number group on the Home tab.

11 Click on the slide outside the embedded object to end editing and restore PowerPoint's ribbon.

12 Switch to Excel and select and copy C3:D11.

13 Switch to PowerPoint, make Slide 4 the active slide, and embed, format, resize, and position the copied cells by completing steps similar to those in Steps 5 through 11.

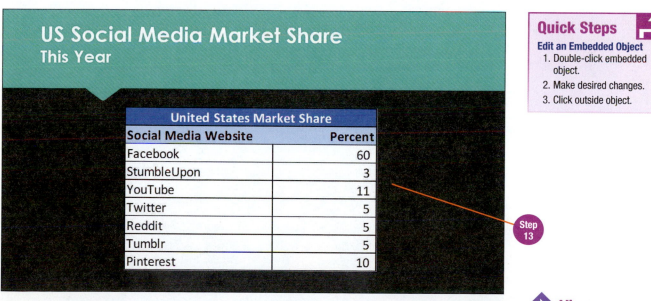

14 Save the revised presentation using the same name (**SocialMediaPres-YourName**) and then close PowerPoint. Leave Excel and the **SocialMediaStats** workbook open for the next topic.

Quick Steps

Edit an Embedded Object
1. Double-click embedded object.
2. Make desired changes.
3. Click outside object.

View

Model Answer
Compare your completed file with the model answer.

Beyond Basics   **Embedding an Entire File**

Embed an entire document or worksheet using the Object button on the Insert tab. At the Object dialog box, click the Create from File tab and then use the Browse button to navigate to the desired file name. Note that this method embeds the entire file contents at the insertion point, active cell, or active slide.

14.5 Linking an Excel Chart with a Presentation and Updating the Link

Skills

Link an Excel chart with a presentation

Turn on automatic link updates

Edit a linked chart

Update links

Tutorials
Linking Data
Linking Objects

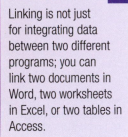
Good to Know

Linking is not just for integrating data between two different programs; you can link two documents in Word, two worksheets in Excel, or two tables in Access.

If the data that you want to integrate between two programs is continuously updated, copy and link the data instead of copying and pasting or copying and embedding. When copied data is linked, changes made to the source data can be automatically updated in any other document, worksheet, or presentation to which the data is linked. Linking avoids duplicating work and ensures that errors are not made when the same data is changed in more than one location. Linked objects are managed using the **Links dialog box**, where a link can be set to update automatically, a link can be broken, or the source location for a linked object can be changed.

1. With Excel active and the **SocialMediaStats** workbook open, use Save As to save a copy of the workbook as **LinkedSocialMediaStats-YourName** in the Ch14 folder within CompletedTopicsByChapter.

2. Start PowerPoint 2016 and then open **SocialMediaPres**.

3. Use Save As to save a copy of the presentation as **LinkedSocialMediaPres-YourName** in the Ch14 folder within CompletedTopicsByChapter.

4. Switch to Excel, select and then copy the *Global Market Share* pie chart.

5. Switch to PowerPoint and then make Slide 3 the active slide.

6. Click the Paste button arrow.

7. Click *Use Destination Theme & Link Data* (third button in *Paste Options* section).

8. Resize and move the chart to the approximate size and position shown in the image below.

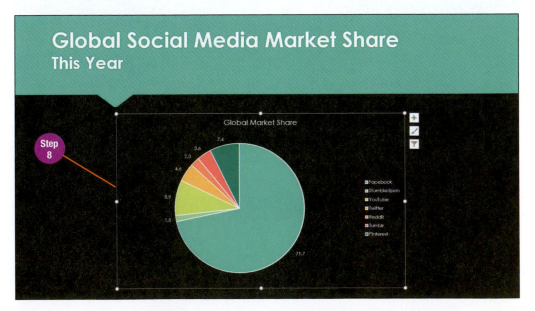

9. Switch to Excel, select and then copy the *United States Market Share* pie chart.

10 Switch to PowerPoint, make Slide 4 the active slide, and link, resize, and move the chart by completing steps similar to those in Steps 6 through 8.

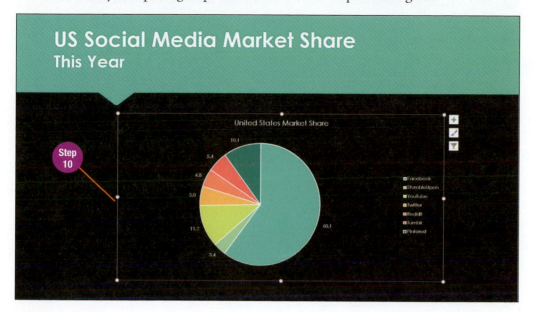

11 Click the File tab and then click *Edit Links to Files* in the *Related Documents* section at the bottom of the Properties panel (bottom right) at the Info backstage area.

12 At the Links dialog box, click to select the first link in the *Links* list box and then click to insert a check mark in the *Automatic Update* check box.

13 Click to select the second link and then click to insert a check mark in the *Automatic Update* check box.

14 Click Close.

15 Click the Back button at the Info backstage area to return to the presentation.

16 Save the revised presentation using the same name (**LinkedSocialMediaPres-YourName**) and then close the presentation.

17 Switch to Excel.

18 Change the value in D6 from *3.4* to *2.9*.

19 Change the value in D7 from *11.2* to *19.3*.

20 Change the value in D11 from *10.1* to *2.5*.

Notice the pie chart updated after each change in value. The revised chart is noticeably different from the original pie chart.

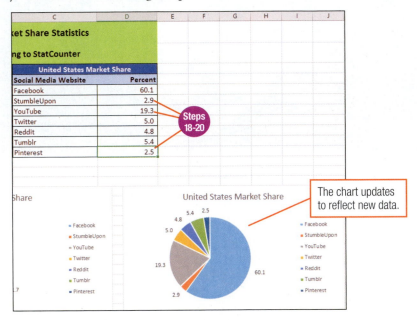

21 Save the revised worksheet using the same name (**LinkedSocialMediaStats-YourName**) and then close Excel.

22 Switch to PowerPoint if PowerPoint is not already active and then open **LinkedSocialMediaPres-YourName**.

Because the presentation contains linked data that is set to automatically update, you are prompted to update links. A dialog box with a security notice and the **Update Links** button appears when you open a file that has links to objects outside the document, worksheet, or presentation.

23 Click the Update Links button at the Microsoft PowerPoint Security Notice dialog box.

App Tip

If the source and destination files are both open at the same time, changes made to the source reflect in the destination file immediately.

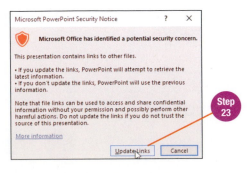

24   Make Slide 4 the active slide. Notice that the chart is updated to reflect the same data as the revised Excel chart.

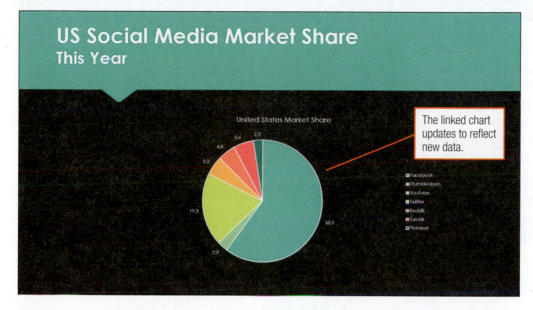

US Social Media Market Share
This Year

United States Market Share

The linked chart updates to reflect new data.

25   Make Slide 3 the active slide and then delete the title inside the chart above the pie.

26   Make Slide 4 the active slide and then delete the title inside the chart above the pie.

27   Save the revised presentation using the same name (**LinkedSocialMediaPres–YourName**) and then close PowerPoint.

Quick Steps

Link Data
1. Open source program and file.
2. Open destination program and file.
3. With source program active, select and copy data.
4. Switch to destination program.
5. Activate destination location.
6. Click Paste button arrow.
7. Click *Use Destination Theme & Link Data*.

Turn on Automatic Link Updates
1. Make destination file active.
2. Click File tab.
3. Click *Edit Links to Files*.
4. Select link.
5. Click *Automatic Update* check box.
6. Click Close.

View

Model Answer

Compare your completed file with the model answer.

Topics Review

Topic	Key Concepts	Key Terms
14.1 Importing Excel Worksheet Data into Access	The Import Spreadsheet Wizard starts when you click the Excel button in the Import & Link group on the External Data tab. Five dialog boxes in the Import Spreadsheet Wizard guide you through the steps to create a new table using data in an Excel worksheet. You can save the Excel import settings to repeat the import later using the same settings. Open an imported table in Access in Design view to modify the table design after the import is complete.	Import Spreadsheet Wizard
14.2 Exporting an Access Query to Excel	Export table or query data to Excel using the Excel button in the Export group on the External Data tab. Specify the file name, drive and/or folder, and export options at the Export – Excel Spreadsheet dialog box. You can elect to export the data with formatting and layout options in the datasheet, and automatically open Excel with the worksheet displayed when the export is complete. Export specifications can be saved to repeat the export later.	
14.3 Embedding an Excel Chart into a Word Document	Embedding inserts a copy of selected data as an object in a document, worksheet, or presentation. The program from which data is copied is called the *source program*. The data that is copied is referred to as the *source data*. The destination program is the program that receives the copied data. The destination document refers to the document, worksheet, or presentation into which copied data is pasted as an object. Click the Paste button arrow and then choose the desired embed option to embed copied data as an object in the destination document.	source program source data destination program destination document
14.4 Embedding Excel Data into PowerPoint and Editing the Embedded Data	Double-click an embedded object to open the object data for editing using the source program's ribbon and tools. Click outside the embedded object to end editing and restore the source program's ribbon. An entire document can be embedded using the Object button on the Insert tab.	
14.5 Linking an Excel Chart with a Presentation and Updating the Link	Source data that is continuously updated should be linked instead of copied and pasted or copied and embedded, to avoid duplication of work and reduce errors made when data is entered more than once. Click the Paste button arrow in the destination document and then choose the desired link option to link copied data. Click *Edit Links to Files* in the Properties panel at the Info backstage area to open the Links dialog box in which you manage links to source data. Select a link in the Links dialog box and insert a check mark in the *Automatic Update* check box to turn on automatic updates for the link. Click the Update Links button in the Security Warning dialog box that appears when you open a file with a linked object to update data from the source.	Links dialog box Update Links

Recheck

Recheck your understanding of the topics covered in this chapter.

Workbook

Chapter review and assessment resources are available in the *Workbook* ebook.

Chapter 15

Using OneDrive and Other Cloud Computing Technologies

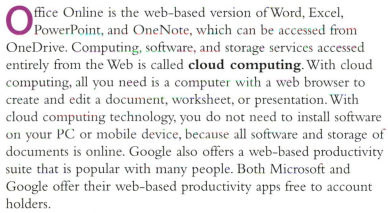

Precheck

Check your understanding of the topics covered in this chapter.

Office Online is the web-based version of Word, Excel, PowerPoint, and OneNote, which can be accessed from OneDrive. Computing, software, and storage services accessed entirely from the Web is called **cloud computing**. With cloud computing, all you need is a computer with a web browser to create and edit a document, worksheet, or presentation. With cloud computing technology, you do not need to install software on your PC or mobile device, because all software and storage of documents is online. Google also offers a web-based productivity suite that is popular with many people. Both Microsoft and Google offer their web-based productivity apps free to account holders.

In Chapter 3, you learned how to save a presentation to OneDrive within PowerPoint. In this chapter, you will learn to create and edit files using Office Online from OneDrive; upload, download, and share files in OneDrive; and create a document using Google Docs from Google Drive.

Note: If you are using a tablet, consider connecting it to a USB or wireless keyboard because parts of this chapter involve a fair amount of typing. In this chapter, you will need to sign in with a Microsoft account and a Google account. If necessary, create a new account at each website. You may wish to check with your instructor before completing this chapter to confirm the required topics.

Learning Objectives

15.1 Create a document in Word Online

15.2 Create a worksheet in Excel Online

15.3 Create a presentation in PowerPoint Online

15.4 Edit a presentation in PowerPoint Online

15.5 Upload and download files to and from OneDrive

15.6 Share a document in OneDrive

15.7 Create a document using Google Docs

SNAP If you are a SNAP user, go to your SNAP Assignments page to complete the Precheck, Tutorials, and Recheck.

Data Files

Before beginning this chapter, be sure you have copied the student data files for this course to your storage medium. Steps on downloading and extracting the data files are provided in Chapter 1, Topic 1.8, on pages 22–23.

15.1 Creating a Document Using Word Online

With Word Online, you create and edit documents within a web browser. **Word Online** is similar to the desktop version of Word that you used in Chapters 6 and 7; however, Word Online has fewer features than the desktop version. The program looks similar to the full-featured Word, but you will notice that for some features, functionality within a browser environment is slightly different than the desktop version.

Note: Microsoft may update Office Online after publication of this textbook, in which case the information, steps, and/or screens shown here may vary.

1 Open a browser window.

2 If necessary, click in the Address bar or select the current web address. Skip this step if you are using a browser where the insertion point is already positioned in a blank address bar.

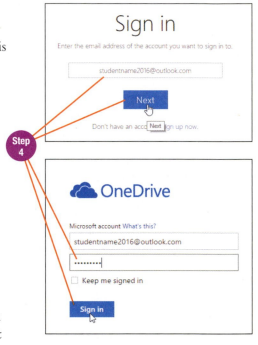

3 Type https://onedrive.live.com and then press Enter.

4 If necessary, click Sign in, type the email address or phone number for your Microsoft account and click Next, then type your password and click the Sign in button at the second sign-in screen. Skip this step if you are already signed in to OneDrive, which occurs with the Microsoft Edge browser when you are already signed in to your Microsoft Account for Windows.

> **Oops!**
>
> Don't know your Microsoft account? If you have a hotmail.com, live.com, or outlook.com email address, your email login is your Microsoft account; otherwise, click the Sign up button or link and create a new account.

Once signed in, the OneDrive window appears similar to the one shown in Figure 15.1.

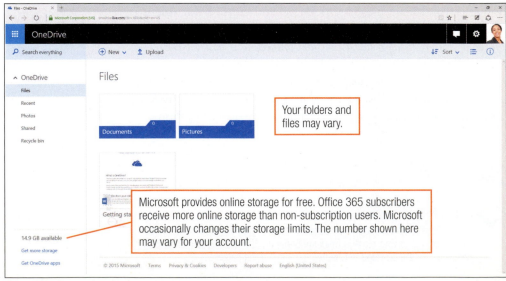

Your folders and files may vary.

Microsoft provides online storage for free. Office 365 subscribers receive more online storage than non-subscription users. Microsoft occasionally changes their storage limits. The number shown here may vary for your account.

Figure 15.1

This is how the OneDrive window appears for a signed-in user.

⑤ Click the Microsoft Apps button in the upper left of the screen next to OneDrive (it displays as a waffle icon and is sometimes called the waffle button) and then click the Word tile.

⑥ Click Got it! if a screen displays with information about new features in Office Online; this message appears the first time you use Office Online, or when changes have been made to the software.

⑦ Click New blank document at the Word Online Start screen.

Word Online launches, as shown in Figure 15.2.

Check This Out ✔

http://CA2.Paradigm College.net/Office

Go here to start Word Online without first starting OneDrive. You can access all the same web-based apps from office.com. Click the desired tile and then sign in with a Microsoft account or a school account.

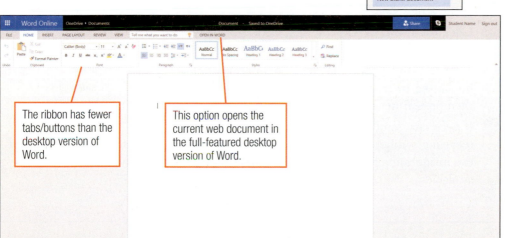

The ribbon has fewer tabs/buttons than the desktop version of Word.

This option opens the current web document in the full-featured desktop version of Word.

Figure 15.2
The Word Online new document window is shown.

⑧ Type the following text in the document window using all the default settings:

What Is Green Computing?

Green computing refers to the use of computers and other electronic devices in an environmentally responsible manner. Green computing can encompass new or modified computing practices, policies, and procedures. This trend is growing with more individuals and businesses adopting green computing strategies every year.

Strategies include the reduction of energy consumption by computers and other devices; reduction in use of paper, ink, and toner; and reuse, recycling, or proper disposal of electronic waste.

⑨ Proofread carefully and correct any typing errors that you find. If necessary, use the Spelling button in the Spelling group on the REVIEW tab to spell check the document.

App Tip
Use the same editing and formatting techniques in Word Online as you learned in the desktop edition of Word.

10 Click to place the insertion point within the title *What Is Green Computing?* and then click the Center button in the Paragraph group on the HOME tab.

11 Select all the text in the document and then change the font size to 12 using the *Font Size* option box arrow in the Font group on the HOME tab.

12 Select the two paragraphs of text below the title and then change the line spacing to 1.5 using the Line Spacing button in the Paragraph group on the HOME tab.

Steps 8-12

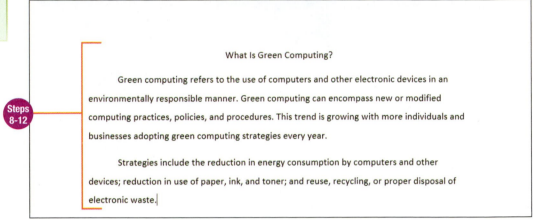

What Is Green Computing?

Green computing refers to the use of computers and other electronic devices in an environmentally responsible manner. Green computing can encompass new or modified computing practices, policies, and procedures. This trend is growing with more individuals and businesses adopting green computing strategies every year.

Strategies include the reduction in energy consumption by computers and other devices; reduction in use of paper, ink, and toner; and reuse, recycling, or proper disposal of electronic waste.

13 Position the insertion point at the end of the last paragraph and press Enter to create a new line.

14 Click the INSERT tab and then click the Online Pictures button in the Pictures group.

15 Type recycling in the *Bing Image Search* text box at the Insert Pictures dialog box and then press Enter or click the Search button.

16 If necessary, scroll down, click the green recycling symbol picture shown in the image below, and then click Insert. Insert the image using the student data file *recyclinglogo* if the picture shown is not available.

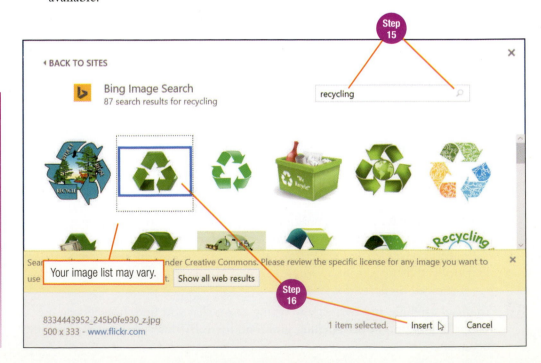

17 If necessary, click to select the green recycling symbol image and then click the PICTURE TOOLS FORMAT tab.

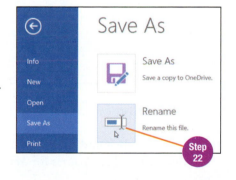

18 Select the current value in the *Scale* text box in the Image Size group, type 35, and then press Enter.

19 With the image still selected, click the HOME tab and then click the Center button in the Paragraph group.

20 Click within the document text to deselect the image.

21 Click the FILE tab and then click *Save As*.

22 At the Save As backstage area, click Rename.

23 Type GreenComputing-YourName in the *Enter a name for this file* text box at the Rename dialog box and then press Enter or click OK.

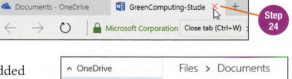

24 Close the document browser tab. Leave OneDrive open for the next topic.

A Word document thumbnail is added to the *Files > Documents* list in OneDrive. Additional options appear in the bar along the top of the OneDrive window when a document is selected. You will use some of these options in later topics.

App Tip

Switch to the desktop version of Word if you need access to the full set of picture formatting and editing tools.

App Tip

Create a printable PDF of the document at the Save As backstage area using the Download as PDF option.

Oops!

OneDrive closed when you closed the Word Online document tab? Sometimes, OneDrive operates within a single browser tab, and closing the document tab also exits OneDrive. Restart OneDrive and sign back in to your OneDrive account. Next time, click the Microsoft Apps button and then click the OneDrive tile instead of closing the browser tab when you finish a topic.

15.2 Creating a Worksheet Using Excel Online

Skills

Create a worksheet in Excel Online

Excel Online looks the same as the full-featured desktop edition of Excel; however, the ribbon contains fewer options, and functionality for some features will vary. You can create a basic worksheet in the web-based version of Excel, but for worksheets that need advanced formulas or editing, the desktop version of Excel is preferred.

1. With OneDrive open, click the Microsoft Apps button (displays as a waffle icon), click the Excel tile, and then click New blank workbook at the Excel Online Start screen.

Excel Online launches and opens a window similar to the window shown in Figure 15.3. Like Word Online, Excel Online workbooks are saved in the same file format as the desktop version of Excel and are transferable between software editions.

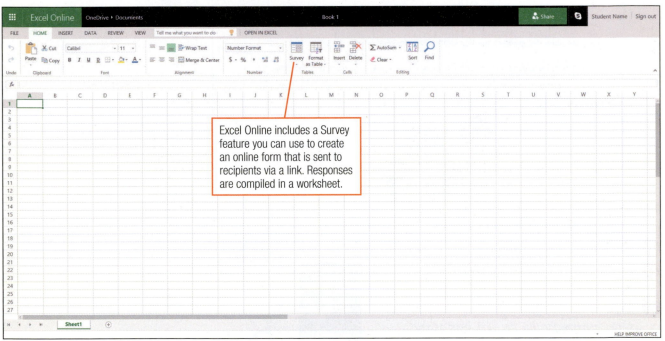

Excel Online includes a Survey feature you can use to create an online form that is sent to recipients via a link. Responses are compiled in a worksheet.

Figure 15.3

As you can see from the Excel Online window, the ribbon has fewer tabs and buttons than the full-featured desktop version of Excel.

2. Type the labels and values in the cells as shown in the image below, substituting your first and last names for *Your Name* in A1.

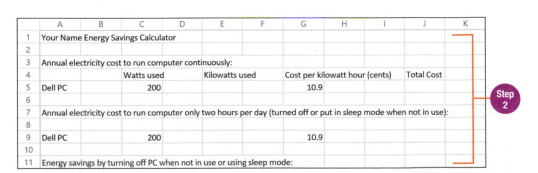

	A	B	C	D	E	F	G	H	I	J	K
1	Your Name Energy Savings Calculator										
2											
3	Annual electricity cost to run computer continuously:										
4			Watts used		Kilowatts used		Cost per kilowatt hour (cents)		Total Cost		
5	Dell PC		200				10.9				
6											
7	Annual electricity cost to run computer only two hours per day (turned off or put in sleep mode when not in use):										
8											
9	Dell PC		200				10.9				
10											
11	Energy savings by turning off PC when not in use or using sleep mode:										

Step 2

3 Click E5 to make it the active cell and then type the formula =(c5*24*365)/1000.

The formula multiplies the watts used by a PC running continuously 24 hours per day, 365 days per year and then divides the result by 1000 to convert watts to kilowatts.

4 Click J5 to make it the active cell and then type the formula: =e5*(g5/100).

The cost per kilowatt hour in G5 is divided by 100 to convert 10.9 to a decimal value representing cents.

5 Type the remaining formulas in the cells indicated:

E9 =(c9*2*365)/1000
J9 =e9*(g9/100)
J11 =j5-j9

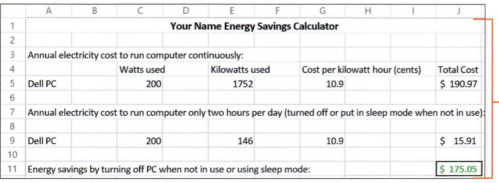

 Steps 3–5

6 Select A1:J1 and then click the Merge & Center button in the Alignment group on the HOME tab.

7 With A1:J1 still selected, apply bold formatting and change the font size to 12.

8 Click J5 to make it the active cell, click the *Number Format* option box in the Number group on the HOME tab, and then click *Accounting* at the drop-down list.

9 Apply the Accounting Number Format to J9 and J11.

10 With J11 the active cell, click the Borders button in the Font group on the HOME tab and then click *Outside Borders* at the drop-down list.

Notice that fewer border options exist in Excel Online.

11 With J11 still the active cell, apply bold formatting and the *Dark Green* font color (sixth color in *Standard Colors* section).

12 Click in a blank cell to view the border in J11 and then proofread carefully and correct any typing errors that you find.

	A	B	C	D	E	F	G	H	I	J
1					Your Name Energy Savings Calculator					
2										
3	Annual electricity cost to run computer continuously:									
4			Watts used		Kilowatts used		Cost per kilowatt hour (cents)			Total Cost
5	Dell PC		200		1752		10.9			$ 190.97
6										
7	Annual electricity cost to run computer only two hours per day (turned off or put in sleep mode when not in use):									
8										
9	Dell PC		200		146		10.9			$ 15.91
10										
11	Energy savings by turning off PC when not in use or using sleep mode:									$ 175.05

This is how the worksheet should be formatted after Steps 6 to 11 are completed.

13 Rename the workbook EnergySavings-YourName and then close the workbook tab by completing steps similar to Steps 21 to 24 in the previous topic. Leave OneDrive open for the next topic.

An Excel workbook thumbnail is added to the *Files > Documents* list in OneDrive.

Quick Steps

Create a Workbook in Excel Online
1. Open a browser window.
2. Navigate to https://onedrive.live.com.
3. If necessary, sign in with Microsoft account.
4. Click Microsoft Apps button.
5. Click Excel tile and then New blank workbook.
6. Type cell entries and format worksheet.
7. Use Save As to rename workbook
8. Close workbook tab.

App Tip

Changes to the worksheet are saved automatically to the workbook in OneDrive.

App Tip

Display the worksheet in a printer-friendly format in a separate browser window and then click the Print button at the Print backstage area.

15.3 Creating a Presentation Using PowerPoint Online

A basic presentation that does not need to incorporate a chart or audio can be created using **PowerPoint Online**. Other PowerPoint Online features that vary from the full-featured desktop version include fewer animation and transition options, the inability to customize a slide show with timings, set up slide show options, and fewer views.

1 With OneDrive open and with the Documents folder displayed, click the Microsoft Apps button (displays as a waffle icon), click the PowerPoint tile, and then click New blank presentation at the PowerPoint Online Start screen.

2 Click the DESIGN tab, click the More Themes button located at the right end of the Themes gallery, and then click *Banded*.

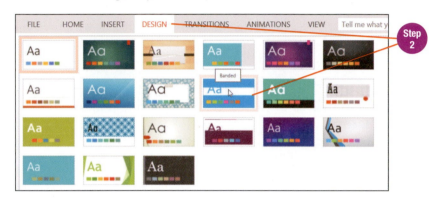

3 Click *Variant 4* in the Variants gallery.

A new presentation with the selected theme is started in the PowerPoint Online window, as shown in Figure 15.4. Similar to Word and Excel, presentations are saved in the same file format and are transferable between PowerPoint Online and the desktop version of PowerPoint.

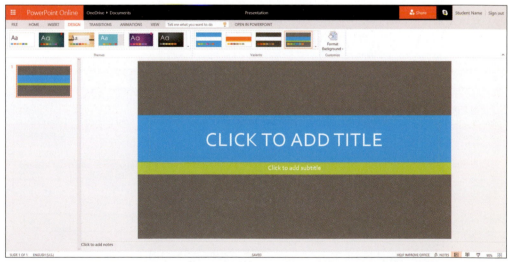

Figure 15.4

The PowerPoint Online window is shown. PowerPoint Online does not have the Slide Show or Review tab found in the desktop version of PowerPoint.

4 Click anywhere in *CLICK TO ADD TITLE* in the title slide and type Green Computing. Click anywhere in *Click to add subtitle* and then type your name.

Note that the font for this theme converts titles to all uppercase text.

5 Click the HOME tab and then click the New Slide button in the Slides group.

6 At the New Slide dialog box with the *Title and Content* layout selected, click Add Slide.

7 Type the text on Slide 2 as follows:

Slide title   What Is Green Computing?
Bulleted list Use of computers and other electronic devices in an environmentally responsible manner including new or modified:
       computing practices
       computing policies
       computing procedures

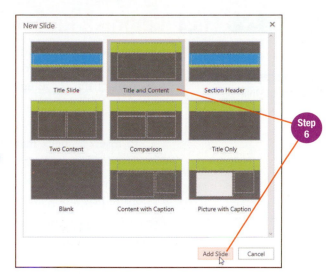

Step 6

8 Add another new slide with the *Title and Content* layout and then type the text on Slide 3 as follows:

Slide title   Green Computing Strategies
Bulleted list Reduction in energy consumption
        Reduction in use of paper, ink, and toner
        Reuse or recycling of devices
        Proper disposal of e-waste

> **App Tip**
> A slight delay may occur after typing or clicking outside a placeholder as the screen refreshes.

9 Make Slide 2 the active slide, click the INSERT tab, and then click the Online Pictures button in the Images group.

10 Type computer in the *Bing Image Search* text box at the Insert Pictures dialog box and then press Enter or click the Search button.

11 Scroll down to the image shown at the right and then double-click to insert it on the slide. Insert the image using the student data file **computer** if the image shown is not available.

12 Move the image to the approximate position as shown in the slide at the right.

Steps 9-12

13 Make Slide 3 the active slide. Insert and position the image as shown in the slide at the right. Search for the image by typing recycling in the *Bing Image Search* text box. If necessary, insert the image using the student data file **worldrecycling**.

14 Rename the presentation GreenComputingPres-YourName and then close the presentation tab. Leave OneDrive open for the next topic.

Step 13

15.4 Editing a Presentation in PowerPoint Online

Skills

Edit a presentation in PowerPoint Online

Opening a presentation from OneDrive in PowerPoint Online displays the slides in Reading view. Switch to editing mode using the Edit Presentation button near the top right of the Reading view window to make changes to slide content or to add new slides.

1 With OneDrive open and with the Documents folder displayed, click to insert a check mark in the check circle at the top right corner of the **GreenComputingPres–YourName** thumbnail if the check circle is empty.

A check mark in the check circle indicates the presentation is selected. Additional options display along the top of the OneDrive window when a presentation is selected.

2 Click <u>Open</u> and then click <u>Open in PowerPoint Online</u>.

The presentation opens in Reading view, as shown in Figure 15.5.

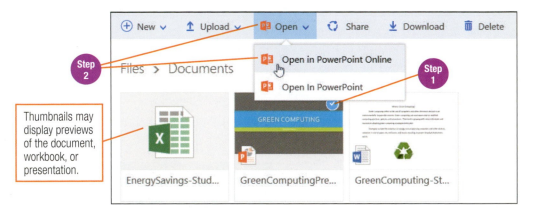

3 Click the Next Slide button (displays as a right-pointing arrow) in the Status bar to view Slide 2 in the window.

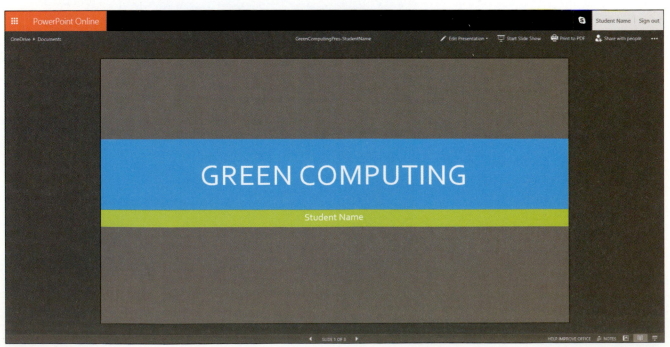

Figure 15.5
A presentation displays in Reading view in PowerPoint Online when opened from OneDrive.

④ Click the Next Slide button again to view Slide 3.

⑤ Click the Edit Presentation button and then click *Edit in PowerPoint Online*.

⑥ If necessary, make Slide 3 the active slide.

⑦ Add a new slide with the *Title and Content* layout and then type the following text on Slide 4:

Slide title   Green Computing Example

Bulleted list A desktop PC can use up to 1700 kilowatt hours per year if left on continuously

Turning off or putting the PC in sleep mode when not in use can save over 1600 kilowatt hours per year for average use of 2 hours per day

This strategy can save $175 per year for electricity cost at 10.9 cents per kilowatt hour

⑧ Search for the image shown below using the key phrase *power on*. Insert and move the image to the approximate position as shown. Insert the image using the student data file **poweron** if the image shown is not available.

⑨ Close the presentation tab. Leave OneDrive open for the next topic.

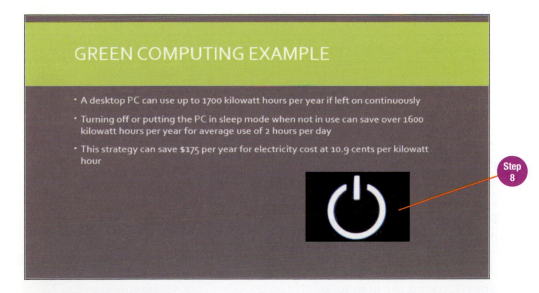

Alternative Method	**Editing a Document, Workbook, or Presentation in Desktop Version of Word, Excel, or PowerPoint**

Select the document, workbook, or presentation thumbnail for the file to be edited, click <u>Open</u>, and then click the option to open the file in the full-featured desktop version of Word, Excel, or PowerPoint.

Quick Steps

Edit a Presentation in PowerPoint Online
1. Open a browser window.
2. Navigate to <u>https://onedrive.live.com</u>.
3. If necessary, sign in with Microsoft account.
4. Select presentation thumbnail.
5. Click <u>Open</u>.
6. Click <u>Open in PowerPoint Online</u>.
7. Click Edit Presentation button.
8. Click *Edit in PowerPoint Online*.
9. Edit as required.
10. Close presentation tab.

15.5 Downloading Files from and Uploading Files to OneDrive

You can copy files from your PC or mobile device to OneDrive for backup storage purposes; to access the files from another device instead of copying the files to a USB flash drive; or to share the files with other people. Conversely, you can download a file from OneDrive to your local PC or mobile device to view or edit the file offline.

1 With OneDrive open, click to clear the check mark in the check circle for the **GreenComputingPres–YourName** thumbnail, and then click to insert a check mark in the check circle for the **GreenComputing–YourName** Word document.

2 Click Download at the top of the OneDrive window.

3 Click the Close button at the right end of the pop-up message at the bottom of Microsoft Edge with the options to Open or View downloads. Skip this step if you are using a different browser and the message shown does not display.

By default, files downloaded from OneDrive are saved in the Downloads folder for the signed in user when you are using OneDrive in the Microsoft Edge browser. Another browser may display a Save As dialog box when you click the Download option. In that case you can select to save the file in the Downloads folder on your computer.

| GreenComputing-StudentName.docx finished downloading. | Open | Open folder | View downloads | ✕ |

Step 3

4 Click to clear the check mark in the check circle for the **GreenComputing–YourName** thumbnail to deselect the document.

5 Select and download the **EnergySavings–YourName** workbook to the Downloads folder by completing steps similar to those in Steps 1 through 3.

6 Clear the check mark for the **EnergySavings–YourName** thumbnail to deselect the workbook and then select and download the **GreenComputingPres–YourName** presentation to the Downloads folder by completing steps similar to those in Steps 1 through 3.

7 Click the File Explorer button on the taskbar to open File Explorer.

8 Click *Downloads* in the Navigation pane.

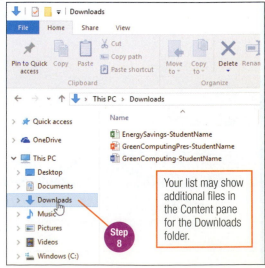

Your list may show additional files in the Content pane for the Downloads folder.

Step 8

The three files downloaded from OneDrive are shown in the Content pane.

9 Select and copy the three downloaded files to a new folder *Ch15* in the CompletedTopicsByChapter folder on your storage medium. Refer to Chapter 1, Topic 1.9 if you need assistance with this step.

10 Close the File Explorer window and return to the browser window with OneDrive active. Clear the check mark in the check circle for the GreenComputingPres-YourName thumbnail if the file is still selected.

In the next steps you will copy three pictures from your storage medium to your account storage at OneDrive.

11 Click Upload in the bar at the top of the OneDrive window, and then click Files at the drop-down list.

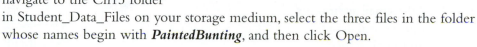

12 At the Open dialog box, navigate to the Ch15 folder in Student_Data_Files on your storage medium, select the three files in the folder whose names begin with *PaintedBunting*, and then click Open.

The three files are uploaded to the current folder in your account storage in OneDrive. A progress message appears near the right end of the bar at the top of the window as the files are uploaded.

13 Select the three painted bunting picture thumbnails and deselect the thumbnail from Step 6 if the presentation is still selected.

14 Click Move to in the bar at the top of the window.

15 Click *Pictures* in the Move items to panel at the right side of the window and then click Move at the top of the panel. Leave OneDrive open for the next topic.

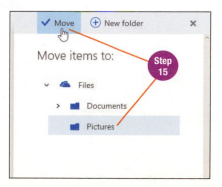

15.6 Sharing a File on OneDrive

OneDrive is an excellent tool for collaborating on documents when working with a team. A team leader can create or upload documents to OneDrive and then share the files with the team members who need them. An individual with shared access to a document receives an email with a link to the file. Changes to the file are made to the copy in OneDrive so that only one document, worksheet, or presentation has to be managed. Collaborating by sharing a file on OneDrive is less cumbersome than sending a file as an email attachment and then trying to manage multiple versions of the same document.

Note: In this topic, you will share a Word Online document with a classmate. Check with your instructor for instructions on with whom you should share the Word document. If necessary, share the document with yourself by using an email address other than your Microsoft account.

1. With OneDrive open, select the **GreenComputing-YourName** document.

2. Click Share in the bar at the top of the window.

3. Type the email address for a classmate in the *To* text box in the Share dialog box with the Invite people panel active.

   More than one email address can be entered at the *To* text box. As with email messages, use a semicolon to separate email addresses.

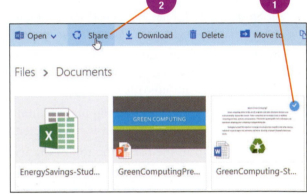

App Tip

You can share files with anyone with a valid email address—the recipient does not have to have a Microsoft account.

Oops!

Sharing from your account is blocked? Sometimes, for new accounts, sharing is blocked until the account is verified. Click the link to verify your account below the message. You may need to give a mobile number that can receive a text message. Enter the code received on your phone, click Next, and then click Done. If necessary, use the Microsoft Apps button to switch from the Outlook tab back to OneDrive when verification is complete and repeat Steps 1 through 5 a second time.

4. Click in the message box and then type Please make your changes to the file accessed from this link..

5. Click Share.

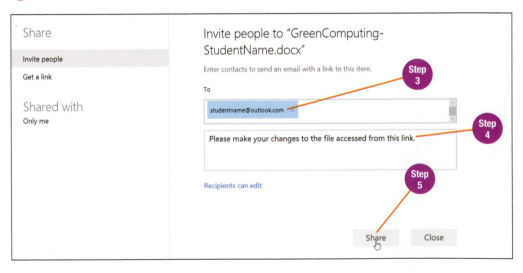

6 Click <u>Close</u> when the classmate's name appears in the *Shared with* section.

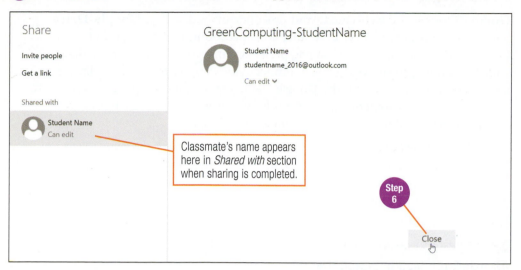

Quick Steps

Share a File on OneDrive
1. Sign in to OneDrive.
2. Select file.
3. Click <u>Share</u>.
4. Type recipient's email address in *To* box.
5. Type message in message box.
6. Click <u>Share</u>.
7. Click <u>Close</u>.

7 Click the Microsoft Apps button next to OneDrive.

8 Click the Mail tile.

9 With *Inbox* the active mail folder, open the message received from a classmate with the subject line informing you the **GreenComputing-StudentName.docx** file has been shared with you in OneDrive.

10 Click the link to the file in the message window.

The file opens in Word Online in Reading view. Notice the Edit Document button in the bar at the top of the window.

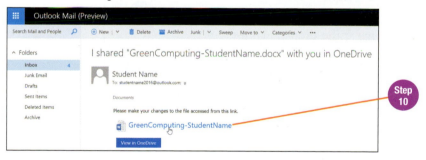

Oops!

No message? Check the email address that the classmate used to ensure the correct address was typed. If an address other than Outlook, Hotmail, or Live was used, you need to go to another mail program to find the message with the link. In that case, sign out of OneDrive, launch your other mail program, and complete Steps 9 and 10. Note also that some mail programs may flag the message as Junk Mail. Check your Junk Mail folder if the message is not in your Inbox.

11 Close the document tab.

12 Use the Microsoft Apps button to switch back to OneDrive.

13 Click Shared in the Navigation pane (left pane) of OneDrive to view the file details of files shared by you and with you in the Content pane.

You will have two entries in the Shared panel because you shared the document with a classmate and a classmate shared his or her document with you.

14 Click the Account button that displays your account picture or the generic user silhouette at the right end of the OneDrive bar and then click <u>Sign out</u>.

15 Close the browser window.

15.7 Creating a Document Using Google Docs

Google Docs is the web-based word processor offered within **Google Drive** (Google's cloud storage service). With a Gmail account, you can sign in to Google Drive and create a document, presentation, spreadsheet, form, or drawing. Gmail accounts and the Google web productivity apps are free to use. Having an introduction to using one of the Google cloud applications is a good idea, because you may encounter someone at your workplace or volunteer organization who prefers to use the cloud technologies offered by Google.

Note: This is an optional topic. Check with your instructor or check the course syllabus to see if you are required to complete this topic. Also note that Google may update Google Drive and/or Google Docs after publication of this textbook, in which case the information, steps, and/or screens shown here may vary.

1. Open a browser window.

2. Navigate to google.com.

3. Click the Sign in button near the top right of the window. If you are automatically signed in when you go to the Google page, skip to Step 5.

4. Type your Google account information and click Sign in. If multiple Gmail accounts are set up on the computer you are using, click to select the account you want to use, type your password, and click Sign in.

5. Click the Google Apps button located near the top right of the page (displays as a waffle icon).

6. Click <u>Drive</u> at the drop-down list.

7. Click Got it or a Close button to close a message if a message appears about new features in Google Drive. A message generally displays the first time you access Drive after creating a new account.

8. Click the NEW button below *Drive* at the left side of the page and then click <u>Google Docs</u> at the drop-down list.

9. Click <u>NO THANKS</u> if a message box appears next to the Docs home button with a message about using templates.

A document window opens similar to the one shown in Figure 15.6 on page 371.

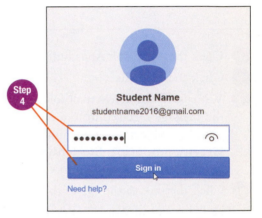

Other applications are available from Google Drive.

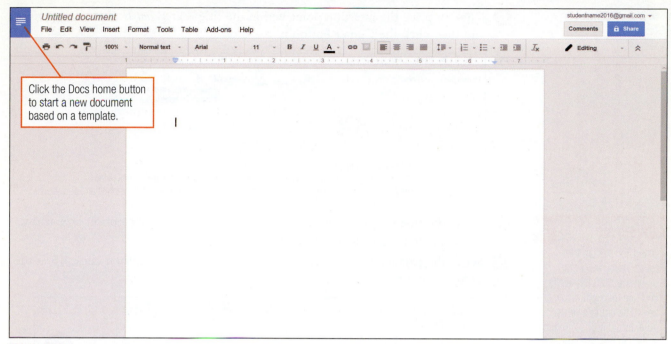

Click the Docs home button to start a new document based on a template.

Figure 15.6

A Google Docs document window is shown. Google Docs automatically saves changes every few seconds to a document named *Untitled document*.

10 Type the following text in the document window using all the default settings:

What Is Cloud Computing?

Cloud computing refers to a delivery model of software and file management using web-based service providers where all resources are online. Consumers of cloud computing services access software and files via a web browser. Some cloud-based services are free, with fees charged to access more storage or software features. [Press Enter twice after the period.]

11 Click the File menu and then click <u>Rename</u> at the drop-down list.

> **App Tip**
>
> You can upload and view a Word document in Google Docs.

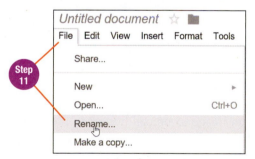

Step 11

12 Type CloudComputing-YourName in the file name text box above the menu bar and then press Enter.

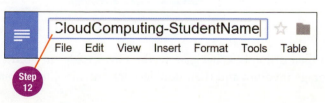

Step 12

13 Click to place the insertion point within the title text *What Is Cloud Computing?* and then click the Center button in the toolbar.

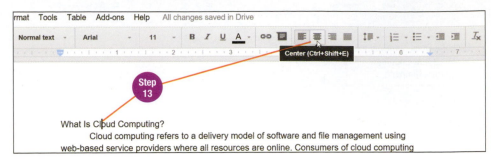

14 Select the title text, click the Bold button, click the *Font Size* option box arrow, and then click <u>14</u> at the drop-down list.

15 Select the paragraph text, click the Line spacing button, and then click <u>1.5</u> at the drop-down list.

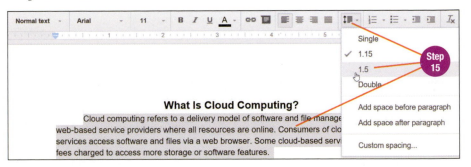

16 Deselect the text and then position the insertion point on the blank line at the bottom of the document.

17 Click the Insert menu and then click <u>Image</u> at the drop-down list.

18 Click the Choose an image to upload button in the middle of the Insert image dialog box.

19 At the Open dialog box, navigate to the Ch15 folder in Student_Data_Files on your storage medium and then double-click the file ***Cloud-computing***.

20 Click to select the image, resize the image using the resizing handles to approximately 2 inches wide by 1.5 inches tall, and then click the Center button.

21 With the image still selected, click the Insert menu and then click <u>Footnote</u> at the drop-down list.

22 Type Cloud computing image courtesy of Wikimedia Commons. in the Footnote pane.

¹ Cloud computing image courtesy of Wikimedia Commons.

Step 22

23 Scroll up to the top of the page.

24 If necessary, click at the end of the paragraph text to deselect the image.

25 Close the tabbed window for the document to return to Google Drive.

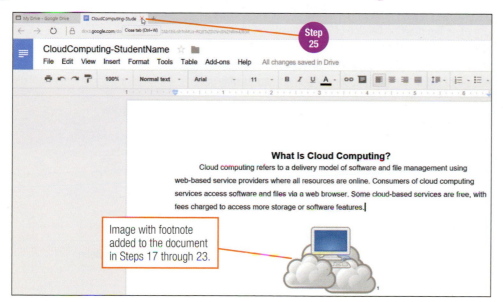

Step 25

What Is Cloud Computing?

Cloud computing refers to a delivery model of software and file management using web-based service providers where all resources are online. Consumers of cloud computing services access software and files via a web browser. Some cloud-based services are free, with fees charged to access more storage or software features.

Image with footnote added to the document in Steps 17 through 23.

A document in *My Drive* list is shown with a thumbnail.

26 Click your account icon near the top right of the window and then click Sign out.

27 Close the browser window.

Alternative Method   **Navigating directly to Google Drive or Google Docs**

Type the URL drive.google.com to navigate directly to Google Drive or docs.google.com to navigate directly to Google Docs.

Quick Steps

Create a Document Using Google Docs
1. Open a browser window.
2. Navigate to google.com.
3. Sign in with Google account.
4. Click Google Apps button, then <u>Drive</u>.
5. Click NEW button, then <u>Google Docs</u>.
6. Type, edit, and format document.
7. Click File.
8. Click <u>Rename</u>.
9. Type document name.
10. Press Enter.
11. Close window tab.

App Tip

Google Docs can be shared with others by selecting the file in the Google Drive list and using the Share button in the bar above the file list.

Check This Out

http://CA2.Paradigm College.net/Zoho

Go here to check out another popular web-based productivity suite. Register for a free account at Zoho to access several web-based applications including Writer (word processor), Show (presentation), and Sheet (spreadsheet).

View

Model Answer
Compare your completed file with the model answer.

Topics Review

Topic	Key Concepts	Key Terms
15.1 Creating a Document Using Word Online	Cloud computing is a service in which computer resources, software, and storage are provided online. Cloud computing applications and files are accessed from a web browser. Word Online is the web-based version of Word accessed from OneDrive. Sign in to OneDrive with a Microsoft account, click the Apps button, click the Word tile, and then click New blank document to start a new document. Documents created in Word Online are saved in the same file format as the desktop version of Word, meaning files can be transferred between editions. Word Online has fewer ribbon tabs and options than the desktop version of Word, and the functions of some features may vary.	cloud computing Word Online
15.2 Creating a Worksheet Using Excel Online	Excel Online is best suited for basic worksheets; use the desktop version of Excel for worksheets that need advanced formulas or editing. Worksheets created in Excel Online are saved in the same file format as the desktop version of Excel, meaning files can be transferred between editions. Like Word Online, Excel Online has fewer features than the desktop version, and some functionality may vary.	Excel Online
15.3 Creating a Presentation Using PowerPoint Online	Use PowerPoint Online to create a presentation that does not need charts, audio, advanced animation or transition effects, or advanced slide show set up options. Presentations created in PowerPoint Online are saved in the same file format as the desktop version of PowerPoint, meaning files can be transferred between editions.	PowerPoint Online
15.4 Editing a Presentation in PowerPoint Online	To open a presentation, select a presentation file thumbnail and use Open from OneDrive. A presentation opens in Reading view from OneDrive. Use the Edit Presentation button and choose to open the presentation in either the desktop version of PowerPoint or PowerPoint Online to make changes to the presentation.	
15.5 Downloading Files from and Uploading Files to OneDrive	Select a file by clicking in the check circle at the top right of a tile to insert a check mark. Click Download to copy a selected file from OneDrive to the Downloads folder on your computer. Open a File Explorer window to copy and paste the file to the desired location. Click Upload to select and copy a file from your computer to your OneDrive storage.	

continued…

Topic	Key Concepts	Key Terms
15.6 Sharing a File on OneDrive	OneDrive can be used to collaborate with team members by sharing one copy of a file among several users. Select a file and choose Share to type the email address of any individual you want to share a file with. Individuals receive an email message with a link to the shared file on OneDrive.	
15.7 Creating a Document Using Google Docs	The web-based word processor offered by Google is called *Google Docs*. Sign in to Google with a Gmail account, click the Google Apps button, click Drive in the drop-down list, and then click the NEW button to start a document, presentation, spreadsheet, form, or drawing. Google Drive is the online file-storage service from Google. Google Docs saves changes automatically every few seconds to an untitled document. Use the Rename option from the File menu to assign a name to the untitled document. Use options from the Menu bar drop-down lists and toolbar to add elements, edit, and format a document.	Google Docs Google Drive

Recheck

Recheck your understanding of the topics covered in this chapter.

Workbook

Chapter review and assessment resources are available in the *Workbook* ebook.

Glossary/Index

A

absolute address Excel address with a dollar symbol before the column letter and/or row number so that the address will not change when the formula is copied to another column or row, 214

Accept button used to send a message to a meeting organizer indicating acceptance of a meeting request, 125

Access 2016 database management application in Microsoft Office suite used to organize, store, and manage related data such as customers, vendors, employees, or products, 64, 66, 288–311, 315–335
  Best Fit in, 299
  color scheme in, 66
  columns
    adjusting widths, 299
    inserting and deleting, 310–311
  control objects in, 331
  database in
    compacting, repairing and backing up, 334–335
    encrypting, with password, 335
    one open at a time, 291
  data in
    finding and replacing, 298–299
    importing Excel worksheet into, 340–343
  datasheets in
    adding records using, 292–293
    editing and deleting records in, 294–295
    previewing, 310–311
  data types in, 318
  Design view in, 304
    creating new tables in, 320–321
    creating queries using, 304–305
    modifying field properties in, 324–325
  Enforce Referential Integrity in, 329
  exporting data from, 311
  field properties in, 324
    formatting and data validation, 325
    modifying, in Design View, 324–325
  Field Properties pane in, 324
  fields in, 289
    adding, to existing tables, 322–323
    deleting, 323
    sorting by more than one, 301
  field values in, 290
    creating lookup list with, in another table, 327
  files in, creating new, 316–317
  finding and replacing data and adjusting column widths in, 298–299
  foreign key in, 329
  forms in, 288, 290
    adding, editing and deleting records in, 296–297
    creating, using the Form Wizard, 331
    creating and editing, 330–331
  Form view in, 330
  Form Wizard in, creating forms using, 331
  identifying database objects, 288–289
  Layout View in, 330
  lookup list in, 326–327
  Lookup Wizard in, 326–327
  opening, 66
  primary key in, 300
    assigning, 320–321
  primary tables in, 329
  queries in, 288, 291
    calculated field in, 310–311
    Design view in creating, 304–305
    entering criteria to select records in, 306–307
    entering multiple to sort, 308–309
    exporting to Excel, 344–345
    Simple Query Wizard in creating, 302–303
  records in, 290
    adding, editing, and deleting in a form, 296–297
    best practices for deleting, 295
    datasheets in adding, 292–293
    editing and deleting using datasheets, 294–295
    entering criteria to select, in queries, 306–307
    entering multiple criteria to select, 308–309
    sorting and filtering, 300–301
    using find to move to, 297
  relationships in
    creating, 329
    displaying and editing, 328–329
    one-to-many, 329
    one-to-one, 328
  reports in, 288, 290
    creating, editing, and viewing, 332–333
    grouping and sorting, 333
  Simple Query Wizard in, creating queries using, 302–303
  tables in, 288, 289
    adding fields to existing, 322–323
    creating lookup list with field values in, 327
    creating new, 318–319
    in Design View, 320–321
    planning new, 316–317
    primary, 329
  terminology in, 288–291
  wildcard characters in, 309

Accounting Number Format button in the Number group on the Home tab in Excel adds a dollar symbol, a comma in the thousands place, and displays two digits after the decimal point, 191–192

Action Buttons, 273

Action Center panel in Windows 10 opened from the Notification icon on the taskbar that displays notifications from settings, apps, applications, and connected accounts and shows quick action tiles to perform system changes, 15

active cell the cell with the green border around its edges and in which the next entry will be stored or that will be affected by a command, 186

Add Account dialog box, 112

Add Contact Picture dialog box, 126–127

adding
  in Access
    records in forms, 296–297
  in Outlook
    contacts in, 126–127
    tasks in, 128–129

Add Page control in OneNote that displays as a plus symbol inside a circle used to add a new page in a section; pages organize note content within a section, 96

Address bar the area in a web browser in which the web address (also called a URL) for a web page is viewed or typed, 41, 43

Adobe Flash Media, 274

advanced search options, 55–56

Alarms & Clock windows, 12–13

Align Left paragraph alignment in which lines of text are aligned at the left margin; left edge of page appears even, 147

alignment
  cells, in Excel, 194–195
  in Excel
    indenting cells, 207

alignment guide colored vertical and/or horizontal line that appears while dragging an object to help you align and place the object; also called smart guides, 77, 249

Align Right paragraph alignment in which text is aligned at the right margin; right edge of page appears even, 147

Angle Counterclockwise option from Orientation button in Alignment group of Home tab in Excel that rotates text within the cell boundaries by 45 degrees, 209

animation involves adding a special effect to an object on a slide that causes the object to move or change in some way during a PowerPoint slide show, 278
  adding, to slide show
    applying to individual objects, 280–281
    using slide master, 278–279
  categories of, 279
  changing sequence of, 281
  presentations, 278–281

Animation Painter button used in PowerPoint to copy animation effects from one object to another object, 280

APA (American Psychological Association) formatting and page layout guidelines, 172
  Works Cited page, 176

app a small program used on a smartphone, tablet, or PC designed usually for one purpose, to work on a touchscreen, use less power than an application, and fewer network resources, 4

adding to the Lock screen, 12
Calendar app, 10
launching, 10–11
Microsoft Excel, 360
   creating worksheet using, 360–361
Microsoft PowerPoint, 362
   creating presentation using, 362–363
   editing presentation in, 364–365
Microsoft Word, 357
   creating document using, 356–359
other, 11
Photos app, 10
Store app, 11
switching between, 11
application a program used on a PC by individuals to do a task that is more powerful than an app, has a wide range of functions, makes use of a full-size keyboard, mouse, larger screen, and more powerful computing resources, 4
appointment any activity in your schedule that you want to keep track of, or be reminded of, including the day and time the activity begins and ends, the subject, and the location, 120
   editing, 122–123
   keyboard shortcuts for, 123
   reminders and tags for, 121
   scheduling, 120–121
      recurring, 122–123
argument parameters for a function formula that appear in parentheses, 216
Arrow keys in Excel, 187, 190
Ask, 54
Attach File button used to attach a file (such as a photo) to an email message in Outlook, 116
audience handouts, in PowerPoint, 258–259
Audio button in Media group on Insert tab in PowerPoint used to add a sound clip to a slide in a presentation, 276
audio file formats, PowerPoint recognition of, 276
AutoComplete feature in software programs in which entries that match the first few characters that you type are displayed in a list so that you can enter text by accepting a suggestion; suggested entries are derived from prior entries that have been typed, 116
   in Excel, 187
AutoCorrect software feature that automatically corrects commonly misspelled words when you press the spacebar, 137
   in Excel, 189
   in Word, 136, 137–138
Auto Fill Excel feature that enters data automatically based on a pattern or series that exists in an adjacent cell or selected range in Excel, 196–197
AutoFit column width setting that adjusts the width of the column to accommodate the length of the active cell, or longest entry in the column, 194
   in Excel, 194–195
   in PowerPoint, 249
AutoFormat software feature that automatically formats text or replaces text with symbols or special characters, 138
   in Word, 136, 138, 145
AutoSum button in Excel used to enter SUM formula or access other functions, 198
AVERAGE function, 217

B

Back button in Excel used to enter SUM formula or access other functions, 43
backing up, database, 334–335
backspace in correcting typing errors, 68
Backspace key, 136
backstage area view accessed from the File tab in Microsoft Office applications that is used to perform file management; start a new document; display file information; manage document properties; and customize application options, 68
   Export tab, 105
   to manage documents, 68–71
   New Notebook, 106
   Notebook Information, 105
   printing document, 70
   Share Notebook, 106
banded row shading or other formatting applied to a row so that every other row is easier to read in a table, 166
Battery saver, 15
Best Fit column width dialog box option in Access that adjusts column width to accommodate the length of the longest field value in the column, 299
Bibliography Word button used to generate a Works Cited or References page in an academic paper, 176
   Works Cited page creation, 176
Bing, 54, 55
Bing Image Search, 158
Blinds option, 278
bookmark button, 145
bookmarking pages
   with Google Chrome, 47
   with Mozilla Firefox, 51–52
Bookmarks bar in Google Chrome, the bar below the Address bar that displays buttons for bookmarked web pages, 47
borders
   in Excel, 192
   in Word, 162–163
Borders gallery feature that adds a border to a paragraph or cell, 162
browsing, tabbed, 44
bullets
   changing symbol, 247
   in PowerPoint, 246, 247, 268
   in Word, 144–145
Bullets button used to format text as a list with a bullet symbol at the beginning of each line; bullets are used for a list of items that are in no particular order, 144

C

calendar, setting in notification center, 14
Calendar app Windows 10 app used to view your appointments and reminders stored in your Microsoft account calendar, 10
Calendar component in Microsoft Outlook, used to schedule, manage, and store appointments, meetings, and events, 120
Calendar Tools Appointment tab, 122
Caption property Access field property used to store a descriptive title for a field; caption text displays as the column heading or control label in a datasheet or form, 319

category axis horizontal axis in a column chart with the names or other labels associated with each bar; also called the x-axis, 226
cell the intersection of a column with a row into which you type text, a value, or a formula in Excel, 186
   absolute addressing, 214–215
   active, 186, 187
   applying styles, 202–203
   changing alignment, 194–195
   copying, 206–207
   currency option, 192
   deleting contents of, 189
   display cell formulas, 205
   editing, 189
   entering text and values, 187–188
   fill handle in copying, 197
   formatting, 190–193
      applying font, 190–191
      numbers, 191–192
   identifying, 186
   indenting, 207
   inserting and deleting, 201
   merging, 193
   naming, 215
   pointers for, 186, 187
   range, 190
   selecting
      mouse in, 190
      touch in, 190
   shading, wrapping, and rotating, 209
   sorting, 202–203
   in Word
      merging and splitting, 168–169
cell formulas, displaying in Excel, 205
Cell Styles set of predefined formatting options that can be applied to selected cells in a worksheet, 202–203
Center paragraph alignment in which text is aligned centered between the left and right margins, 147
character formatting changing the appearance of characters, 146
charts
   in Excel
      column, 226–227
      embedding, into Word document, 346–347
      line, 228–229
      pie, 224–225
      recommended, 227
      sparkline, 232–233
   in PowerPoint
      creating, on slides, 270–271
      formatting, 271
      using data from Excel in creating, 271
Check Full Name dialog box, 127
chime sounds, setting, 14
Chrome free web browser from Google that runs on PCs or Macs, 46
Chrome menu in Google Chrome, the button near the top right of the window that displays three short bars (often called a hamburger button) used to access the menu system. Displays the ScreenTip *Customize and control Google Chrome*, 47

citation reference to the source used for a quotation or for paraphrased text in an academic paper, 174
    inserting and editing, 174–175

Clear button used in Excel to clear contents, formats, or both contents and formats in a cell or range, 189

click and type feature in Word in which you can double-click anywhere on a page and start typing; paragraph alignment in effect depends on location within the line at which you double-click, 137

clip art, inserting, 265

Close the red button with white X that displays at the top right of a window and is used to close the window, 12, 13

cloud computing Software, storage, and computing services accessed entirely on the Internet, 355

cloud storage, 84

color scheme
    in Access, 66
    changing background, 19
    in Excel, 66
    in PowerPoint, 66
    in Word, 66

column
    in Access
        adding fields to, in query design grid, 305
        adjusting width, 299
        inserting and deleting, 310–311
    in Excel
        adjusting width, 194–195
        deleting, 200–201
        inserting, 200–201
    in Word
        inserting and deleting, 167
        modifying column width, 168–169

column chart a chart in which each data point is represented by a colored bar extending upward from the category axis with the bar height extending to its corresponding value on the value axis, 226
    category axis in, 226
    value axis in, 226
    creating and modifying, 226–227

comma, in email addresses, 114

Comma Style button in the Number group on the Home tab in Excel that adds a comma in thousands and two decimal places, 191

comment a short note associated with text that provides explanatory information, poses a question, or provides feedback, 178
    in Excel
        inserting, 232–233
        printing, 233
    in PowerPoint, 240, 241
        adding, 254–255
        deleting or hiding, 255
    in Word, 178–179
        inserting, 178–179
        replying to, 178–179

comment balloon, 178

Comments pane at right side of slide pane in Normal view of PowerPoint in which comments are added to a presentation, 240, 254

Compact & Repair Database button at Info backstage area used to perform a compact and repair routine for the current database, 334

computer. *See* personal computer

conditional formatting in Excel, 235

contacts in Outlook, adding and editing, 126–127

Content pane, 113

contextual tab a tab that appears when an object is selected that contains commands or options related to the type of selected object, 76

control object a rectangular content placeholder in a form or report, 331

Convert to SmartArt button in Paragraph group on Home tab in PowerPoint used to convert existing text into a SmartArt graphic object, 268

Copy button or menu option used to make a duplicate copy of a file or selected text, 28
    files, 28–29
    folders, 28–29

copyright-free multimedia, 277

Cortana, 12

COUNTA function, 217

COUNT function, 217

cover letters in Word, creating from templates, 180–181

crawler a program that reads web pages and other information to generate index entries; also called a spider, 54

Create a New Section tab control used to add a new section to a OneNote notebook; sections are used to organize notes by category, topic, or subject, 96

Creative Commons license, 158

curly quotes, 138

currency option, 192

custom sort in Excel, 203

Cut button or menu option used to move a file or selected text to another location, 30

D

data
    in Access, finding and replacing, 298–299
    downloading and extracting data files, 23
    embedding copies data using Paste Special dialog box, 347
    in Excel
        embedding and editing, into PowerPoint presentation, 348–349
        entering or copying with Fill command, 196–197
        importing, into Access, 340–343
    exporting from Access, 311
    linking copied data using, 353
    in PowerPoint, editing, 271

database data stored in an organized manner to provide information for a variety of purposes, 287
    in Access
        compacting, repairing and backing up, 334–335
        encrypting, with password, 335

database management system (DBMS) software that organizes and keeps track of data, 287

datasheet
    adding records using, 292–293
    editing and deleting records in, 294–295
    modifying field properties using, 325
    previewing, 310–311
    records in
        sorting and filtering, 300–301

data types in Access, 318

DATE function, 219

Date Navigator the calendars displayed above the Folder pane with which you can change the day that is displayed in the Appointment area in the Calendar, 120

dates in Excel, entering, 218–219

decimal places, adjusting, in Excel, 192

Decrease Decimal button used in Excel to remove one decimal place from each selected value each time the button is clicked, 192

Decrease Indent button that moves a paragraph closer to the left margin in a document or left edge of a cell in a worksheet each time the button is clicked, 148

Decrease List Level button used to move text left to the previous indent position within a bullet list placeholder in PowerPoint, 246

Delete button or menu option used to remove a file, folder, selected text, or other object from storage or a document, 32
    in Access
        columns, 310–311
        fields, in, 323
        records in datasheets in, 294–295
        records in forms in, 296–297
    emails, 116–117
    in Excel
        cells in, 201
        columns in, 200–201
        contents of cell, 189
        rows in, 200–201
    files, 32–33
    folders, 32–33
    notes, in OneNote, 95
    in PowerPoint
        comments in, 255
        slides in, 251
    in Word
        columns and rows, 167

Delete button in the Cells group on the Home tab of the Excel ribbon used for deleting cells, rows, or columns, 201

Deleted Items mail folder to which messages that have been deleted are moved; messages are not permanently deleted until Deleted Items folder is emptied, 117

Delete key, 95

Design tab, 149

Design view Access view for an object in which the structure and/or layout of a table, form, query, or report is defined, 304
    adding new field to table in, 323
    creating new tables in, 320–321
    creating queries using, 304–305
    modifying field properties in, 324–325

desktop the display with icons that launch programs and a taskbar used to start programs, switch between open programs, and view system notifications, 9
    element on, 9–10
    personalizing, 20–21

destination document worksheet, or presentation into which copied data is pasted, embedded, or linked, 346

destination program program into which copied data is pasted, embedded, or linked, 346

dialog box a box that opens in a separate window with additional options or commands for the related ribbon group as buttons, lists, sliders, check boxes, text boxes, and option buttons, 77

dialog box launcher diagonal downward-pointing button located at bottom right of a group on the ribbon that opens a task pane or dialog box when clicked, 77

digital notebook, 93

documents
 Backstage area to manage, 68–71
 creating
  Google Docs in, 370–373
  using Word Online, 356–359
 pinning, to *Recent* option list, 71
 printing, 70
 saving to OneDrive, 84–86
 sending, as PDF document, 70–71
 tracking changes made to, 179
 in Word
  creating and editing new, 136–139
  creating from template, 152–153
  cropping and removal picture background, 161
  editing pictures for, 160–161
  inserting captions with picture, 161
  inserting pictures from computer, 160
  inserting pictures from online sources, 158–159
  Save As to save copy of document, 68–69
  saving new, 138

Dogpile, 54

domain name, 41

downloading the practice of copying content (image, audio clip, video clip, music file) from a web page to your PC or mobile device, 58
 content, from a web page, 58–59
 files from OneDrive, 366–367

Draft view displays a document in Word without print elements such as headers or footers, 177

DuckDuckGo, 54

E

editing the practice of making changes to a document after the document has been typed, 138
 in Access
  forms, 330–331
  records in datasheet, 294–295
  records in forms, 296–297
  reports, 332–333
 in Excel, cells, 189
 in Outlook
  contacts in, 126–127
  tasks in, 128–129
 in PowerPoint
  Excel data in, 348–349
  images, 265
  online, 364–365
  text on slides, 243
 in Word
  citations, 174–175
  documents, 138–141
  pictures, 160–161
  sources, 175

Effects Options button, 278

electronic mail (email) the sending and receiving of digital messages, sometimes with documents or photos attached, 112

 in Outlook
  attaching files to, 116–117
  creating, 113–114
  deleting, 116–117
  forwarding, 115
  meeting request, 124
  previewing attachments, 118–119
  replying to, 114–115
  sending, 113–114
  signatures for, 115

endnotes in Word, 177

Enforce Referential Integrity Access relationship option that verifies as a new record is added to a related table that a record with the matching field value in the joined field already exists in the primary table in the relationship, 329

event an activity in the Calendar that lasts an entire day or longer, 121

Excel 2016 spreadsheet application in Microsoft Office suite used to calculate, format, and analyze primarily numerical data, 64, 66, 185–209, 213–235
 absolute addresses in, 214–215
 accounting number format in, 191–192
 adding borders in, 192
 adjusting decimal places in, 192
 alignment of text entries in, 187
 Angle Coutnerclockwise, 209
 arguments in, 216
 Arrow keys in, 187, 190
 AutoComplete in, 187
 AutoCorrect in, 189
 Auto Fill in, 196–197
 AutoFit in, 194–195
 AutoSum in, 198–199
 cells in, 186
  active, 186, 187
  changing alignment in, 194–195
  copying, 206–207
  currency option, 192
  deleting contents of, 189
  displaying formulas, 205
  editing, 189
  entering text and values, 187–188
  font formatting, 190–191
  formatting, 190–193
  identifying, 186
  indenting, 207
  inserting and deleting, 201
  merging, 193
  mouse in selecting, 190
  number formatting, 191–192
  pointers in, 186, 187
  shading, wrapping, and rotating, 209
  sorting and applying styles in, 202–203
  touch in selecting, 190
  using fill handle to copy, 197
 Cell Styles, 202–203
 changing margins in, 231
 changing orientation in, 204
 charts in
  column, 226–227
  line, 228–229
  pie, 224–225
  recommended, 227
  sparkline, 232–233
 color scheme in, 66
 columns in
  adjusting width, 194–195

  deleting, 200–201
  inserting, 200–201
 comma style format in, 191
 comments in
  inserting, 232–233
  printing, 233
 conditional formatting in, 235
 copying cells, 197
 custom sort in, 203
 data in
  entering or copying with Fill command, 196–197
  using, in creating PowerPoint chart, 271
 dates in
  entering, 218–219
  Region setting and, 219
 Decrease Decimal button in, 192
 default setting
  column width, row height, 194
 editing of cells in, 189
 embedding and editing data in PowerPoint presentation, 348–349
 embedding charts into Word document, 346–347
 exporting Access query to, 344–345
 Fill Color button, 209
 Fill command, 196–197, 199
 fill handle in, 197
 Fill Right, 196–197
 filters in, 234
 Filter Slicer panes in, 235
 Find & Select button in, 208
 Flash Fill in, 196–197
 footers in, 230–231
 format cells dialog box in, 193
 formula bar in, 186, 187, 188
 formulas in
  absolute addressing and range names in, 214–215
  creating, to perform calculations, 188
  equals sign in, 188
  mathematical operators in, 188
  order of operations in, 189
  statistical functions in entering, 216–217
 Freeze Panes in, 208
 function library in, 216, 219
 functions in
  AVERAGE function, 217
  COUNTA function in, 217
  COUNT function, 217
  DATE function, 219
  FV function, 223
  IF function in, 220–221
  MAX function, 217
  MIN function, 217
  TODAY function, 218–219
 Go To command in, 208
 Go To Special command in, 208
 headers in, 230–231
 horizontal scroll bar in, 186, 187
 importing worksheet data into Access, 340–343
 Increase Decimal button in, 192
 linking charts with PowerPoint presentation and updating links, 350–353
 linking worksheets to Access table, 343
 mixed addresses in, 214–215
 name box in, 186, 187
 new sheet button, 186, 187
 opening, 66

Page Layout view in, 230
Percent Style in, 221
PMT function in, 222–223
pound symbols in, 193
printing
    changing options, 205
quick analysis in, 190
range in, 190, 202
    names for, 215
    sorting, 203
relative addresses in, 214–215
rows in
    adjusting height, 194–195
    deleting, 200–201
    inserting, 200–201
scaling in
    changing, 204–205
    options in, 204–205
sheet tab bar in, 186, 187, 206
sheet tab in, 186, 187
Slicer pane in, 235
status bar in, 186, 187
SUM function in, 198–199
tables in, 234–235
on touch device, 190, 197
vertical scroll bar in, 186, 187
view button in, 186, 187
View tab in, 88
Web App, 360–361
workbooks in, 185
    themes of, 203
worksheets in, 185
    centering, 231
    creating new, 186–188
    creating using Excel Online, 360–361
    inserting, 206–207
    renaming, 206–207
Wrap Text button, 209
zoom button in, 186, 187
Excel Online web-based version of Microsoft
    Excel accessed from OneDrive that is similar
    to the desktop version of Excel but has fewer
    features; some functionality within features
    may also vary from the desktop version, 360
    creating worksheet using, 360–361
Exchange server, use of Outlook with, 111
exporting, document as PDF file, 70–71
expression, calculating in OneNote, 97

F

Facebook, inserting pictures from, 158
Favorites web addresses for pages that you visit
    frequently saved to the Favorites list, 44
    adding web page to, 44–45
field one characteristic about a person, place,
    event, or item in an Access table; for example,
    Birthdate is a field in a table about students, 289
    adding, to existing tables, 322–323
    adding new field, using Design View, 323
    adding to columns, in query design grid, 305
    data types, 318
    deleting, 323
    sorting by more than one, 301
Field Properties pane lower half of Design
    view window for an Access table that contains
    field properties for the active field, 324
field property a characteristic or attribute for
    an Access field that defines the field's data,
    format, behavior, or other feature, 324
    formatting and data validation, 325
    modifying, in Design View, 324–325

modifying, using datasheet, 325
field value data in Access that is stored within
    one field in a record; for example, Jane is the
    field value for a FirstName field in a record for
    Jane Smith, 290
        creating lookup list with, in another table,
        327
        replacing, 298–299
file a document, spreadsheet, presentation,
    picture, or any text and/or image that is saved
    as digital data, 22
    browsing with File Explorer, 24–25
    copying, 28–29
    creating new, in Access, 316–317
    deleting, 30–31
    downloading and extracting data files, 21
    downloading and uploading from and to
        OneDrive, 366–367
    email attachments
        attaching to email messages, 116–117
        opening, 118–119
        previewing, 118–119
        saving, 119
    inserting, into notebook, 100–101
    moving, 28–29
    name of, 20
    organization of, in folders, 20
    renaming, 29–30
    saving to OneDrive, 84–86
    scrolling within, 87
    sharing, on OneDrive, 368–369
file attachments
    cautiousness in opening, 119
    for Email messages, 116–117
    previewing, 118–119
File Explorer utility used to browse files and
    folders and perform file management routines,
    21
    browsing files with, 22–23
    copying files and folders in, 27
    create new folder, 25
    move, rename and delete files and folder,
        28–31, 30
file name a series of characters you assign to a
    document, spreadsheet, presentation, picture, or
    other text or image that allows you to identify
    and retrieve the file later, 20
File tab, 67, 68
Fill button in Editing group of Home tab in
    Excel used to access fill options, 196–197, 199
Fill Color button located in Font group of the
    Home tab in Excel used to add shading to the
    background of a cell, 209
fill handle small green square at bottom right
    corner of active cell or selected range used to
    extend the pattern or series in adjacent cells, 197
    copying cells, 197
    with mouse, 197
    on touch device, 197
Fill Right Excel feature that automatically
    copies entry in first cell of a range to the
    adjacent cells to the right, 196
Filter Email button, 131
filtering of records, in Access, 300–301
Filter Slicer panes in Excel, 235
filter temporarily hides data that does not meet
    a criterion, 234
Find feature that moves insertion point or cell
    to each occurrence of a word or phrase, 142
    finding and replacing data in database, 299

to move to record in database, 297
    in Word, 142–143
Find & Select button in Editing group on the
    Excel Home tab with Find, Replace, and Go
    To options, 208
Find bar in a web browser, the area used to
    locate words or phrases on a web page, 48
Find Tags OneNote button used to open the
    Tags Summary pane in which you view tagged
    notes and can click a link to jump to a tagged
    note, 103
Firefox free web browser from Mozilla
    foundation that runs on PCs or Macs, 50
Fit Sheet on One Page Excel Scaling option
    in Print tab backstage view that shrinks the
    size of text to fit all columns and rows on one
    page, 205
Flash Fill Excel feature that automatically fills
    data in adjacent cells as soon as a pattern is
    recognized, 196–197
Flickr
    Commons, 58
    inserting pictures from, 158
folder a name assigned to a placeholder or
    container in which you store a group of
    related files, 22
    copying, 28–29
    creating, 26–27
    deleting, 32–33
    New folder, 26
    organizing files with, 22
    pinning, to *Recent* option list, 71
    renaming, 31–32
    subfolder, 25
Folder pane, 113
font a typeface that includes the design and
    shape of the letters, numbers, and special
    characters, 146
    in Excel, 190–191, 193
    in PowerPoint, 246–247
    in Word, 146–147
Font Color button, 146
Font dialog box, 146
footer text that appears at the bottom of each
    page, 172
    in Excel, 230–231
    in PowerPoint, 259
    in Word, 172–173
footnotes in Word, 177
foreign key Access field added to a related table
    that is not the primary key and is included for
    the purpose of creating a relationship, 329
form Access object used to enter, update, or
    view records generally one record at a time,
    288, 290
    adding, editing and deleting records in,
        296–297
    creating and editing, 330–331
    Form Wizard in creating, 405
Format Painter clipboard option used to copy
    formatting options from selected text or an
    object to other text or another object, 78
    for copying formatting options, 80–81
formatting changing the appearance of text, 146
    conditional, in Excel, 235
    copy and paste options, 81
    in Excel
        cells, 190–193
        numbers, 191–192

Format Painter to copy formatting options, 80–81
in PowerPoint
with font and paragraph options, 246–247
in Word
character formatting, 146–147
paragraph formatting, 147
tables, 166–167
Theme, 151
using styles, 150–151
Formula bar in Excel, 186, 187, 188
formula cell entry beginning with an equals sign (=) and followed by a statement that is calculated to determine the value displayed in the cell, 188
absolute addressing and range names in, 214–215
creating, to perform calculations, 188
equals sign in, 188
mathematical operators in, 188
order of operations in, 189
statistical functions in entering, 216–217
Form view Access view in which data is viewed, entered, and updated using a form and is the view that is active when a form is opened, 330
Form Wizard in Access, creating forms using, 331
Forward button used to send a copy of an email message to someone else, 115
Forward control or button in web browser or other software that moves forward or to the next screen, 43
forwarding Email messages, 115
Fractions, recognition of, in AutoFormat, 138
Freeze Panes Excel option that fixes column and/or row headings in place so that headings do not scroll off the screen, 208
From Beginning button in PowerPoint Slide Show tab used to start a slide show from slide 3, 256
function library in Excel, 216, 219
functions, in Excel
AVERAGE function, 217
COUNTA function in, 217
COUNT function, 217
DATE function, 219
FV function, 223
IF function, 220–221
MAX function, 217
MIN function, 217
PMT function, 222–223
statistical functions in entering, 216–217
TODAY function, 218–219
FV function, 223

G

gallery in a drop-down list or grid, visual representations of options that can be applied to a selection, 76
gesture an action or motion you perform with your finger, thumb, stylus, or mouse on a touchscreen device, 4
press and hold, 5
slide, 5
swipe, 5
tap or double-tap, 5
zoom out, 5
Google account, 370
Google Chrome
navigating Web using, 46–49

advanced search options, 57
bookmarking pages in, 47–48
customizing and controlling, 47
displaying web pages with, 46–47
multiple, 47–48
finding text on web page, 48–49
hamburger button, 47
incognito browsing, 48–49
keyboard shortcuts, 49
searching Web from address bar, 48
starting, 46
Google Docs web-based free word processor offered by Google, 370
creating document using, 370–373
sharing, 373
Google Drive cloud file storage service and the place where you access free cloud applications offered by Google, 370
Go To feature to move active cell to a specific location in the worksheet, 208
Go To Date feature in Outlook with which you go directly to a specific date to display in the Appointment area of the Calendar, 120
Go To Special dialog box with options for moving the active cell by various cell attributes, 208
grammar check in Word, 140–141

H

hanging indent a paragraph in which the first line remains at the left margin but subsequent lines are indented, 148
hard page break a page break that you insert before the maximum number of lines that can fit on the page has been reached, 171
hard return, 139
header text that appears at the top of each page, 172
in Excel, 230–231
in PowerPoint, 259
in Word, 172–173
help, in Office program, 82–83
hidden formatting in Word, 144
home page the first page that displays for an organization when the user navigates to the web address of the organization, 43
Home tab, 67
horizontal scroll bar
in Excel, 186, 187
using, 86–87
Hub panel in Microsoft Edge in which you view Favorites, Reading list, browsing history, or downloads, 45
hyperlink (link) text, image, or other object on a web page that when clicked takes you to a related Web page, 40
Hypertext Transfer Protocol, 41

I

I-beam pointer in Word, 136
IF function, 220–221
images, editing, in PowerPoint presentation, 265
Import Spreadsheet Wizard Access wizard used to perform an import of Excel data into a new table in Access, or to append data to the bottom of an existing Access table, 340
using, 340–343
Inbox Outlook mail folder into which received email messages are placed and is the active folder displayed by default when you start Outlook, 113

incognito browsing, 49
Increase Decimal button used in Excel to add one decimal place to each selected value each time the button is clicked, 192
Increase Indent button that moves a paragraph away from the left margin in a document or left edge of a cell in a worksheet each time the button is clicked, 148
Increase List Level button used to move text right to the next indent position within a bullet list placeholder in PowerPoint, 246
information, using search engine, to locate on the Web, 54–55
inline reply typing a reply to the sender of an email message from the Reading pane in Outlook, 114
Insert button in Cells group on Home tab of Excel ribbon used to insert new cells, rows, or columns, 200
Insert Caption feature used to add text above or below an image to label the image or add other descriptive text, 161
inserting
in Access
columns, 310–311
in Excel
cells, 201
columns, 200–201
commands, 232–233
header and footer, 230–231
rows, 200–201
worksheets, 206–207
in PowerPoint
pictures, 264–265
slides, 242–243
SmartArt graphics, 266–267
tables, 244–245
WordArt, 268–269
in Word
caption with picture in documents, 161
citations in research paper formatting, 174–175
columns and rows in tables, 197
pictures from computer in documents, 160
pictures from online sources, 158–159
table, 164–165
insertion point in Word, 136, 138–139
Insert Options button that appears in worksheet area after a new row or column has been inserted with options for formatting the new row or column, 201
Insert Picture dialog box, 59
Instant Search, 130–131
International Standards Organization (ISO) date format, 219
Internet (net) a global network that links other networks such as government departments, businesses, nonprofit organizations, educational and research institutions, and individuals, 40
Internet service provider (ISP) a company that provides access to the Internet's infrastructure for a fee, 40
Investment, using FV to calculate future value of, 223

J

Justify paragraph alignment in which extra space is added between words so that the text is evenly distributed between the left and right margins; both sides of the page appear even with this alignment, 147

K

keyboard. *See also* touch keyboard
  navigating Windows 10 with, 6–7
keyboard command a key or combination of keys (such as Ctrl plus a letter) that performs a command, 6
  adding new records in datasheet, 293
  for appointments, 123
  basics for Window 10, 6
  in Chrome, 49
  copy, 28
  cut, 30
  in Excel
    editing cell, 189
    formatting, 193
    selecting cells, 190
  paste, 28
  Word
    page breaks, 171
    paragraph alignment, 147
kiosk show, setting up, in PowerPoint presentation, 282–283

L

landscape layout orientation in which the text is rotated to the wider side of the page with a 9-inch line length at the default margin setting, 170, 204
Layout Options gallery that provides options to control how an image and surrounding text interact with each other, 158
Layout tab, 149
Layout view Access view in which you edit the structure and appearance of a form or report, 330
Line and Paragraph Spacing button in Word's Paragraph group used to change the spacing of lines between text within a paragraph and the spacing before and after a paragraph, 148
line breaks, 139
  versus new paragraphs in Word, 139
line chart a chart in which the values are graphed in a continuous line that allows a reader to easily spot trends, growth spurts, or dips in values by the line's shape and trajectory, 228
  creating and modifying, 228–229
line spacing in Word, 148–149
Links dialog box a dialog box opened from Info backstage area in which linked objects can be updated or otherwise managed, 350
link text, image, or other object on a web page that when clicked takes you to a related web page, 40
  Automatic Link Updates, 350–353
  linking Excel chart with PowerPoint presentation, 350–353
  managing, 353
list
  bulleted list in PowerPoint, 246
    converting to SmartArt, 268
live preview displays a preview of text or an object if the active option from a gallery is applied, 76

live updates, turning on an doff, 19
loan payment, PMT function and, 222–223
local account user name and password or PIN used to sign in to Windows that is stored on the local PC or device (not a Hotmail, live. com, or outlook.com email address), 8
  signing in with, 8–9
Lock screen the screen that displays when a Windows computer is locked to prevent other people from seeing programs or documents you have open, 8
  adding app to the, 21
  changing background, 21
  locking screen, 16
  personalizing, 20–21
lookup list a drop-down list in an Access field in which the list entries are field values from a field in another table or a fixed list of items, 326
  creating, 326–327
    with field values in another table, 327
    with Lookup Wizard, 326–327
Lookup Wizard Access wizard used to assist with creating a lookup list for a field, 326

M

Mail component within Microsoft Outlook used to send, receive, organize, and store email messages, 113
margins in Excel, changing, 231
Mark Complete, Remove from List versus, 129
Markup Area area at the right side of the screen in which comments and other changes made to a document are shown, 178
mathematical operators in Excel, 188
MAX function, 217
Maximize button near top right of a window used to expand the window size to fill the entire desktop, 13
meeting an appointment in your Calendar to which people have been invited to attend via email messages, 124
  accept request for, 125
  canceling, 125
  requesting in email message, 124
  scheduling, 124–125
  updating, 125
meeting request email message sent to an invitee of a meeting that you have scheduled in the Outlook Calendar, 124
Merge & Center button used in Excel that combines a group of selected cells into one and centers the content within the combined cell, 193
metasearch search engines, 54
Microsoft account email address from Hotmail.com, live.com, or outlook.com used to sign in to Windows, 8
  signing in with, 8
Microsoft Edge the web browser included with Windows 10, 42
  navigating Web using
    adding page to Favorites, 44–45
    annotating web page, 45
    displaying web pages, 42–44
      multiple, 44–45
    Hub, 45
    starting, 42
Microsoft Excel Web App, 360
  creating worksheet using, 360–361

Microsoft Office 2016, 63–89
  alignment guide in, 77
  backstage area to manage documents, 68–71
    exporting document as PDF files, 70–71
    printing documents, 70
  changing display options
    screen resolution, 88–89
    zoom options, 88
  color schemes in, 66
  contextual tabs, 76
  customizing and using Quick Access toolbar, 72–73
  dialog box, selecting options in, 77
  editions of, 64
  exploring ribbon interface, 76
  finding help in program, 82–83
  Format Painter to copy formatting options, 80–81
  objects, 76
  Office clipboard in, 78–80
  popularity of, 63
  scroll bars, using, 86–87
  selecting text and objects using, 74
  Smart Lookup, 82–83
  starting new presentation in, 66–67
  starting program in, 64–66
  switching between programs in, 66
  system requirements for, 64, 65
  task panes, 77
  Tell Me feature, 82–83
  Undo and Redo, 87
  using Mini toolbar and ribbon, 74
Microsoft Office suite, 339
Microsoft PowerPoint Web App
  creating presentation using, 362–363
  editing presentation in, 364–365
Microsoft Word Web App, 357
Middle Align alignment button in Excel that centers text vertically between the top and bottom edges of the cell, 195
MIDI files, 276
MIN function, 217
Minimize button near top right of a window used to reduce window to a button on the taskbar, 12
Mini toolbar toolbar that appears next to selected text or with the shortcut menu that contains frequently used formatting commands, 75, 146
  using, 75
mixed address cell addresses in a formula that have a combination of relative and absolute referencing, 214
MLA (Modern Language Association)
  formatting and page layout guidelines, 172
  source without page number, 175
  web addresses and URLs, 174
  Works Cited page, 176
mobile devices, shutting down, 17
mouse pointer, 75
mouse pointing device that is used for computer input, 6
  in Excel
    changing column width or row height, 195
    selecting cells using, 190
    using fill handle, 197
  movements and actions, 7
  navigating Windows 10 with, 6–7

peek and, 120
selecting text and objects using, 74
Move Chart button used to move a selected
chart to a new chart sheet in a workbook, 214
Mozilla Firefox
navigating Web using, 50–53
bookmarking pages, 51–52
displaying web pages, 50–51
multiple, 51–52
finding text on a page, 52–53
searching Web using Search bar, 53
starting Firefox, 50
MP3 audio files, 276
MP4 audio files, 276
MP4 videos, 274
MPEG 4 (MP4) movie file, 274
creating, 283

N

Name box in Excel, 186, 187
Network panel, 14
New Appointment button used to add a new
activity into the appointment area in the
Calendar, 120
New Contact button used to add a new contact
to the People list, 126
New Email button used to create a new email
message in Outlook, 113
New folder button in New group on Home
tab in File Explorer window used to create a
new folder on a device, 26
New Meeting button used to schedule a
meeting in Outlook Calendar for which
others are invited to attend, 124
New Notebook backstage area panel used in
OneNote to create a new notebook, 106
New quick note OneNote button with scissors
on the taskbar, or revealed by clicking Show
hidden icons in the Notification area on the
taskbar, 98
New (blank) record Access button located in
Record Navigation and Search bar used to add
a new record in a table or form, 292
New sheet button that displays as a plus
symbol inside a circle in the Sheet tab bar
used to add a new worksheet to an Excel
workbook, 186, 187, 206
New Slide button used in PowerPoint to
insert a new slide after the active slide in a
presentation, 242
New tab a control in a web browser that is
clicked to open a new tab in which a web
page can be displayed, 44
New Task button used to create a new task in
Outlook using the Task window, 128
Normal view in PowerPoint, 240
Notebook Information backstage area panel
that displays when the File tab is clicked with
a OneNote notebook open, 105
notebooks in OneNote
creating new, 106–107
opening, 94–95
printing and exporting sections, 105
sharing, 106–107
note container a box on a OneNote page that
contains note content, 95
notes
in OneNote
creating table in, and calculating
expressions, 97
deleting, 95

highlighting, with color, 96
jumping to tagged, 103
searching, 104–105
tagging, 102–103
typing, 97
in PowerPoint, 240, 241
adding, 254–255
Notes Pages handout option for printing slides
in which one slide is printed per page with
the slide printed in the top half of the page,
and the speaker notes or blank space when
no notes are present in the bottom half of the
page, 259
notes pane below slide pane in Normal view of
PowerPoint in which speaker notes are typed
and edited, 240, 254
Note-taking software, 93
notification area area at right end of taskbar
with icons to view or change system settings
or resources such as the date and time or
speaker volume, 10
using, 14–15
numbered list, creation of, in Word, 144–145
Numbering button used to format text as a list
with sequential numbers or letters used at the
beginning of each line, 144
numbers, formatting in Excel, 191–192, 193

O

object a picture, shape, chart, or other item that
can be manipulated separately from text or
other objects around it, 76
applying animation effects to individual,
280–281
opening and closing, 289–291
selecting, using mouse and touch, 74
Office 365 subscription-based edition of
Microsoft Office 2016, 64
advantage of, 64
Office 365 University, 64
Office Clipboard, using the, 78–80
Office Home and Business 2016, 64
Office Home and Student 2016, 64
Office Presentation Service, in
delivering presentation online, 281
Office Professional 2016, 64
OneDrive secure online storage provided by
Microsoft that is available to users signed in
with a Microsoft account, 84
creating document using Word Online,
356–359
creating presentations in using PowerPoint
Online, 362–363
creating worksheet using Excel Online,
360–361
downloading and uploading files from and
to, 366–367
editing presentations in PowerPoint Online,
364–365
save a file to, 84–86
sharing files on, 368–369
storage of OneNote notebooks on, 93
OneNote 2016 note-taking software application
included in the Microsoft Office 2016 suite,
64, 93–107, 94
notebooks in
adding sections and pages to, 96–97
closing, 105
creating new, 106–107
creating table in, and calculating
expressions, 97

inserting files in, 100–101
inserting pictures in, 100–101
inserting Web content into, 98–99
opening, 94–95
printing and exporting sections, 105
sharing, 106–107
notes in
deleting, 95
highlighting, with color, 96
jumping to tagged notes, 103
searching, 104–105
active page only, 105
tagging, 102–103
typing, 97
pages in, adding, 96–97
search features in, 104–105
sections in, 96
adding, 96–97
tags in, 102–103
viewing, 103
OneNote notebook electronic notebook
created in OneNote organized into sections
and pages in which you store, organize, search,
and share notes of various types, 94
one-to-many relationship an Access
relationship in which the common field used
to join two tables is the primary key in only
one table, 329
one-to-one relationship an Access relationship
in which two tables are joined on the primary
key field in each table, 328
Online Pictures button used to search for an
image to insert into a document from the Web,
Flickr, Facebook, or OneDrive, 158
Open a new tab in Mozilla Firefox, the control
that displays with a plus symbol used to open a
new tab for displaying a web page, 51
operating systems, purposes of, 3
orientation
in Excel, 204
in PowerPoint, 259
in Word, 170–171
Outline view displays content as bullet points,
177
Outlook 2016 personal information
management (PIM) software application
included in the Microsoft Office suite, 64,
111–131, 112
appearance of window, 113
appointments in
editing, 122–123
scheduling, 120–121
scheduling recurring, 122–123
Calendar tool in, 120–125
contacts in, adding and editing, 126–127
email messages in
attaching files to, 116–117
creating and sending, 113–114
deleting, 116–117
previewing attachments, 118–119
replying to, 114–115
signatures for, 115
meetings in
accept request for, 125
canceling, 125
scheduling, 124–125
updating, 125
People tool in, 126–127
scheduling events in, 120–121
searching feature in, 130–131

setting up, 112–113
tasks
    adding and editing, 128–129
    Remove from List versus Mark
      Complete, 129

P

Page Borders button used in Word to choose a border that is drawn around the perimeter of a page, 162
page breaks in Word
    hard, 181
    soft, 171
page layout, in Word, 170–171
Page Layout view Excel view in which you can add or modify print options while also viewing the worksheet, 230
Page Number button used to insert page numbering in a header or footer, 172
pages, adding in OneNote, 96–97
paragraph formatting changing the appearance of paragraphs, 146
paragraphs in Word
    alignment of, 147
    formatting, 146–147
    line breaks versus new in, 139
    spacing of, 148–149
paragraph symbol, 144
Password text box, 9
Paste button or menu option used to insert the copied or cut text or file, 26
Paste Options, 79
Paste Special dialog box, 347, 353
path to the web page, 41
PDF document a document saved in Portable Document Format that is an open standard for exchanging electronic documents developed by Adobe systems, 70
    creating printable, at Save As backstage area, 359
    exporting documents as, 70–71
    exporting notebook sections as, 105
    opening, 71
peek, 120
Pen and Touch dialog box, 4
People card contact information for an individual displays in a People card in the Reading pane, 126
People component in Microsoft Outlook used to add, maintain, organize, and store contact information for people with whom you communicate, 126
Percent Style in Excel, 221
personal computer
    adding video from file on, 275
    inserting pictures from, into PowerPoint presentation, 264
    shutting down, 17
personal information management (PIM) program a software program to organize messages, schedules, contacts, and tasks, 111
Photos app Windows 10 app used to view pictures from a local device and OneDrive, 10
Photo window, 13
picture password, 9
pictures
    adding for contacts, 126–127
    adding of yourself, to Microsoft account, 21
    inserting
        into OneNote page, 100–101

        into PowerPoint presentation
            from computer, 264
            from web resource, 264
    saving, from web pages, 58–59
    in Word 2016
        cropping and removing background, 161
        editing, 160–161
        inserting, from computer, 160
        inserting, from online sources, 158–159
        inserting caption with, 161
Pictures button used to insert an image stored on your PC or mobile device into a document, workbook, or presentation, 160
Picture Tools Format tab, 76, 265
pie chart a chart in which each data point is sized to show its proportion to the total in a pie shape, 224
    creating and modifying, 224–225
Pin to Start Windows option to add a tile for the selected app or application to the Start menu, 19
placeholder rectangular container on a slide in which text or other content is added, 240, 241
    selecting, resizing, aligning, and moving, 248–249
PMT function, 222–223
pointer the white arrow or other icon that moves on the screen as you move a pointing device, 6
Point explosion, 225
points, 139
portrait layout orientation in which the text on the page is vertically oriented with a 6.5-inch line length as the default margin setting, 170, 204
pound symbols in Excel, 194
PowerPoint 2016 presentation application in Microsoft Office suite used to create multimedia slides for an oral or kiosk-style presentation, 64, 66, 239–259
    audience handouts in, 258–259
    AutoFit in, 249
    bullets/bulleted list on, 246, 247
        changing symbol, 247
        converting to SmartArt, 268
    color scheme in, 66
    comments in, 241
        adding, 254–255
        deleting or hiding, 255
    Comments pane in, 240, 254
    editing of text on slides, 243
    embedding and editing Excel data into presentation in, 348–349
    embedding entire file in, 349
    linking Excel chart with, and updating links, 350–353
    Normal view in, 240
    notes in, 241
        adding, 254–255
    Notes Pages in, 259
    Notes pane in, 240, 254
    opening, 66
    placeholders in, 240, 241
        selecting, resizing, aligning, and moving, 248–249
    presentations in, 263–283
        adding sound to, 276–277
        adding timing for, 282–283
        adding video to, 274–275

    animation
        applying, to individual objects, 280–281
        applying, using the slide master, 278–279
        changing sequence of, 281
    categories of, 279
    charts in
        creating, on slides, 270–271
        formatting, 271
    creating new, 240–243
    creating using PowerPoint Online, 362–363
    editing data in, 271
    editing images in, 265
    editing in PowerPoint Online, 364–365
    pictures in
        inserting, from your computer in, 264
        inserting from web resource, 265
    publishing, to a web service, 259
    rehearse timings in, 283
    setting up a self-running, 282–283
    setting up kiosk show in, 282–283
    shapes in
        drawing, 272–273
        transforming, 269
    SmartArt graphics
        converting text to, 268–269
        inserting, 266–267
        modifying, 257
    text boxes in
        adding, 272–273
    transitions
        adding, to slide show, 278
    WordArt in
        inserting, 268–269
        transforming, 269
Presenter view in, 256
recognition of audio file formats, 276
recognition of video file formats, 274
slide layouts in, 242
slide master in, 252
    modifying, 252–253
Slide Master view in, 252–253
slide pane in, 240, 241
slide show in
    displaying, 256–257
    toolbar buttons for, 256
slides in
    adding text to bottom of, 253
    closing, 257
    deleting, 251
    duplicating, 250
    editing text on, 243
    footers on, 259
    headers on, 259
    hiding, 251
    inserting, 242–243
    inserting table on, 244–245
    moving, 250, 251
    orientation of, 259
    previewing, 259
    size of, 259
slide size in, 242
Slide Sorter view in, 250
Slide Thumbnail pane in, 240, 241
Smart guides in, 249
speaker notes in, 259
status bar in, 241

tables in
  inserting, on slides, 244–245
  modifying, 245
  text in
    editing, 243
    formatting, with font and paragraph options, 246–247
  theme, changing, 244–245
  title slides in, 240
  variants in, 240
  view button in, 241
  View tab in, 88
  Web App, 362–365
  zoom button in, 241
PowerPoint Online web-based version of Microsoft PowerPoint accessed from OneDrive that is similar to the desktop version of PowerPoint but has fewer features; some functionality within features may also vary from the desktop version, 362
  creating presentation using, 362–363
Power & sleep settings, 15, 17
presentation application software used to create slides for an electronic slideshow that may be projected on a screen, 239
Presenter view PowerPoint view for a second monitor in a slide show that displays the slide show along with a preview of the next slide, speaker notes, timer, and slide show toolbar, 256
previewing file attachments, 118–119
primary key the Access field that contains the data that uniquely identifies each record in the table, 300
  assigning, 320–321
primary table a table in an Access relationship in which the joined field is the primary key and in which new records should be added first, 329
printing
  audience handouts and speaker notes, 258–259
  comments, 233
  current page only, 70
  Excel, changing options, 205
  notebook sections, 105
  web pages, 56–57
Print Layout view default view in Word that shows the document as it will appear when printed, 177
Print Preview right panel in Print backstage area that shows how a document will look when printed, 70
private browsing, 49
Protected View View in which a file is opened when the file is opened as an attachment from an email message or otherwise downloaded from an Internet source; file contents can be read but editing is prevented until Enable Editing is performed, 119
public domain multimedia, 277
Publisher, 64
pull quote a quote placed inside a text box in a document, 162

Q
query Access object used to display information in a datasheet from one or more tables, 288, 291
  calculated field in creating, 310–311
  criteria to select records in, 306–307

Design view in creating, 304–305
entering multiple to sort, 308–309
exporting to Excel, 344–345
Simple Query Wizard in creating, 302–303
Quick Access Toolbar toolbar with frequently used commands located at the top left corner of each Office application window, 72
  add button to, 72–73
  customizing and using, 72–73
  moving below ribbon, 73
  Redo button on, 87
  Save button on, 84
  Undo button on, 87
Quick access view in which File Explorer opens by default in Windows 10 with frequently-used and recently-used files displayed in the content pane, 24
quick action tile A quick action tile displays in Notification panels and is used to turn on or turn off a setting or open a setting window, 14
Quick analysis in Excel, 190
quick notes, 99
Quick Print button, 73
Quick Table a predefined and formatted table with sample data, 164
QuickTime movies, 274
quotation marks, 138

R
range a rectangular group of cells referenced with the top left cell address, a colon, and the bottom right cell address (e.g., A1:B9), 190, 202
  names for, 215
  sorting, 203
Reading pane, 113, 115
Read Mode view displays a document full screen in columns or pages without editing tools such as the QAT and ribbon, 177
Real Time Presence, 178
Recent option list
  Open backstage area, 68
  opening document from, 65
  pinning documents and folder to, 71
Recommended Charts button, 227
record all the fields for one person, place, event, or item within an Access table, 290
  adding, editing, and deleting in a form, 296–297
  adding, using datasheets, 292–293
  best practices for deleting, 295
  editing and deleting using datasheets, 294–295
  entering criteria to select, in queries, 306–307
  entering multiple criteria to select, 308–309
  finding, 298–299
  sorting and filtering, 300–301
  using find to move to, 297
Recurrence button in Outlook that opens the Appointment Recurrence dialog box in which you set up particulars of an appointment that repeats at the same day and time each week for a set period of time, 122
Recycle Bin, 18
  empty, 33
  maximum size for, 32
  window, 33
Redo button, 87
red wavy lines, 136–137
Refresh, 44

Region setting, Excel and, 219
rehearse timings in PowerPoint presentations, 283
relationships in Access
  creating, 329
  displaying and editing, 328–329
  one-to-many, 329
  one-to-one, 328
relative address Excel default addressing method used in formulas in which column letters and row numbers update when a formula is copied relative to the destination, 214
reminders, for appointments, 121
Remove from List, versus Mark Complete, 129
Rename button or menu option used to change the name of a file, folder, or tab, 31
  files and folders, 31–32
Replace feature that automatically changes each occurrence of a word or phrase to something else, 142
Reply All, 115
Reply button used to compose a reply to the sender of an email message, 114
replying to email messages, 114–115
report Access object used to display or print data from one or more tables or queries in a customized layout and with summary totals, 288, 290
  creating, editing, and viewing, 332–333
  grouping and sorting, 333
Report view Access view that displays a report's data without editing tools and is the view that is active when the report is opened, 333
research paper, formatting in Word
  adding text and page numbers in headers, 172–173
  citations, inserting and editing, 174–175
  editing sources, 175
  footnotes and endnotes in, 177
  MLA and APA guidelines, 172
  removing page numbers from first page, 173
  word views in, 176–177
  Works Cited page, creating, 176–177
Restart, 17
Restore Down button that replaces Maximize button when a window has been maximized used to return the window to its previous size before being maximized, 13
resume, creation from templates in Word, 180–181
Ribbon Display Options button used to change the ribbon display to show tabs only or auto-hide the ribbon, 67
ribbon interface that displays buttons for commands organized within groups and tabs, 26
  buttons in, 78
  exploring, 76
  moving Quick Access toolbar below, 73
  rotating, in Excel, 209
rows
  in Excel
    adjusting height, 194–195
    deleting, 200–201
    inserting, 200–201
  in Word
    inserting and deleting, 167
Run button in Access Query Tools Design tab used to instruct Access to perform the query instructions and display the query results datasheet, 305

S

Safely Remove Hardware and Eject Media option used to eject a USB flash drive from a PC or mobile device, 33

Save As backstage area
    creating printable PDF of document at, 359
    to save copy of document, 68–69
    saving to OneDrive, 84–86

Save As command, 68–69

Save As dialog box, 105

scaling in Excel, 204–205

screen
    locking the, 16
    Start screen, 9

screen clipping
    into OneNote, 98–99

screen resolution display setting that refers to the number of picture elements, called pixels, that make up the image shown on a display device, 88
    viewing and changing, 88–89

scroll bar a horizontal or vertical bar with arrow buttons and a scroll box for navigating a larger file when a document exceeds the viewing space within the current window, 86
    using, 86–87

scroll box a box between the two arrow buttons on a scroll bar that is dragged up, down, left, or right to navigate a larger file that cannot fit within the viewing area, 86

Search bar in Mozilla Firefox, the area used to type search phrases to find web pages, 53
    searching Web using, in Mozilla Firefox, 53

search engine a company that searches Web pages to index the pages by keyword or subject and provides search tools to find pages, 54
    advanced search options, 55–56
    metasearch, 54
    using, to locate information on the Web, 54–55

search options, using advanced, 55–56

Search People text box, 131

search text box box next to Start button in Windows 10 used to search for apps, files, settings, or web links, 8, 104
    Cortana, 12
    using, 12

section break in Word, 171

sections, adding in OneNote, 96–97

selection handle empty circles that appear around a selected object, or at the beginning and end of text on touch-enabled devices, that are used to manipulate the object or define a text selection area, 74, 75

self-running presentation, setting up a, 282–283

semicolons in email addresses, 114

Send to OneNote 2016 printer, 98

Set Up Slide Show button used in PowerPoint to configure options for a slide show such as setting up a self-running presentation, 282

shading color added behind text, 162
    in Excel, 209
    in Word, 162–163

shapes
    drawing, in Word, 160
    in PowerPoint presentation
        drawing, 272–273
        transforming, 269

Shapes button used to select the type of shape to be drawn on a slide, in a document, or in a worksheet, 272

Share Notebook backstage area view used in OneNote to share the current notebook with another individual by providing an email address, 106

Sheet tab, 186, 187

Sheet tab bar bar above Status bar at bottom left of Excel window where sheet tabs are displayed, 186, 187, 206

shortcut menu
    modifying table using in Word, 168–169

Show Formulas button in Formula Auditing group of the Formulas tab used to turn on or turn off the display of formulas in cells, 205

Show/Hide, 144

Show your bookmarks button in Mozilla Firefox used to open the bookmarks menu, 51

shut down process to turn the power off to the PC or mobile device to ensure all Windows files are properly closed, 17

signatures, in email, 115

signing in
    with local account, 8–9
    with Microsoft account, 8

sign out action that closes all apps, applications, and files and displays the lock screen; also called logging off, 16

Simple Query Wizard Access wizard that assists with creating a new query by making prompting the user to make selections in a series of dialog boxes, 302
    creating queries using, 302–303

sleep setting, 15, 17

Slicer pane in Excel, 235

slide
    adding
        Action Button, 273
        text to bottom of, 253
        transitions, 278
        video to existing, 275
    closing, 257
    creating charts on, 270–271
    deleting, 251
    duplicating, 250
    editing text on, 243
    footers on, 259
    headers on, 259
    hiding, 251
    inserting, 242–243
    inserting table on, 244–245
    moving, 250, 251
    orientation of, 259
    previewing, 259
    size of, 242, 259
    timing settings for, 282–283

slide deck a collection of slides in a presentation, 239

slide layout an arrangement of content placeholders that determine the number, position, and type of content for a slide, 242

slide master a slide master in PowerPoint is included for each presentation and slide layout and determines the default formatting of placeholders on each new slide, 252
    applying animation effects using, 278–279
    modifying, 252–253

Slide Master view PowerPoint view in which global changes to the formatting options for slides in a presentation are made, 252

slide pane PowerPoint pane that displays the current slide in Normal view, 240, 241

SlideShare, 259

slide show, in PowerPoint
    displaying, 256–257
    toolbar buttons for, 256

Slide Show button in PowerPoint Status bar used to start a slide show from the active slide, 257

Slide Show view PowerPoint view in which you preview slides full screen as they will appear to an audience, 256

Slide Sorter view PowerPoint view in which all of the slides in the presentation are displayed as slide thumbnails; view is often used to rearrange the order of slides, 250

slide thumbnail pane PowerPoint pane at left side of Normal view in which numbered thumbnails of the slides are displayed, 240, 241

SmartArt a graphic object used to visually communicate a relationship in a list, process, cycle, or hierarchy, or some other object diagram, 266
    in PowerPoint
        converting text to, 268–269
        inserting, 266–267
        layout categories of, 266
        modifying, 257
    in Word, 160

smart guide a colored vertical and/or horizontal line that appears to help you align and place objects; also called an alignment guide, 249

Smart Lookup option on Tell Me drop-down list that opens a task pane at the right side of the window when clicked populated with web links and a definition for the term typed in the Tell Me text box, 82–83

smart quotes, 138

Snap Assist Windows 10 feature that pops up when a portion of the screen is empty after snapping a window. Thumbnails for the remaining open windows are shown so that you can click the window you want to fill the remaining empty space., 13

Snap Windows feature that lets you dock a window to a half or quadrant of the screen without having to move the window and manually resize it, 13

soft page break a page break inserted automatically by Word when the maximum number of lines for the page has been reached with the current page size and margins, 171

Sort & Filter button in Editing group of Home tab in Excel used to sort and filter a worksheet, 202

sorting in Access
    of queries, 308–309
    of records, 300–301
    reports, 333

sound, adding, to PowerPoint presentation, 276–277

source data data that is selected for copying to be integrated into another program, 346

source program program in which data resides that is being copied for integration into another program, 346

spacing, default in Word, 139

sparkline chart a miniature chart embedded into a cell in an Excel worksheet, 232
  creating and modifying, 232–233

speaker notes in PowerPoint, 259

Spell Check in Word, 140–141

Spelling & Grammar feature in software applications that flags potential errors, displays suggestions for correction, as well as other options for responding to the potential error, 140

spider a program that reads web pages and other information to generate index entries; also known as a crawler, 54

spreadsheet application software in which you work primarily with numbers that you organize within a grid-like structure called a worksheet, perform calculations, and create charts, 185

Start button at lower left corner of Windows display used to display the Start menu, 4

Start menu Windows 10 user interface that displays a pop-up menu with program names in the left pane and tiles in the right pane, which are used to launch apps or other programs, 4
  customizing, 18–19
  pinning and unpinning tiles to and from, 19
  rearranging tiles, 19
  resizing, 18

Start screen the opening screen that displays when you start an application in the Microsoft Office suite such as Word, Excel, PowerPoint, or Access. In the left pane are recently opened files and the right pane displays the templates gallery., 9, 65
  changing design or color scheme, 21
  personalizing, 20–21

Status bar
  in Excel, 186, 187
  in PowerPoint, 241
  in Word, 136

Store app Windows app used to download or buy new apps for your device, 11

style guide set of rules (such as MLA or APA) for formatting academic essays or research papers, 172

style set a set of formatting options for each style based upon the theme of the document, which is changed with options in the Document Formatting group on the Design tab. Each style set has different formatting options associated with it., 150

style set of predefined formatting options that can be applied to selected text with one step, 150
  formatting using, 150–151

subfolder a folder within a folder, 25

SUM function in Excel, 198–199

Symbol gallery used to insert a special character or symbol such as a copyright symbol or fraction character, 140

symbol insertion in Word, 140–141

T

tabbed browsing a feature in web browsers that allows you to view multiple web pages within the same browser window by displaying each page in a separate tab, 44

table in an Access database, all of the data for one topic or subject; for example, Customers is one table in a database that tracks invoices, 289
  in Access, 288, 289
    adding fields to existing, 322–323
    creating lookup list with field values in, 327
    creating new, 318–319
      in Design View, 320–321
    linking Excel worksheet to, 343
    planning new, 316–317
    primary, 329
  in Excel, 234–235
  in OneNote, 97
  in PowerPoint
    inserting, on slides, 244–245
    modifying, 245
  in Word
    banded rows in, 166
    creating, 164–165
    formatting, 166–167
    inserting and deleting columns and rows, 167
    merging and splitting cells, 168–169
    modifying column width and alignment, 168–169

table cell a box that is the intersection of a column with a row in which you type text in a table in a Word document or PowerPoint presentation, 164

Table Styles gallery of predefined table formatting options that include borders, shading, and color, 166

tag a short category or other label attached to an item such as a note in a OneNote page that allows you to categorize or otherwise identify the item, 102
  appointment, in Outlook, 121
  in OneNote, 102–103
    jumping to tagged notes, 103
    To-Do, 103
    viewing, 103

Tags Summary pane in OneNote, the pane opened from the Find Tags button that is used to navigate to the location of a tagged item, 102

taskbar bar along the bottom of the desktop that has the Start button, search box, icons to start programs and the notification area; buttons on the taskbar are used to switch between open programs, 10

task pane a pane that opens at the right or left side of an application window with additional options or commands for the related ribbon group as buttons, lists, sliders, check boxes, text boxes, and option buttons, 77

Tasks component in Microsoft Outlook used to maintain a to-do list, 128
  adding and editing, 128–129
  Remove from List versus Mark Complete, 129

Task View button on taskbar that displays all open apps, programs, or other windows with large thumbnails in the center of the screen, 11

Tell Me text box in Office 2016 that displays *Tell me what you want to do* in which you type a feature name or option and press Enter to find help resources or locate an option rather than searching through the ribbon tabs, 82

finding help using, 82–83

template a document with formatting options already applied, 152
  in Word
    creating new document from, 152–153
    creating resumes and cover letters from, 180–181

text
  in Excel
    alignment of, 187
    finding on web page
      using Google Chrome, 48–49
      using Mozilla Firefox, 52–53
  in PowerPoint
    editing, 243
    formatting, with font and paragraph options, 246–247
  selecting, using mouse and touch, 74
  in Word
    changing line and paragraph spacing, 148–149
    finding and replacing, 142–143
    formatting, with font and paragraph alignment options, 146–147
    indenting, 148–149
    moving, 144–145

Text Box button used to create a text box in a document, workbook, or presentation, 162
  in PowerPoint presentations
    adding, 272–273
    resizing, 269
  in Word
    editing, 163
    inserting, 162–163

Text Highlight Color tool, 96

Theme a set of colors, fonts, and font effects that alter the appearance of a document, spreadsheet, or presentation, 151
  in PowerPoint, changing, 244–245
  for workbook, 203

Thesaurus in Word, 143

tile square or rectangular shapes on Windows Start menu used to launch apps or programs, 4
  pinning and unpinning, to and from the Start menu, 19
  quick action tile, 14
  rearranging, 19
  resizing, 19

timing, adding, for slide, 282–283

Title bar, 12

title slide first slide in a presentation that generally contains a title and subtitle, 240

TODAY function, 218–219

To-Do List view in Outlook that displays a list of tasks to be done, 128

To-Do tag, 103

toggle button a button that operates in an on or off state, 80

touch device
  in Excel
    Paste Options button, 206
    selecting cells, 190
    using fill handle, 197
  in navigating Windows 10, 4–6
  selecting text and objects using, 74
  signing in, 6–7
  Touch/Mouse Mode, 72

touch keyboard onscreen keyboard that displays for touch- enabled devices, 4, 372
  in handwriting mode, 6
  in thumb mode, 5
    using, 4
Touch/Mouse Mode, 72
touchpad a flat surface on a laptop or notebook operated with your finger(s) as a pointing device, 6
trackball a mouse in which the user manipulates a ball to move the on-screen pointer, 6
track changes, 179
trademark symbol, 140
transition a special effect that plays during a slide show as one slide is removed from the screen and the next slide appears, 278
  adding to slide show, 278
Trim Video button used to modify a video clip to show only a portion of the clip by changing the start and/or end times in the Trim Video dialog box, 274
Turn live tile off Windows option to stop displaying headlines or other notifications from online services on a tile on the Start menu, 19
Turn live tile on Windows option to display headlines or other notifications from online services on a tile on the Start menu, 19
typeface, 146

U

Undo command that restores a document to its state before the last action was performed, 87, 164
uniform resource locator (URL) a text-based address used to navigate to a website; also called a web address, 40
Unpin from Start Windows option to remove selected tile from the Start menu, 19
Update Links button that when clicked causes linked objects to be updated with new data; button appears inside Security Notice dialog box when destination document with linked objects set to automatically update is opened, 352
uploading, files from OneDrive, 366–367
user account the user name and password or PIN that you type to gain access to a PC or mobile device, 8
user interface (UI) the icons, menus, and other means with which a person interacts with the OS, software application, or hardware device, 4

V

value axis vertical axis in a column chart scaled for the values that are being charted; also called the y- or z-axis, 226
values in Excel, 191–192
variant a style and color scheme in the active PowerPoint theme family, 240
vertical scroll bar
  in Excel, 186, 187
  using, 86–87
  in Word, 136
video
  adding, to existing PowerPoint slides, 275
  adding, to PowerPoint presentations, 274–275
  linking to YouTube, 275
video file formats, PowerPoint recognition of, 274
video playback, options for, 275

Video Tools Playback tab, 275
View buttons
  in Excel, 186, 187
  in PowerPoint, 241
  in Word, 136
View Printable link, 57
View tab
  in Excel, 88
  in PowerPoint, 88
  in Word, 88

W

waffle button, 357
web
  navigating, using Google Chrome, 46–49
    bookmarking pages in, 47
    customizing and controlling, 47
    displaying web pages with, 46–47
    finding text on web page, 48–49
    hamburger button, 47
    incognito browsing, 49
    multiple, 47–48
    searching Web from address bar, 48
    starting, 46
  navigating, using Microsoft Edge, 42–45
    displaying web pages, 42–44
      multiple, 44–45
    starting, 42
  navigating, using Mozilla Firefox, 50–53
    bookmarking pages, 51–52
    displaying web pages, 50–51
      multiple, 51–52
    finding text on a page, 52–53
    searching Web using Search bar, 53
    starting Firefox, 50
  publishing presentations to, 259
  using search engine to locate information on the, 54–55
    advanced options in, 55–56
    metasearch, 54–55
web address a text-based address to navigate to a website; also called uniform resource locator (URL), 40
  parts of, 41
web browser a software program used to view web pages, 40. *See also* specific by name
web content, inserting, into notebook, 98–99
Web Layout view displays document in Word as a web page, 177
web page a document that contains text and multimedia content such as images, video, sound, and animation, 40
  displaying
    with Google Chrome, 46–47
      multiple, 47–48
    with Microsoft Edge, 42–44, 42–45
      annotating in, 45
      multiple, 44–45
    with Mozilla Firefox, 50–51
      multiple, 51–52
  downloading content from, 58–59
  file name for, 41
  finding text on
    with Google Chrome, 48–49
    with Mozilla Firefox, 52–53
  path to, 41
  printing, 56–57
  refresh, 44
  saving pictures from, 58–59
web servers, 40

websites, 40
  copyright-free or public domain multimedia on, 277
what-if analysis, 185
Wi-Fi network, connecting, 14
Wikimedia Commons, 50–51
Wildcard characters in Access, 309
window
  close, 13
  maximize, 13
  minimize, 12
  resize, 12–13
  restore down, 13
  Snap feature, 13
  standard features of, 12
Windows 10 Microsoft's Operating System for PCs released in July 2015, 3, 4
  locking screen, 16
  navigating
    with keyboard, 6–7
    with mouse, 6–7
    with touch, 4–6
  shutting down, 17
  signing out of, 16–17
  starting, 8–10
  touch gestures, 4–5
  user accounts at, 8
  user interface of, 4
  using and customizing Start menu, 18–19
Windows audio files (.wav), 276
Windows Live ID, 8
Windows Media Audio (.wma) files, 276
Windows Media Video (.wmv) files, 274
Word 2016 Word processing application in Microsoft Office suite used to create, edit, and format text-based documents, 64, 65, 136–153, 146, 157–181
  AutoCorrect in, 136, 137–138
  AutoFormat in, 136, 138, 145
  blank document screen, 136
  borders in, 162–163
  bullet insertion in, 144–145
  changing page layout options, 170–171
  character formatting in, 146–147
  color scheme in, 66
  comments in
    inserting, 178–179
    replying to, 178–179
  cover letter creation from templates, 180–181
  default setting
    font, 146
    spacing, 139
    view for new documents, 177
  Design tab, 149
  documents in
    creating and editing new, 136–139
    creating from template, 152–153
    creating using Word Online, 356–359
    cropping and removal picture background, 161
    editing pictures for, 160–161
    inserting caption with picture, 161
    inserting pictures from computer, 160
    inserting pictures from online sources, 158–159
    opening, 68–69
    saving new, 138
    sending, as PDF document, 70–71
    tracking changes made to, 179

editing in
    of citations, 174–175
    documents, 138–141
    pictures, 160–161
    of sources, 175
embedding Excel chart into document, 346–347
endnotes in, 177
fonts in, 146
    changing, 146
footnotes in, 177
formatting
    font options, 146–147
    paragraphs, 147
    Theme, 151
    using styles, 150–151
hanging indent in, 148
hidden formatting in, 144
I-beam pointer in, 136
insertion point in, 136, 138–139
landscape orientation in, 171
Layout tab, 149
line and paragraph spacing in, 148–149
maximize screen, 65
numbering in, 144–145
opening, 64–66
    document, 68–69
orientation in, 170–171
page breaks, 171
page layout in, 170–171
paragraph
    alignment in, 147
    formatting in, 147
pictures in
    cropping and removing background, 161
    editing, 160–161
    inserting, from computer, 160
    inserting, from online sources, 158–159
    inserting caption with, 161
portrait orientation in, 171
pull quote in, 162–163
research paper formatting
    adding text and page numbers in headers, 172–173
    citations, inserting and editing, 174–175
    editing sources, 175
    footnotes and endnotes in, 177
    MLA and APA guidelines, 172
    removing page numbers from first page, 173
    Works Cited page, creating, 176–177
resume creation from templates in, 180–181
section break in, 171
shading in, 162–163
shapes, drawing, 160

SmartArt in, 160
spelling and grammar check in, 140–141
status bar in, 136
symbol insertion in, 140–141
tables in
    banded rows in, 166
    create with dialog box, 165
    create with Quick Table, 164–165
    formatting, 166–167
    inserting and deleting columns and rows, 167
    merging and splitting cells, 168–169
    modifying column width and alignment, 168–169
templates in, 152–153
    resume and cover letter creation, 180–181
text box, inserting, 162–163
text in
    finding and replacing, 142–143
    formatting, with font and paragraph alignment options, 146–147
    indenting, 148–149
    moving, 144–145
thesaurus in word replacement in, 143
vertical scroll bar in, 136
View buttons in, 136
views in, 177
View tab in, 88
Web App, 357
Welcome back! balloon in, 177
WordArt in, 160
word search in, 142
wordwrap in, 136
Zoom buttons in, 136
WordArt text that is created and formatted as a graphic object, 268
    in PowerPoint
        changing font size, 269
        inserting, 268–269
        transforming, 269
    in Word, 160
Word Online web-based version of Microsoft Word accessed from OneDrive that is similar to the desktop version of Word but has fewer features; some functionality within features may also vary from the desktop version, 356
    creating document using, 356–359
word processing application software used to create documents containing mostly text, 135
    See also **Word 2016**
word search in Word, 142
wordwrap feature in word processing applications in which text is automatically moved to the next line when the end of the current line is reached, 136

workbook a spreadsheet file saved in Excel, 185
    opening blank, 186
    themes of, 203
Works Cited page at the end of an MLA paper or report that provides the references for the sources used in the paper, 176
    APA and MLA style guide for, 176
    creating, 176
worksheet the grid-like structure of columns and rows into which you enter and edit text, values, and formulas in Excel, 185
    centering, 231
    creating new, 186–188
    creating using Excel Online, 360–361
    entering text and values, 187–188
    importing data into Access, 340–343
    inserting, 206–207
    linking to Access table, 343
    renaming, 206–207
    scaling, 204–205
World Wide Web (Web) the global collection of electronic documents circulated on the Internet in the form of web pages, 40
wrapping
    in Excel, 209
Wrap Text button in Alignment group on Home tab in Excel wraps text within the cell's column width, 209

Y

Yahoo!, 54, 55
YouTube, linking video to, 275

Z

ZIP file a file with the extension .zip that is used to bundle and compress a group of files together. A ZIP file is sometimes referred to as an archive file, 22
Zoho, 373
Zoom buttons
    in PowerPoint, 241
    in Word, 136
Zoom In control that displays as a plus symbol at bottom right corner of application window that increases magnification by 10 percent each time button is clicked, 88
Zoom Out control that displays as a minus symbol at bottom right corner of application window that decreases magnification by 10 percent each time button is clicked, 88
Zoom slider bar located near bottom right corner of an application window that is used to change the magnification option, 88

Photo Credits: Microsoft images used with permission from Microsoft.